Cybersecurity Protocols in Industrial Automation

Cybersecurity Protocols in Industrial Automation

Dr. K. Aruna Bhaskar

ARCLER PRESS

www.arclerpress.com

Cybersecurity Protocols in Industrial Automation

Dr. K. Aruna Bhaskar

Arcler Press
4164 Lakeshore Road
Burlington, ON L7L 1A4
Canada
www.arclerpress.com
Email: orders@arclereducation.com

ISBN: 978-1-77956-975-2 (Hardcover)

ABOUT THE AUTHOR

Dr. K. Aruna Bhaskar, currently working as an Associate Professor at KLEF, KL Deemed to be University, Vaddeswaram, Andhra Pradesh, in the Department of Computer Science & Engineering. He received Ph.D degree from Krishna University, Machilipatnam, and M. Tech in Computer Science from JNTUK, Kakinada.His academic background, which includes over fifteen years of teaching experience, demonstrates a commitment to both established and developing areas of technology.Dr. K. Aruna Bhaskar has advanced understanding in fields like biomedical image processing, computer networks, blockchain technology, Agile Technologies, artificial intelligence (AI), and machine learning (ML) while his long career. His multidisciplinary experience connects intelligence-driven approaches, secure network systems, decentralized computing (via blockchain), and imaging techniques with a healthcare focus.

TABLE OF CONTENTS

LIST OF FIGURES

PREFACE

The modern world relies on automation for efficiency, consistency, and productivity in industries such as energy, manufacturing, transport, and utilities. Industrial automation transformed the way machines, processes, and humans work together. Isolated and mechanical control systems in the past are now highly integrated with digital networks, offering fresh opportunities for real-time monitoring, predictive maintenance, and operation control. However, this digital expansion brings with it a critical challenge: security. Whereas traditional IT systems can accept downtime or loss of data, industrial facilities cannot afford this because even small breaches will have catastrophic consequences like financial loss, safety vulnerabilities, and breaches of national security.

As industries come online with cloud services, add Industrial Internet of Things (IIoT) devices, and build smarter infrastructure, the line between information technology and operational technology blurs further. This convergence demands an equally severe focus on cybersecurity. Industrial control systems, which were once thought to be too specialized to target, are now considered one of the prime targets for cybercriminals, hackers, and even state-sponsored attackers. Malware attacks against industrial facilities, attacks on critical infrastructure, and disruptions caused by exploits against industrial devices show that there is a pressing need for strong cybersecurity solutions tailored to automation environments.

Cybersecurity Protocols in Industrial Automation meets this growing demand with a structured and comprehensive guide that blends technical depth with practical application. The book discusses significant concepts, security controls, and defensive mechanisms that empower industrial networks and introduce safeguarding against new cyber-threats. With a focus on both core knowledge and advanced practices, it supports students, researchers, practitioners, and policy-makers to build safe security systems in critical systems. Each chapter presents core issues, explains technologies, and connects theory and practice to prepare the reader for the unique challenges of safeguarding industrial environments.

Chapter 1 presents industrial automation and its cyber security needs. It outlines the role of automation within modern industries, the industrial control systems architecture, and why cyber security takes center stage in their operations. It recognizes the IT-OT disparity in security needs and presents the foundation pillars that guide protection activities in industrial settings.

Chapter 2 deals with network designs in industrial control systems. The chapter addresses the architecture of ICS networks, including SCADA systems, PLCs, and DCS. There is also focus on segregating networks, controlling communication channels, and realizing differences between conventional enterprise networks and industrial-oriented architectures. Readers understand how design affects both efficiency and security.

Chapter 3 provides information on common threats and vulnerabilities industrial systems commonly experience. The chapter offers details on attack vectors like malware, ransomware, denial-of-service attacks, and insider attacks. The chapter further addresses vulnerabilities in protocols, obsolete devices, and unpatched systems. Incident case studies of real-world incidents are used to illustrate how threats happen and the threat they pose to critical operations.

Chapter 4 focuses on authentication and access control mechanisms. It explains how to ensure that only authorized users and devices access industrial systems. From role-based access control to multi-factor authentication, the chapter lists measures to enhance identity management within automation environments. Practical scenarios such as controlling contractor or remote employee access are addressed.

Chapter 5 covers safe industrial network communication protocols. As industrial environments require secure and trustworthy transfer of data, the chapter covers encryption methods, tunneling protocols, and safe versions of industrial communication protocols. It emphasizes that low latency should be ensured while still ensuring confidentiality, integrity, and availability of data across industrial networks.

Chapter 6 covers intrusion detection and prevention in ICS. The chapter explains an overview of industrial traffic monitoring tools and industrial anomaly detection techniques. Readers are instructed how to distinguish between normal process activity and malicious traffic, and how to implement intrusion prevention systems that do not cause disruption while preventing threats.

Chapter 7 addresses risk assessment and security policies. The chapter illustrates how industrial systems can be identified, prioritized, and mitigated in risk by organizations. It introduces risk management frameworks, policy development criteria, and the necessity of harmonizing security policies with technical requirements and regulatory compliance. Hands-on guidelines for conducting risk assessments are given.

Chapter 8 covers endpoint and device security. Industrial sites usually have numerous devices such as PLCs, sensors, and actuators. In this chapter, endpoint hardening, firmware update deployment, physical access management, and device configuration security are emphasized. Legacy device-specific vulnerabilities are also discussed along with compensating control measures.

Chapter 9 explores industrial IoT and cloud integration's security. As industries rely increasingly on interconnected devices and the cloud, this chapter examines the advantages and challenges of IIoT deployment. This chapter depicts the challenge of securing device-to-device communication, secure storage on the cloud, and data privacy in highly connected industrial systems.

Chapter 10 concludes with incident response, forensics, and future trends. It focuses on strategies for handling industrial security breaches through detection, containment, eradication, and recovery. Digital forensics for ICS is discussed as an incident investigation tool and better defenses. The chapter also looks to the future challenges and emerging technologies that will impact industrial cybersecurity.

This book is meant for a general audience interested in defending industrial automation environments. Undergraduate students pursuing cybersecurity, information technology, or industrial engineering will find it a simple-to-read introduction plus in-depth explanations of protocols and security models. Control engineers, IT security specialists, and plant managers in industry can use it as a reference to hire and sustain strong security measures in their firms. Research experts will value its formatted presentation of problems and solutions, while policymakers and regulators will be able to take ideas from it to inform successful standards and compliance mandates.

Bringing together the world of cybersecurity technicalities and industrial operation's needs, *Cybersecurity Protocols in Industrial Automation* provides readers with the knowledge and hands-on tools to push defenses forward, reduce vulnerabilities, and build resilience in one of the most important sectors of infrastructure today.

CHAPTER 1

INTRODUCTION TO INDUSTRIAL AUTOMATION AND CYBERSECURITY

Contents

INTRODUCTION

Industrial automation encompasses a wide range of technologies, systems, and processes that maximize manufacturing efficiency through equipment and process automation. Through applications such as robotics, sensors, and software, business industries are able to achieve productivity and precision levels that were unthinkable before, reducing human error as well as operation costs. Such a shift to automation does not only streamline processes but allows organizations to allocate resources more effectively. As industries increasingly rely on automated processes, the need for highly evolved cybersecurity solutions takes precedence. With intelligent systems being incorporated into manufacturing settings, data integrity and cybersecurity protection concerns rise, particularly in industries where operational technology is pivotal. As this transition occurs, awareness of the critical infrastructures associated with industrial automation ends up leading to protective solutions receiving more attention. Institutions like the European Reference Network for Critical Infrastructure Protection (ERNCIP) play a significant part in ensuring that there are guidelines and frameworks to safeguard industrial systems from potential cyberattacks. The importance of carrying out such collaborative efforts cannot be overstated since they facilitate the sharing of knowledge among stakeholders in the effort to avoid weaknesses in automated surroundings more effectively. By focusing on technical protective security solutions, ERNCIP underscores the relevance of a robust cybersecurity strategy to supplement industrial automation development. A security strategy that covers the entire asset lifecycle not only safeguards assets but promotes system-wide resilience against possible threats, leading to enhanced operational continuity. In industries with high-sophistication applications such as oil and gas, the role of cybersecurity in automated systems is particularly critical. The deployment of operational technologies is bringing unprecedented efficiencies but also brings advanced cybersecurity concerns that require thorough risk analysis. For example, a model based on Adversarial Risk Analysis can be applied for systematic evaluation of cyber threats in offshore drilling operations. This process offers a systematic framework for addressing potential vulnerabilities in industrial automation, which makes it more suitable for practical applications. By implementing such rigorous cybersecurity tests, organizations can make informed decisions weighing operational effectiveness against the demands of security and hence foster a culture of resilience. The interface between industrial automation and cybersecurity effectively represents the crux of modern manufacturing, and thus it calls for ongoing research and cooperation to confront emerging threats and strengthen defenses against cyber-attacks.

Definition and Scope of Industrial Automation

Industrial automation refers to a broad span of technology and methods designed to enhance the processes of manufacturing by reducing the touch of people and enhancing operation efficiency. Essentially, it combines sensors, actuators, and control systems to automate operations entailing material handling, production processes, and quality inspection. With this combination, real-time data collection and analysis are achievable, and businesses can make decisions based on data, which optimizes productivity and reduces expenses. Industrial automation is not only limited to the traditional manufacturing industries; it also encompasses industries such as pharmaceuticals, food and beverage, and automobiles. With an increasing number of industries adopting automated technology, it is increasingly necessary for the overall industry to comprehend operation requirements as well as safety protocols, particularly against the background of cyber threats that accompany such technologies. The rapid advancement of technologies in the fourth industrial revolution, termed Industry 4.0, has significantly expanded the potential and challenges associated with industrial automation.

Figure 1.1. Industrial automation.

Source: https://www.surecontrols.com/wp-content/uploads/2013/01/Industrial-Automation-1.jpg.

The age is characterized by the merging of information technology and operational technology to support intelligent factories based on data analysis, machine learning, and the Internet of Things (IoT). Institutions such as the Technical University of Catalonia have found the importance of building pertinence skills through programs such as the Industry 4.0 Technologies Laboratory (I4Tech Lab) (Prieto et al., 2019). These organizations have a tremendous responsibility to equip the workforce with appropriate skills to adapt to the complexities arising from these emerging technologies to facilitate graduates to work in tandem with automated surroundings and handle cybersecurity challenges inherent to these technologies. Moreover, the convergence of operational technology (OT) and information technology (IT) in industrial automation has introduced new vulnerabilities that require adequate countermeasures. When firms implement automation by elevating connectivity between enterprise systems and plant floors, the ICS becomes much more complicated. As complexity rises, it gets challenging to have a clear understanding of end-to-end data flow and operational requirements. When the operators lack this awareness, it poses a higher risk of security vulnerabilities and system crashes, necessitating the setup of robust cybersecurity paradigms. By adhering to set standards and guidelines, organizations are better placed to analyze risks associated with automated environments, hence enhancing resilience to potential threats as automation continues to advance further (Green et al., 2016) (Figure 1.1).

Historical Development and Evolution

While attempting to follow the historical development and evolution of industrial automation, it is worth observing the revolutionary impact of technological innovations over the years. The transformation from mechanization to automation began during the late 18th century with the Industrial Revolution, which laid the groundwork for modern automation. Innovations such as the steam engine and power looms revolutionized the production capabilities so that it became possible to manufacture commodities on an unprecedented scale. As industrial sectors evolved, the application of computers and programmable logic controllers (PLCs) in the latter half of the 20th century further enhanced automation processes. The machines added accuracy and efficiency, enabling industries to streamline processes and reduce human error, thereby ushering in today's fully automated settings. The later 20th and early 21st centuries were a formative period in the trend towards integrated systems and smart technologies, assisted in very large part by progress in artificial intelligence. It was a period that catapulted big data analytics to the forefront as a key component in industrial process management, particularly in the case of energy management and smart grids. Big data analysis not only enhances operational efficiency but also supports decision-making by separating large sets of data to yield insights, thus overcoming the limitations of conventional approaches that struggled with numbers of variables and complex computation. This is critical in situating industrial automation within a broader context of technological convergence, encompassing IoT devices and cloud computing, to facilitate real-time processing and analysis of data across various industrial functions (Gharehpetian et al., 2019). In addition, the co-evolution of cybersecurity within industrial automation is crucial in meeting the challenges in the implementation of such technologies by industries currently. As automated systems increasingly rely on interconnected networks, cyber-attack vulnerabilities overwhelm them, rendering data integrity and operational resilience ever more uncertain. The prevalence of artificial intelligence both civilian and military further complicates, since AI can be thought of as an enabling technology that will prospectively redefine competitive advantage globally. The exchange between cybersecurity and automation emphasizes developing robust security frameworks to respond to potential adversities in this continuously evolving world. Evaluating the consequences of such technological developments is crucial for organizations that aim to hold a competitive edge while their operations are safe (Horowitz et al., 2018).

Key Technologies in Industrial Automation

The discipline of industrial automation is significantly enhanced through the application of sophisticated Building Automation Systems (BAS). Centralization of all building services within a single system through integrated communications and protocols facilitates automated and effective operation. With the integration of Internet Protocol (IP)-based communications, BAS can also be integrated onto corporate networks for facilitating remote access for monitoring and management. But at this security cost, this ease is bought. Primarily built without security being a priority, most BAS suffer from an ineffective or nonexistent protective arrangements, leaving them exposed to potential attacks. To seal these loopholes is crucial, as they might serve as the entry point for cyberattacks not only on the buildings but also the related corporate networks, indicating the significance of comprehensive cybersecurity measures in current industrial automation systems (Johnstone et al., 2014). In understanding the industrial automation environment, it is impossible not to acknowledge the key position of robotics and artificial intelligence (AI). Conditioned to repeat tasks with accuracy, robots enhance productivity and reduce the amount of errors made by humans in manufacturing activities. Furthermore, AI enhances the ability of robots to make decisions with the assessment of vast amounts of data in real-time, thereby leading to increased efficiency of operations. However, the integration of AI with robotics comes with challenges, particularly in terms of cybersecurity. With the increased automation of systems, they are also exposed to cyber-attacks that can disrupt operations or steal confidential data. Securing such high-tech systems is paramount because a breach would lead to substantial financial sanctions and system downtime. For this purpose, good attention to cybersecurity measures is required to ensure the stability and reliability of automated systems (Burov et al., 2020). The convergence of IoT devices in industrial automation systems is accompanied by threats and opportunities. IoT devices enable smart connectivity among a range of machines, sensors, and systems, enabling real-time data exchange that can facilitate production process simplification and predictive maintenance. Connectivity can lead to increased efficiency and reduced operations cost. It also carries with it the risk of being more vulnerable to cyberattacks. The majority of IoT devices lack robust security controls, and their ubiquitous deployment offers many points of attack to attackers attempting to exploit vulnerabilities within the network. Having robust cybersecurity controls is therefore critical in an effort to protect such systems against attacks that may result in major operational disruptions. In all, the deployment of safe practices together with IoT is essential in safeguarding industrial automation infrastructures against the evolving cyber threats, ensuring that the benefits of automation may be optimized (Johnstone et al., 2014) (Figure 1.2).

Figure 1.2. Key technologies in industrial automation.

Source: https://www.te.com/content/dam/te-com/images/corporate/marketing/global/story-hero-rendition/tablet-control-factory-automation-robots-1024-400.jpg/jcr:content/renditions/hero-story.jpg.

1.1. COMPONENTS OF INDUSTRIAL AUTOMATION SYSTEMS

Industrial automation systems are a complex set of components designed to enhance productivity and operation efficiency across different areas. Inserted into these systems are Programmable Logic Controllers (PLCs), the automation mind that carries out pre-programmed control commands on equipment and operations. PLCs are connected with sensors and actuators, which feed real-time data and provide actionable signals. The inclusion of Human-Machine Interfaces (HMIs) enables operators to observe, monitor, and manage processes in a graphical format, with quicker decision-making and fault-debouncing. With advancements in automation technologies, there is a growing reliance on digital systems and, with it, the necessity for an in-depth understanding of their working nature as well as susceptibilities, particularly of cybersecurity vulnerabilities related to integrated systems (Kabir-Querrec et al., 2016). A growing trend in industrial automation is the drift towards increased connectivity facilitated by the Industrial Internet of Things (IIoT). This development bridges the gap between otherwise standalone components and enterprise-level information systems so that they exchange data in real time.

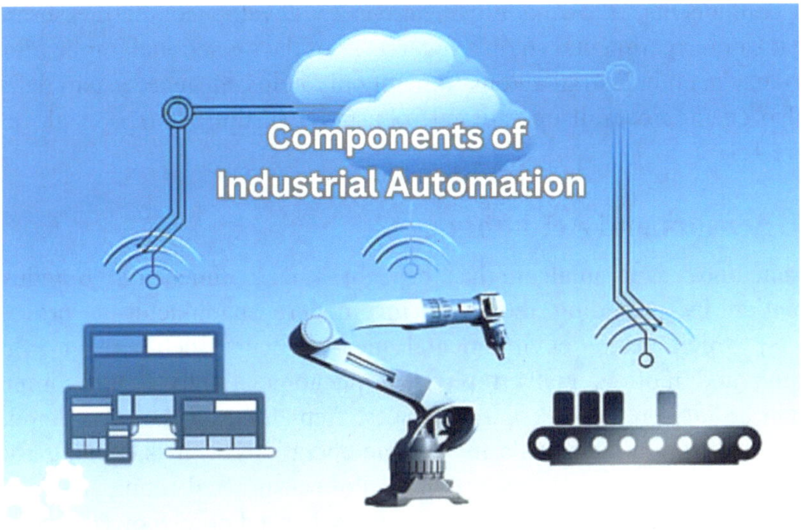

Figure 1.3. Industrial automation systems are a complex assembly of components designed to enhance productivity and operational efficiency across various sectors.

Source: https://i0.wp.com/www.itsoji.com/wp-content/uploads/2024/01/Components-of-industrial-automation.webp?fit=900%2C600&ssl=1.

But as automation components become more intertwined using protocols such as Modbus and Profibus, they unknowingly expose sensitive infrastructure to the possibility of cyber-attacks. The erosion of the "air gap" that once separated control systems is a gradual process, but it allows hackers to exploit such interconnections remotely. Thus, organizations must face the double challenge of maintaining efficiency of operations and safeguarding against the exposures caused by such networked arrangements. The transition towards smart devices and remote monitoring calls for a visionary cybersecurity strategy in order to constrain risk in the expanding industrial automation universe (Evangelopoulou et al., 2017). Besides PLCs and IIoT, another critical aspect is using intelligent electronic devices (IEDs) that enhance industrial system monitoring and control. These IEDs, with advanced sensors, become highly critical in automatic decision-making and data acquisition. Nevertheless, the way in which these IEDs are able to communicate within a network, more so in the core infrastructures of energy and manufacturing, is one of the highest priorities regarding cybersecurity. The recent events highlight the importance of studying the vulnerabilities of such devices. For instance, testing scenarios utilizing the employment of cyber threat simulation can identify weaknesses in IEDs and other components, allowing organizations to develop countermeasures. As evident from experimental studies, awareness of the operational implications of cyber-attacks on industrial automation not only helps enhance system defenses but also ensures overall operational robustness (Kabir-Querrec et al., 2016) (Figure 1.3).

1.1.1. Sensors and Actuators

The actuator-sensor amalgam has brought a new dimension to industrial automation by providing the basis for modern manufacturing processes. Sensors, which detect environmental measurements such as temperature, pressure, and motion, convert physical phenomena into electric impulses that can be interpreted by control devices. Actuators, on the other hand, are commanded from these signals and actuate operations such as opening a valve or swinging a robot arm. This combination of sensing and acting enables real-time control of manufacturing processes for enhanced efficiency and accuracy. Moreover, as they become increasingly linked in the IIoT framework, there is the possibility of sophisticated control and monitoring, enabling industries to optimize production processes and adopt predictive maintenance procedures. This, however, also necessitates robust cybersecurity systems to guard against emerging threats. The development of sensor and actuator has continued to advance with technological progress, increasing the complexity of the social and technical intersections of human-machine interfaces. With more IoT devices, human-machine relationships have been transformed into what can be referred to as Cyber-Physical Social Machines. These systems employ networked digital technologies to manage interaction and knowledge creation that continuously evolves in their operating environment. As debated in literature, merging continuous sensing and actuation functionality introduces new layers and intricacies of handling and interpreting industrial systems, particularly issues of security and privacy. This is an indication that not only must sensor and actuator networks be thought of as independent, one-off systems, but rather as part of large socio-technical systems that must be constantly watched over against security attacks that exploit their connectedness (Creese et al., 2018). Historically, industrial control systems that involved safety maintained some form of protection by being isolated, generally termed as being "air gapped." In the legacy model, sensors and actuators such as Programmable Logic Controllers

(PLCs) operated independent of larger networks and thereby minimized possibilities of external threats. However, the gradual erosion of such isolation has resulted in more connectivity through gateways and switches to facilitate the flow of important process information into enterprise information systems. While the move facilitates efficiency and streamlined decision-making within senior management, it equally facilitates access to security vulnerabilities. Notably, recent geopolitical events, such as cyberattacks on infrastructures, show the pressing need for rigorous cybersecurity protocols protecting such networked systems (Evangelopoulou et al., 2017). As industrial automation increasingly relies on data-dependent information supported by sensors and actuators, the stakes in securing these technologies increase to an urgent level, necessitating a reexamination of dominating security models (Figure 1.4).

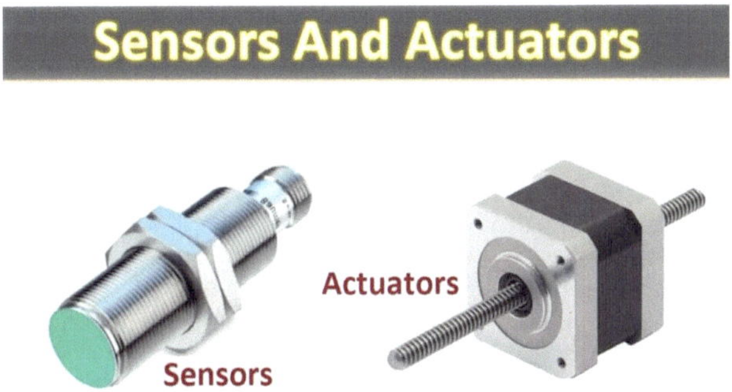

Figure 1.4. The integration of sensors and actuators has transformed industrial automation.

Source: https://cranesvarsity.com/wp-content/uploads/2023/07/csq1.jpg.

1.1.2. Programmable Logic Controllers (PLCs)

PLCs form the core of industrial automation as the basis of modern manufacturing and process control systems. PLCs provide real-time monitoring and automaton, which automatically enhances efficiency and reduces human involvement in production processes. PLC structure is designed to withstand factory environments, so they operate even where other computing systems won't. With industries becoming more digitized every day, integrating PLCs into new technologies is of prime significance; they can talk to sensors, actuators, and even higher control systems, all in one network. This adaptability positions PLCs not only as controllers but as central nodes within the Internet of Things (IoT), where they share critical information that can be used to guide decision-making and rationalize operating processes. The evolution of PLC technology is directly linked with advancements within cybersecurity measures, especially in consideration of their role within Industrial Control Systems (ICSs). As noted by present studies, the threat landscape for such systems is rapidly growing in sophistication and intelligence, making them good targets for cyberattacks. With growing interconnectivity, security vulnerabilities in PLCs have implications that spread beyond localized system failure, with the risk of having an impact on larger infrastructures critical to national security, including energy and water supply infrastructure (Al-Khateeb et al., 2019). This interconnectedness calls for not only robust security measures to shield against intrusions, but also innovative approaches to maintain the integrity and functionality of control systems.

Figure 1.5. Programmable logic controllers (PLCs).

Source: https://www.toshiba.com/tic/product_images/_ASR5089.jpg.

Therefore, integrating security protocols in every step of PLC operation is essential to forestall persistent threats while ensuring reliability in automated processes. Moreover, education is also placing greater emphasis on the importance of PLC training within the Industry 4.0 era, where information technology and operational technology intersect. Academic institutions are discovering their pivotal role in preparing engineers with the capability to grasp the complexity of this new industrial landscape. Initiatives such as the Industry 4.0 Technologies Laboratory at the Technical University of Catalonia demonstrate how forward-thinking engineering education can be facilitated through hands-on experience with PLCs and related technologies (Prieto et al., 2019). These initiatives aim at equipping students with competencies to address both the technical and the cybersecurity challenges encountered in modern automation environments. This emphasis on experiential learning in PLCs produces a new generation of engineers capable of designing and protecting smart manufacturing environments, and this explains the need for a multidimensional education system that responds to evolving industrial demands (Figure 1.5).

1.1.3. Human-Machine Interfaces (HMIs)

In industrial automation, HMIs are significant interfaces offering a mechanism of interaction between operators and systems. These interfaces translate industrial control system data into comprehensible visualizations to facilitate simple monitoring and control of operations. As the times change and industries are improving with advanced technologies, HMI functionality and design come into the limelight. Proper HMI design improves operations and makes them less susceptible to human error, which can be catastrophic in sensitive environments. The emergence of automation calls for a people-oriented strategy centered on clarity, reactivity, and usability, literally setting the tone for how operators engage with technology. The role of HMIs extends beyond utility; they are crucial in controlling cybersecurity threats in industrial environments. As systems become increasingly integrated, the danger of cyberattack is amplified. HMIs must incorporate robust security measures for protecting both operation integrity and sensitive data. The vulnerability in the HMI system can serve as attack sites for malicious actions, and hence it is crucial to implement best practices in cybersecurity. New frameworks that combine traditional security methods with new-era methods such as orthogonality will enhance the overall security position. By utilizing multiple layers of assurance controls at each layer of the system architecture, organizations can better resist adaptive threats to critical national infrastructure, underpinning the necessity for the development of HMIs that consider security as highly as usability (Al-Khateeb et al., 2019). HMIs play significant roles in the successful deployment of automated systems depending on user acceptance. As emerging technologies such as connected and autonomous vehicles become more mainstream, it is paramount to engage potential users in the development process. Users' demands and preferences must be known in order to develop trust and enhance perceived usability of HMIs. Stakeholders will need to re-gauge established beliefs about traditional system interaction and make these new interfaces intuitive and simple to use. Research such as those outlined in the Drive2TheFuture initiative emphasize the need to solve the issues of users towards maximum adoption. With expert advice that centers on affective, persuasive, and trustworthy HMI functionalities, industries are able to obtain seamless implementation of automatic technologies, hence realizing the utmost desired value (Diederichs et al., 2021) (Figure 1.6).

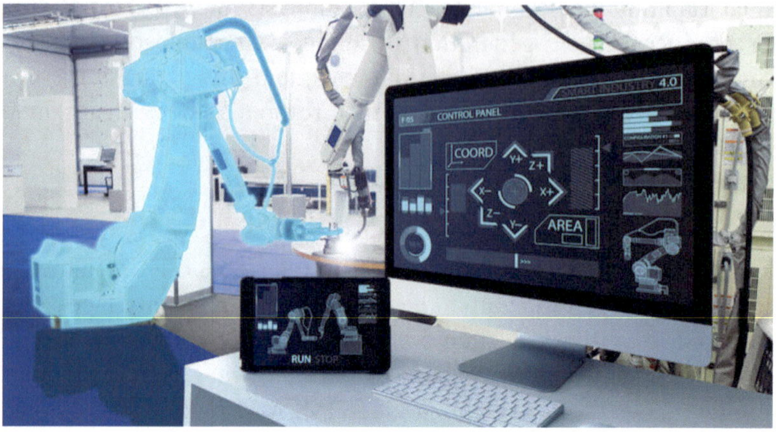

Figure 1.6. Human-machine interfaces (HMIs).

Source: https://www.automationreadypanels.com/wp-content/uploads/2024/01/Human-Machine-Interface-HMI-750x400-1.jpeg.

1.2. BENEFITS OF INDUSTRIAL AUTOMATION

Industrial automation, with its remarkable rate of development, poses an enormous level of benefits which significantly foster operational efficiency, productivity, and competitiveness in various industries. Core to these advantages is the remarkable reduction of production costs, which can be achieved by the reduction of labor costs and minimizing waste. Automation is made possible by the fact that machines can be operated 24/7, thereby producing maximum output without overtime pay, hence rapid return on investment. Moreover, the reliability and precision that automated systems guarantee lead to improved product quality, therefore reducing rework and maximizing customer satisfaction. As increasing numbers of businesses adopt automation technologies, the search for easier processes is at the forefront, and businesses that do not take advantage of these technologies fall behind their industry peers. Industrial automation aside from economic benefits promotes a culture that fosters growth and innovation in technology. A highly qualified staff is needed to operate, maintain, and improve complex systems for the shift to automated processes. This demand drives investment in training and education, thereby elevating the overall level of skills among employees. Furthermore, automation has the potential to liberate employees from routine and harmful physical labor, allowing them to focus on higher-value processes that demand analytical capacity and imagination. This transformation can lead to a culture of innovation within the company, which not only enhances job satisfaction but also positions the company at the forefront of innovation. These are aligned with broad industrial policies for building competitiveness and innovation, including the kind of measures presented in broader international industrial strategy debates such as those represented in Bartsch et al. (2020). In addition, integrating cybersecurity into industrial automation systems becomes increasingly important as increasingly connected devices proliferate throughout manufacturing processes. With the implementation of Industry 4.0, the benefits of enhanced automation must be paired with robust cybersecurity to protect sensitive data and infrastructure. Robust cybersecurity can be a pillar of industrial automation, having the automated systems operate smoothly without any interference and remaining immune to potential cyber-attacks. This balance between industrial automation and cybersecurity encompasses formulating best practices for information asset protection, as outlined in various studies and reports on cyber defense infrastructure (Anglano et al., 2018). With companies operating in the complex environment of automated transformation, it becomes essential to onboard cybersecurity not only for security on investment but also for long-term operational sustainability (Figure 1.7).

Figure 1.7. The rapid evolution of industrial automation.

Source: https://d27athks9omvvg.cloudfront.net/uploads/pageImages/automation-analytics.jpeg.

1.2.1. Increased Efficiency and Productivity

The onset of industrial automation means that there comes with it a new era of increased efficiency and productivity in industries. With the application of emerging technologies such as artificial intelligence, robotics, and Internet of Things (IoT), industries can make their operational processes much more streamlined. These technologies facilitate real-time monitoring and analyzing of data, which allows organizations to reduce waste and streamline workflows. For instance, automation creates time for the monotonous work such that human labor is free to focus on more strategic and creative projects. Furthermore, this approach not only enhances the rate of productivity but also meets with the virtue of sustainability, as process optimization often involves less resource consumption and emission. The concept of Industry 4.0 is key in attaining enhanced efficiency and productivity by combining information technology and operational technology. Digital revolution supports the adoption of smart factories in which systems are interconnected and capable of independent decision-making. It establishes structural stability within vertical and horizontal processes of manufacturing, which is essential for maximizing output. According to existing research, companies adopting Industry 4.0 witness dramatic improvement in their digital management skills, resulting in streamlined operations and less production downtime. The positive effects of the phenomenon extend beyond immediate productivity gains, positioning companies to respond to changes in markets and consumer demands more agilely (Bobák et al., 2017). Companies are increasingly recognizing their central role in building the skills necessary to ride out this industrial revolution. Especially, university courses like the Industry 4.0 Technologies Laboratory, Technical University of Catalonia I4Tech Lab promote active learning activities that combine theoretical understanding and practical application. The laboratories provide room for a space that is conducive to exploring and developing competencies in digital and automation technologies. These education frameworks not only equip learners with the skills needed in their future workplaces but also inspire innovation that can contribute to the higher productivity of industries. Through focus on a balanced education in engineering on the basis of active learning methodologies, these programs aim to prepare the future engineers to respond to the challenges of increasing automation, ultimately fostering ongoing improvement in the industry (Prieto et al., 2019).

1.2.2. Enhanced Quality Control

Enhanced quality control in industrial automation has been identified as a critical ingredient in efficiency in operations and product integrity. With the help of advanced monitoring systems, manufacturers can identify faults in real time, thus improving the overall quality of the output. Application of artificial intelligence and machine learning methods in quality control allows for the possibility that vast data gathered during production can be examined. Application of digitalization allows for predictive maintenance and anomaly detection to be applied in maintaining standards when producing. Above all, vulnerabilities in traditional monitoring systems have heightened the demand for robust cybersecurity systems because cyberattacks can ruin such critical quality control systems and lead to large-scale operations failures (Ginter et al., 2018). Consequently, an interdisciplinary approach that consolidates quality control with cybersecurity methods forms the core of making production lines both efficient and secure. Besides, big data analytics can potentially revolutionize quality control in industry. Since they have the capability to handle and analyze huge quantities of data, businesses can make more efficient conclusions regarding their production activities.

Figure 1.8. The enhanced quality control.

Source: https://www.compliancequest.com/wp-content/uploads/2023/11/automated-quality-control-inspection.jpg.

This approach is beyond traditional quality control techniques, ensuring more efficient understanding of product performance and operational efficiencies. With technological advancements in IoT devices, manufacturers can collect real-time data from production lines, which in turn allows them to respond to inefficiencies and faults in advance. Additionally, big data analytics enhances the quality assurance processes to ensure they meet industry standards as well as remain responsive to the dynamic demands of the market (Gharehpetian et al., 2019). As more and more businesses rely on automated tools, scalability of process for quality control will determine their competitive edge in the digital marketplace. Finally, convergence of high-quality control in an automated system should also address the dangers of cybersecurity. When production processes are integrated, there is potential for cyber-attacks, not only on product quality but also consumer safety. Cyberattacks against quality control systems can have catastrophic implications, including the production of unsafe products, which in turn erodes brand and manufacturer trust. Risk management principles must, therefore, include robust cybersecurity practices that protect quality control systems from external attacks while enabling business continuity. Future automation innovation must be complementary to cyber resilience, where quality control processes are not only efficient but also secure (Ginter et al., 2018). Ultimately, an integrated approach of enhanced quality control and cybersecurity will drive innovation and ensure industrial innovation (Figure 1.8).

1.2.3. Cost Reduction and Resource Management

The integration of industrial automation into practice has been slowly but surely viewed as a prime strategy for cost cutting and maximum utilization of resources. Automated technology reduces the cost of operations by increasing efficiency and minimizing the cost of staff. One such instance of this is when companies use automation technologies such as Supervisory Control and Data Acquisition (SCADA) systems, then they are able to streamline the production process and optimize energy consumption. This technology, besides reducing downtime, also enables predictive maintenance, thus increasing equipment life. By switching to automated systems, companies can save big on costs, concurrently reallocating resources for purposes of innovation and business development initiatives. The balance among these innovations assures that institutions are well positioned to fight the challenges of the dynamic market environment. Cybersecurity becomes an integral element of cost and resource management in industrial environments. As firms embrace automation technology, they mostly introduce complex network systems that require robust cybersecurity to protect assets that are essential to operations from cyber threats. A system without protection is open to a series of cyberattacks, demonstrated by instances that jeopardize operational integrity and lead to costly downtime. The first-ever cyberattack on a power grid in Ukraine is an excellent instance of the devastating effects when cyber protection is weak, and it highlights the necessity for an active approach in addressing such threats. Organizations should invest in comprehensive cybersecurity that is not only defending against attacks but also enhances automated systems' stability so that efficiency can be maintained uninterrupted and the cost of cyberattacks is minimized (Ginter et al., 2018). Also, a new resource management style is exemplified in designing fresh architectures for rural distribution networks, which aid in updating traditional networks into Smart Grids. Through this innovation, there is a better utilization of resources through harnessing synergies between telecommunications and advanced power electronics. Remote and autonomous management empowerment of rural distribution networks optimizes efficiency in integrating distributed energy resources. Through the deployment of this architecture, especially in pilot projects across Europe, the optimization of resource management is achieved, allowing the operators to take advantage of new technologies that can bring massive cost savings (Díaz González et al., 2018). In addition to simply copying current sustainability principles, this transformation also equips organizations to prosper in a progressively competitive marketplace where efficient resource management is paramount (Figure 1.9).

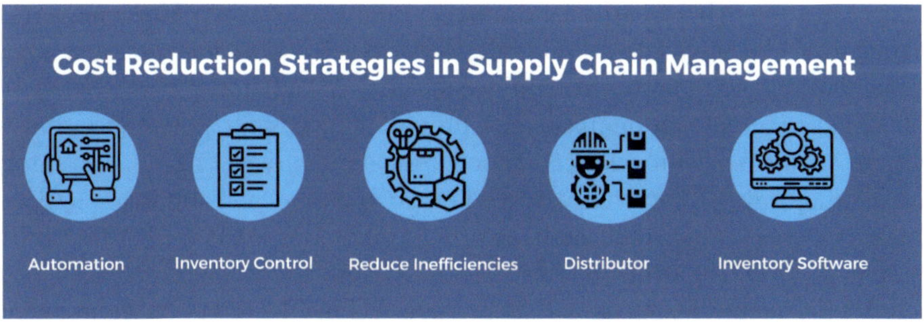

Figure 1.9. Cost reduction and resource management.

Source: https://lh4.googleusercontent.com/6L9aL3JYM9kGa33s0DAFyKKUEn0TpuY-W_QsO0HBWdZnJvNiNciVUdiqSQ_FVKp09IZCCwCBQkJdX13i-HD-4vNLVwLGHhjJWj4ayPcm5wUhZSTF6Kus2PZDgACLCNujXV06uH-Sum9xrigUuyVLjprSU9e0F76-4CwiEyCO81_0DZxHaAVqanUONeUkvizA.

1.3. INTRODUCTION TO CYBERSECURITY IN INDUSTRIAL SETTINGS

As industrial environments increasingly adopt automation technologies, the need for cybersecurity plays a prominent role. Previously, industrial control systems (ICS) operated in isolated systems, but with smart manufacturing and the Internet of Things (IoT), they are now exposed to a wide variety of cyber-attacks. The confluence of operational technology (OT) and IT is a two-edged sword that introduces higher efficiency but opens the attack surface for attackers. Effective cybersecurity solutions in such a setup are not merely significant to protect important data, but also to ensure critical infrastructure availability and integrity. It means that there must be an appropriate understanding of cybersecurity protocols with emphasis on industrial application. As companies are finding it difficult to deal with the complexity of new technologies, including artificial intelligence (AI) in cybersecurity products is offering opportunity, as well as risk. AI has the potential to introduce a paradigm shift in vulnerability management and threat identification in industrial environments. For instance, machine learning algorithms can analyze huge volumes of data in real time, highlighting unusual patterns that may be a sign of a cyber-attack (Diep et al., 2020). However, this technology has a drawback. Enemies may utilize AI to enhance their own offensive plans, creating a never-ending arms race in the cyber world of cybersecurity. Hence, industrial operators must remain vigilant and continue investing in robust AI-based defense mechanisms in order to stay ahead of emerging threats. The creation of Cyber-Physical Social Machines complicates the cybersecurity landscape in industries further. These networks, characterized by their intricate interdependencies between humans, machines, and networked technology, pose unique security risks (Creese et al., 2018). As IoT devices increase in number, they provide avenues for potential intrusion that can have far-reaching effects on operational effectiveness as well as safety. The interdependent character between human operators and automatic systems demands increased consciousness regarding privacy issues as well as the ability for human errors in cyber defense. Thus, organizations must establish end-to-end cybersecurity measures that involve technical protection components and human factors in order to counter threats of such socio-technical systems effectively (Figure 1.10).

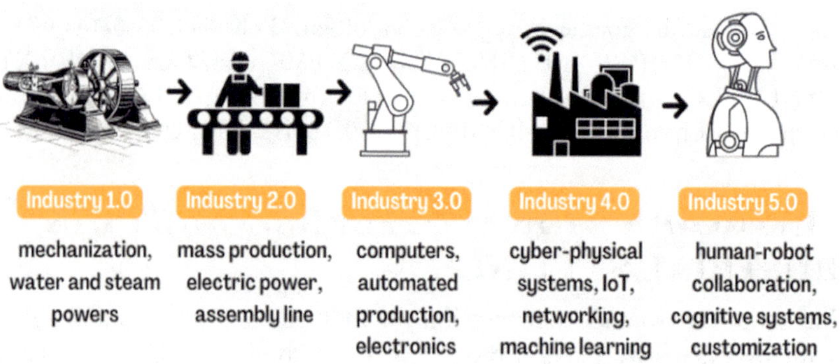

Figure 1.10. Introduction to cybersecurity in industrial settings.

Source: https://pub.mdpi-res.com/sensors/sensors-23-00120/article_deploy/html/images/sensors-23-00120-g001.png?1671779752.

1.3.1. Definition and Importance of Cybersecurity

In the rapidly evolving industrial automation setting, cybersecurity is an important area of study dedicated to protecting systems, networks, and information from cyber-attacks. It addresses a range of practices to preserve the confidentiality, integrity, and availability of information. As the number of devices connected explodes and digital technologies are integrated into industrial environments, never has the need for robust cybersecurity practices been more acute. Against the backdrop of Industry 4.0, which encourages smart factories through the Internet of Things (IoT) and artificial intelligence, cybersecurity protects these crucial technologies from malicious attacks that can be employed by threatening forces. Efficient cybersecurity hence becomes an elementary pillar that not only guards assets but also provides trustworthiness and reliability in industrial operations, leading to better operational efficiencies. The role played by cybersecurity in industrial automation cannot be overemphasized, given the rising cyberattacks on critical infrastructure. The attacks pose real threats that could impede or cripple manufacturing processes, causing financial losses and reputational consequences. Moreover, as organizations keep striving to implement advanced technologies, they unintentionally expand their attack surfaces. Advanced cybersecurity solutions are therefore essential to adequately counteract these threats. Some of such initiatives involve preventive measures like risk analysis, employee training, and the installation of advanced security technologies. With the inclusion of an integrated approach to cybersecurity, organizations are able to ensure the stability of their automated systems and maintain uninterrupted operations even in the wake of potential threats. It is this stability that plays a central role in facilitating competitive advantages in an increasingly digital and interconnected industrial climate. Schools also play a significant role in the development and shaping of the discipline of cybersecurity. Integrating cybersecurity education into engineering curricula is essential in equipping future professionals with the appropriate competencies to address the issues of industrial automation challenges. Industry 4.0 programs concentrate on both the technologies and on competencies that cover the demands of addressing the threat which currently is and will be present. Efforts such as the Industry 4.0 Technologies Laboratory at the Technical University of Catalonia are evidence of the efforts being made in building a qualified workforce that can address the complexities of industrial cybersecurity (Prieto et al., 2019). Secondly, large-scale training and awareness initiatives aimed at both the management and technical staff are necessary to inculcate a cybersecurity culture among organisations, thus contributing to the resilience of industrial automation systems (Anglano et al., 2018) (Figure 1.11).

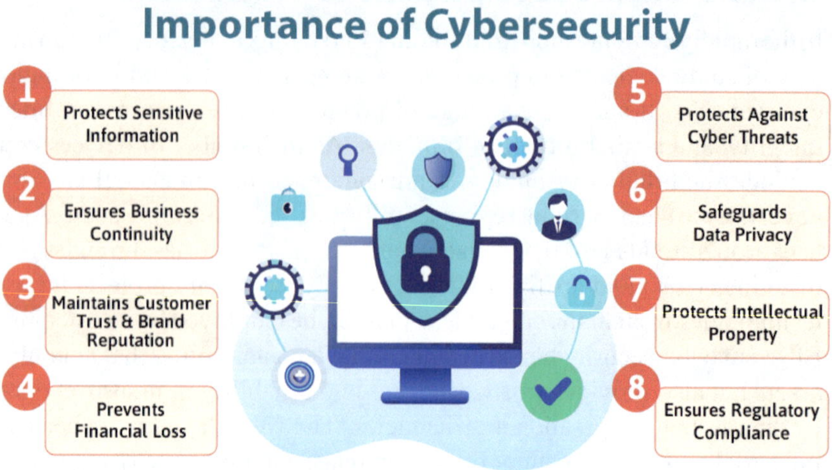

Figure 1.11. Importance of cybersecurity.

Source: https://cdn.educba.com/academy/wp-content/uploads/2024/10/Importance-of-Cybersecurity.png.

1.3.2. Differences between Information Technology (IT) and Operational Technology (OT) Security

The distinction between OT security and IT security is foundational when it comes to industrial automation and cyber security. Whereas IT security is primarily focused on the safety of data networks, devices, and information systems, OT security is focused on the safety of systems that control physical processes in manufacturing and utilities industries, among others. The general purpose of IT security is to ensure confidentiality, integrity, and availability of information, typically by using control mechanisms such as firewalls, encryption, and access controls. OT security, by contrast, aims to protect systems controlling mission-critical infrastructure operations with a focus on system availability and safety, rather than data confidentiality.

This divergence of focus is indicative of the different operational settings and priorities present within IT and OT environments, necessitating tailored strategies and methodologies to most effectively deal with the unique challenges of each field. A second deep distinction lies in the risks and vulnerabilities present within IT and OT environments. IT environments are most typically at risk from malware, phishing, and hacking, which can cause disruption to organizational processes and data breaches. On the other hand, OT systems reveal security risks as a result of physical danger and hardware breakdown, which may result in safety incidents and extensive operating downtime. The widespread application of legacy systems in most OT environments also complicate security processes because they lack robust security features common in modern IT systems. This convergence creates a multi-faceted landscape under which security must not just be robust but also synchronized with the accessible operational capabilities, so there is the requirement for balanced appreciation of both IT and OT environments in an effort to develop efficient cyber security measures (Allodi et al., 2018).

Coupled with this is the divergent world perception of players that function within IT and OT environments. IT professionals would focus more on adherence to cybersecurity frameworks and compliance regulations, while OT personnel may emphasize more on operational continuity as well as safety procedures. Coordinated efforts from both domains are required in order to create an end-to-end security strategy that has the ability to include each domain's specific emphasis. Also, with increased adoption of converged technology by organizations, it is important that IT and OT security be aligned. This convergence needs training and awareness initiatives closing knowledge gaps such that all the parties in both domains gain a detailed comprehension of security vulnerabilities and countermeasures. During this process, creating a common language and global goals for IT and OT professionals is essential to supplement overall cybersecurity resilience in industrial environments (Anglano et al., 2018) (Figure 1.12).

Figure 1.12. The distinction between Information Technology (IT) security and Operational Technology (OT) security.

Source: https://hoanlk.com/wp-content/uploads/2023/04/itvsot.png**.**

1.3.3. Common Cybersecurity Threats in Industrial Automation

In the industrial automation industry, cybersecurity threats have risen sharply, with attackers exploiting the overlap between operational technology (OT) and information technology (IT). One of the biggest threats involves ransomware attacks, which can put manufacturing processes on hold by encrypting confidential data and demanding money for its release. These attacks not only induce significant operational downtime but have the capability of leaking precious trade secrets and intellectual property. With increasing industries interconnected through the Internet of Things (IoT), the potential attack bears greater capacity, in that attackers can target various points of entry through an organization's infrastructure.

This complexity generates the need for strong cybersecurity controls explicitly in place to fight unique vulnerabilities present in industrial environments. Another common industrial automation cybersecurity attack is the susceptibility of human operators to social engineering attacks. The attackers use people to beguile them into revealing confidential information or unknowingly permitting entry to key systems. As most operators are typically asked to deal with evolving technologies, their ignorance of possible phishing activities or other fraud could be perilous. The development of cognitive processes, as seen in the context of the experiences in hybrid warfare, illustrates how vulnerable segments—especially the youth talent joining the workforce—can become victims of sophisticated social engineering maneuvers. Equipping personnel with knowledge of such threats and companies having continuous training programs in place is imperative for building the human side of cybersecurity. CPS integration also advances cybersecurity challenges in industrial sectors. They are built on networked devices and sensors that gather and share data continuously, offering a rich source of entry points for attackers. This kind of complexity introduced by these networks requires a deep understanding of technical and socio-technical systems, which is characteristic of Cyber-Physical Social Machines.

Different exemplars of CPS reveal the intricate human-machine relationship wherein security controls must be responsive to avoid dangers brought about by external and internal threats. Without vigilant observation and comprehensive security controls, industries do not only lose more than operational interruptions but also risk catastrophic failures that can be the consequence of their vulnerable automated processes, thus making their security profile more difficult (Burov et al., 2020; Creese et al., 2018).

1.4. CYBERSECURITY FRAMEWORKS AND STANDARDS

In the context of industrial automation, the use of cybersecurity frameworks and standards is essential for the safeguarding of organizational assets and the maintenance of operational continuity. They provide frameworks with systematic guidelines that enable organizations to delineate risks, implement controls, and maintain compliance. For example, the National Institute of Standards and Technology (NIST) Cybersecurity Framework is now widely accepted across most industries as having a comprehensive approach to the improvement of critical infrastructures. Through offering a common language for cybersecurity risk management, it enables organizations to perform self-assessment of their cybersecurity stance, which is increasingly relevant with the introduction of IoT devices more susceptible to cyber-attacks. This is emphasized in the context of IoT systems, where a substantial discrepancy exists in self-assessment methodologies to approximate cyber risk stance, emphasizing the need for robust frameworks in efficiently fighting emerging vulnerabilities (Ani et al., 2020). Furthermore, it is necessary to take into account particular cybersecurity standards specifically developed for industrial environments, like the International Society of Automation (ISA) 62443 series. These standards present best practices for meeting cybersecurity challenges particular to industrial control systems (ICS) and operational technology (OT). The ISA 62443 framework fosters a risk-based approach with a central focus on the interaction between technological and human factors. This action assists organizations in identifying key vulnerabilities, as seen in the case of the Ukraine power grid cyberattack, which caused tremendous devastation through compromised SCADA systems. A proactive cybersecurity framework based on reliable standards can be utilized to combat threats revealed by such an incident, thus underscoring the importance of constant assessment and enhancement of cybersecurity (Ginter et al., 2018). Moreover, adopting a holistic cybersecurity approach allows organizations not only to defend themselves against current threats but also to prepare against the next set of challenges. With the evolving nature of the cybersecurity landscape, frameworks must be dynamic and incorporate experience gained from previous incidents and newly advent technologies. By focusing on an objective-dependent model of IoT risk assessment, organizations are better able to quantitatively assess and bolster their cybersecurity stance. This adaptability is vital because it enables organizations to manage successfully the escalating levels of cyber threats without compromising the operational integrity of their industrial automation systems. Continuous evaluation, as per guidelines determined, ultimately promotes a culture of being prepared for cybersecurity, which is basic in addressing risks resulting from advanced persistent threats and formulating general resilience within industrial environments (Ani et al., 2020) (Figure 1.13).

Figure 1.13. The implementation of cybersecurity frameworks and standards.

Source: https://www.birlasoft.com/sites/default/files/html_files_migration/images/common/cyber-security-frameworks.jpg.

1.4.1. NIST Cybersecurity Framework (CSF)

The NIST CSF provides an end-to-end solution to organizations, particularly industries, to attain and minimize cybersecurity threats. As operational technology (OT) and information technology (IT) have merged in industrial automation, the framework enables organizations to evaluate their cybersecurity posture using a systematic mechanism. It consists of five essential functions: Identify, Protect, Detect, Respond, and Recover. Each function addresses key areas of cybersecurity so that organizations can identify weaknesses and implement adequate protections. Through constructing these essential elements, the CSF enables industries to align cybersecurity activities with business objectives, thereby establishing overall cyber resilience. In an industrial automation environment, the integration of Building Automation Systems (BAS) is exemplary of the relevance of the NIST CSF. While BAS systems increasingly rely on networked devices using IP-based communication, their inherent weaknesses are the focus of cybersecurity activity. Interestingly, the majority of BAS were designed with security not being a key concern, leading to security mechanisms based on obscurity rather than sound defense (Johnstone et al., 2014).

Figure 1.14. The NIST cybersecurity framework (CSF).

Source: https://sternumiot.com/wp-content/uploads/2024/02/NIST.png.

The NIST CSF serves as a guideline in optimizing security in such systems, enabling organizations to identify vulnerabilities in their systems, strengthen their protective controls, and put entire monitoring in place to be able to identify potential threats effectively. This systematic approach is instrumental in safeguarding critical infrastructure against emerging cyber threats. Furthermore, as the Internet of Things (IoT) continues to gain momentum in industrial settings, NIST CSF highlights that organizations need to continuously re-examine their cybersecurity stance. The convergence of IoT devices introduces more complexities and risks, necessitating proactive risk management strategies (Ani et al., 2020). Currently, most organizations lack sound self-assessing techniques for quantifying their IoT cyber risk stance, thereby reacting rather than acting anticipatorily towards threats. NIST CSF enables organizations to identify a quantitative assessment framework uniquely tailored to their security requirements. By using the CSF in combination with breakthrough approaches, industries can develop targeted solutions that not only address current cybersecurity challenges but also anticipate future threats, thereby ensuring long-term operational resilience (Figure 1.14).

1.4.2. ISA/IEC 62443 Standards

The ISA/IEC 62443 standards are a comprehensive framework designed to address cybersecurity challenges unique to Industrial Automation and Control Systems (IACS). These standards place great importance on adopting a holistic strategy that balances both safety and security factors along the lifecycle of industrial processes. With industries becoming increasingly dependent on interlinked systems, the need for strong cybersecurity frameworks becomes increasingly acute. The 62443 standards offer an established methodology to determine the possible threats, so that security arrangements can be implemented effectively within the existing workings. This convergence not only provides a systematic methodology for managing risk but also provides room for compliance with emerging regulatory requirements and enhancing the overall resiliency of industrial environments. In this case, judicious enforcement of ISA/IEC 62443 standards ensures industrial systems to execute securely in the face of shifting cyber threats, providing operational continuity and safety (Fischer et al., 2019). A core aspect of the ISA/IEC 62443 standard lies in its emphasis on collaboration and definition of safe operating boundaries. By defining responsibilities and roles for various stakeholders, including manufacturers, integrators, and end-users, the standards foster a culture of shared accountability in managing cybersecurity risk. This ethos of cooperation is particularly important in industrial settings where the merging of information technology (IT) and operational technology (OT) brings complex security challenges. The standards highlight the necessity for risk management practices specifically tailored to unique operational environments, ensuring that organizations are adequately equipped to respond to vulnerabilities and enhance the resilience of their cybersecurity position. By placing emphasis on multi-layered defense approaches, ISA/IEC 62443 allows organizations to adopt a more proactive stance in safeguarding their assets, thereby adhering to best practices in the cyber world (Manubolu et al., 2024). In addition, maintained compliance with ISA/IEC 62443 standards not only enhances an organization's security systems but also acts as a bridge to gaining long-term sustainability in industrial processes. As the world economy continues to emphasize digitalization, industries should be able to fit new technologies and potential threats accompanying them. By applying the guidelines prescribed under the 62443 framework, organizations are able to review their security vulnerabilities in a methodical manner, implement appropriate countermeasures, and ensure that their systems are resilient to internal as well as external threats. The continuous development and fine-tuning of security protocols based on such standards enable organizations to prepare for future challenges in the industrial cybersecurity landscape. This adaptability is required to provide operational integrity and build trust with stakeholders, forming a solid foundation for a safe and efficient industrial ecosystem (Fischer et al., 2019) (Figure 1.15).

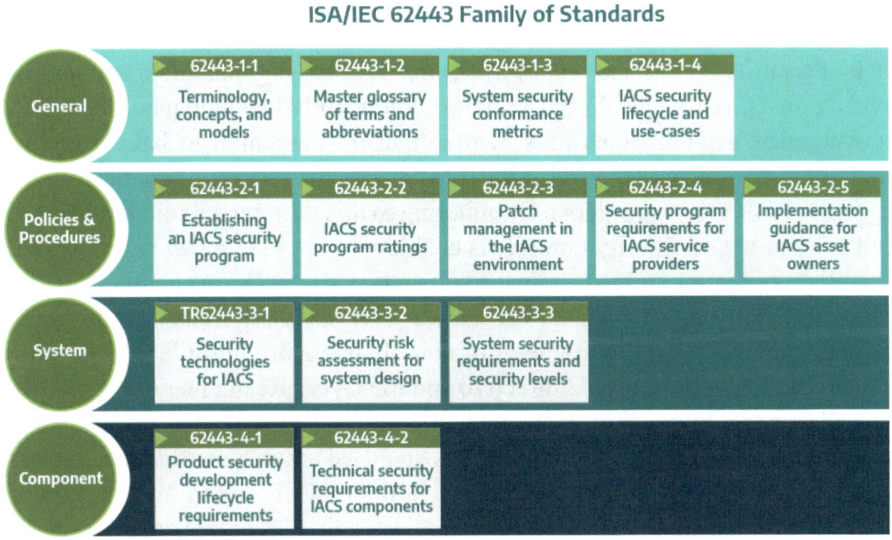

Figure 1.15. The ISA/IEC 62443 standards.

Source: https://www.dragos.com/wp-content/uploads/2023/04/image.png.

1.4.3. ISO/IEC 27001 for Information Security Management

For cybersecurity and industrial automation, ISO/IEC 27001 is a foundation standard to be used for the establishment, implementation, maintenance, and ongoing improvement of an information security management system (ISMS). The standard is essential for organizations that are dedicated to protecting sensitive information from a wide range of potential threats, including cyber-attacks and data breaches, that are typical in operational technology (OT) environments. ISO/IEC 27001 deployment enables a structured risk management process, one that is critical in industries where IT and OT convergence creates increasingly advanced security demands. This is illustrated by the fact that operators risk overlooking the embedded data flows and system dependencies needed for effective risk assessment, and this emphasizes the value of ISO/IEC 27001 in establishing a robust security position. Through integrating these practices in risk management, firms are well positioned to become increasingly resistant to the evolving threat landscape that forms part of today's industrial environments (Green et al., 2016). Moreover, implementing ISO/IEC 27001 is more than compliance; it fosters a culture of security awareness. Human resources become key stakeholders of the security system through training and engagement, limiting the scope for human error, which is the most common target of attack by cyber-attacks. This emphasis on the human factor is particularly essential in OT systems where typically there would be a lack of security controls in legacy systems, exposing organizations to emerging cyber threats. Although digital transformation initiatives enhance IT and OT convergence, organizations must realize that it is essential to have a merged strategy, supported by ISO/IEC 27001, to address the security loopholes created by such convergence problems. By adhering to this time-tested standard, end-to-end assessment of security controls is enabled, hence making an organization's overall defense more robust against advanced persistent threats (Tuliainen et al., 2025). In addition, ISO/IEC 27001 assists with many regulatory requirements and best practices, facilitating ease of compliance with multiple cybersecurity regulations, such as the NIS2 directive and the Cyber Resilience Act (CRA) for critical sectors. In providing stakeholders with the assurance that information security management is functioning, this standard assures confidence and trust in the organization's ability to protect confidential information. The process of certification to ISO/IEC 27001 also indicates that an organization is serious about maintaining better security standards and it is a competitive edge in the market. Since industrial automation businesses are increasingly faced with cyber risks as well as compliance mandates, the role of ISO/IEC 27001 is also more highlighted. Here, adopting the standard as a principal strategy to enhance cybersecurity resilience and readiness for forthcoming regulations is a way of safeguarding organizational interests and public security (Tuliainen et al., 2025).

1.5. CHALLENGES IN IMPLEMENTING CYBERSECURITY

Implementation of cybersecurity in industrial automation comes with numerous intricate challenges from technical as well as organizational factors. One of the biggest issues is the merging of the old systems, which do not possess the high-end security features that their newer counterparts carry and are at times vulnerable in their nature. With the businesses running combinations of new and old technology, protecting these systems from cyber-attacks is becoming increasingly complex. Also, the constant evolution of cyber threats means that organizations need to not only have strong cybersecurity in place but be on constant watch for new vulnerabilities. This dynamic culture demands ongoing training and vigilance across staff, with recognition of the probable impact of operational technology compromises. This complex dynamic of these factors is the reason why detailed cybersecurity measures that are better suited to cater to the specific needs of industrial environments are paramount. The second challenge here is the disconnect between cybersecurity risk assessments and actual operating procedure. Organisations typically find it difficult to quantify the risks posed by cybersecurity threats, and so fail to undertake adequate resource allocation and prioritization. For instance, in sectors such as power grids, cascading failures can be caused by cyber-attacks, as had been experienced in the event of the Ukraine cyber-attack on SCADA systems (Ginter et al., 2018). While asset owners strive to move their investments in cybersecurity to the front, they must tread the thin line between operational efficacy and risk avoidance and reduction. This needs not just a good understanding of the risks but also an efficient framework for their knowledge base and reaction to the rapidly changing patterns of risks that determine the operational environment. The gap between risk perception and fact can lead to vulnerability, increasing the likelihood of serious cyber events. Additionally, the addition of cybersecurity controls poses organizational and human resource challenges. Stakeholders may have varying levels of commitment towards cybersecurity procedures depending on variations in organizational culture and risk tolerance. Such inconsistency may undermine the effectiveness of cybersecurity efforts, given the necessity of cooperation and oversight to shield industrial systems from threats. Moreover, the complexity of cybersecurity in operating technology environments, especially oil and gas sectors, is too complicated to manage with existing frameworks (Houmb et al., 2014). Developing an integrated cybersecurity strategy that brings risk assessment models together at the same time developing a culture of security awareness is fundamental to overcoming such challenges. Proper training and communication across all organizational levels can enable overall cybersecurity postures so that people are adequately trained to identify and respond to threats. This way, addressing the overall challenges of implementing cybersecurity not only secures industrial automation systems but enhances their immunity against future weaknesses.

1.5.1. Legacy Systems and Integration Issues

Legacy systems in industrial automation environments are a major point of integration issues affecting both operational effectiveness and security. Legacy systems, which are typically developed using outdated technology, employ proprietary protocols that limit compatibility with contemporary systems and devices. With the aim to increase productivity through digitalization, these aging components are a hindrance that makes integration with newer technologies difficult. The lack of interoperability not only leads to inefficiencies but also heightens the risk of cyber vulnerabilities because earlier systems may not have the strong security features incorporated in today's solutions. Therefore, evaluating and fixing these integration issues become essential for firms looking to modernize their businesses while maintaining a strong cybersecurity position. Moreover, the shift from air-gapped, standalone industrial control systems to consolidated networks has made legacy systems exposed to higher numbers of threat vectors. In the past, safety-related systems ran in air-gapped environments, where physical access was the threshold to compromising components. In the convergence of technologies such as the Industrial Internet of Things (IIoT), these barriers were dispelled, and new avenues for cyber threats emerged. As organizations connect legacy systems to enterprise networks, they inadvertently increase their attack surface, creating opportunities for malicious players to exploit vulnerabilities.

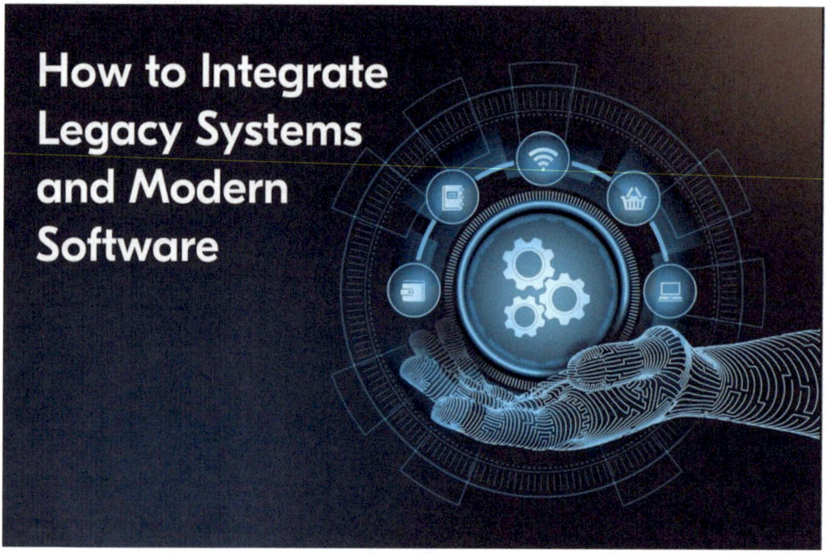

Figure 1.16. Legacy systems and integration issues.

Source: https://acropolium.com/img/articles/legacy-systems-integration/img01.jpg.

Learning from past experiences, such as those seen in attacks on critical infrastructure, underscores the importance of critically assessing legacy systems' security vulnerabilities to pro-actively address potential risks entailed in modernization initiatives (Evangelopoulou et al., 2017). In addition, building automation systems (BAS) are a good example of how legacy technologies can hinder both operational effectiveness and secure integration with contemporary cyber-physical systems. While BAS are now being integrated using IP-based communication protocols, the majority of these systems were not designed with security as a primary concern. As a result, they operate with inadequate security controls, relying on obscurity rather than robust protection. This legacy mentality is a case in point for the challenges of integrating older systems into the current cybersecurity landscape, in which attacks evolve quickly and must be met with responsive security tools. Research on improving security to modernize protection frameworks for such systems is critical, towards delimiting threat profiles and modernizing detection models. By solving these issues, firms will be in a position to protect their operational technologies more effectively from increased risks of cyber terrorism and hacktivism (Johnstone et al., 2014) (Figure 1.16).

1.5.2. Skills Gap and Workforce Training

The increasing inclusion of automation and cybersecurity in industrial processes reveals a large skills gap in the workforce, which calls for extensive training programs. As industries are changing, technology innovation becomes faster than current skills among employees, and most are left unprepared to keep up with the demands of the modern workplace. Skills gaps are felt most significantly in areas such as cybersecurity, for which specific training is needed to safeguard valuable systems from emerging threats. Organizations must invest in tailored workforce training programs that not only address immediate skills gaps but also encourage continuous learning. Investment in developing talented individuals makes industries more responsive and resilient when faced with ever-evolving technology. Beyond this, emerging digital technologies increasingly necessitate rethinking existing paradigms for workforce training. Traditional pedagogical strategies may no longer suffice, and there should be a transition to experiential learning and application. Schools, together with industries, have to develop curricula that are aligned with real-world needs, equipping students with theoretical knowledge coupled with practical skills. The programs ought to aim at the development of flexible learning environments that emphasize cybersecurity practices, automation procedures, and critical thinking. Industry partners and learning organizations working together will enable the creation of a dynamic workforce that can meet the challenges of sophisticated industrial practice, leading to sustainable competitiveness in the global marketplace. In response to these challenges, national and regional policy needs to adapt to address workforce training effectively. Cooperation between government and industry is required in making strategic plans to bridge the skills gap through public investment, incentives to firms to invest in employee training, and the establishment of research institutions for workforce development. For instance, updating industrial strategies, as highlighted in current European debate, is necessary to propel innovation and align the labor market with the requirements of automation and cybersecurity. The aim is to build an educated, skilled workforce capable of driving economic development and technological advancement across industries while also addressing the urgent need for effective cybersecurity measures (Anglano et al., 2018; Bjerkem et al., 2019).

1.5.3. Balancing Security with Operational Efficiency

Security vs. efficiency of operations is increasingly crucial in industrial automation and cybersecurity. As smart grid technology continues to rise, operators are challenged with the requirement to enhance the resilience of the system while ensuring smooth operation. Good security practice is crucial to protecting critical infrastructure from cyber-attacks that have evolved with advancing technology. As these new technologies, such as Internet-of-Things devices, become part of operational deployments, the task of enhancing security without compromising productivity becomes increasingly complex. Computerized security protocols can mitigate the risk of human frailty, a well-known vulnerable point, as well as maintaining operational efficiency kept abreast of safety. This double focus is required for successful defenses as well as building a robust infrastructure in accordance with regulatory standards and best industry practices. Moreover, the smart grid design helps to highlight the necessity of integrating operational efficiency and robust security measures. For example, decentralized management systems, as designed to supplement traditional Supervisory Control and Data Acquisition (SCADA) systems, enable operators to leverage weak rural distribution networks effectively and maximize integration capabilities. By leveraging intelligent management methods based on telecommunications breakthroughs, this architecture facilitates decentralized control of energy resources. All of this allows for ultimately increased efficiency across the system, that reflects on the necessity of adaptive technology to be able to secure vulnerable infrastructures. All of these advancements go not only towards resolving the immediate security issue but also to push the operational efficiency, in order for users and operators to benefit from the synergy among cutting-edge technology as well as rigorous cybersecurity practices in an industrial environment (Díaz González et al., 2018). Finally, the increasing interconnectedness of the cyber and physical infrastructures calls for an ongoing effort at balancing security and efficiency. Notably, the characterizing of the electric grid as "third-world" implies the pressing need for upgrading and integrating smart grid technologies that broaden the range of energy management. Combining these technologies with traditional grid capabilities has the potential to revolutionize on a scale as dramatic as in some pilot efforts aimed at using renewable resources and making grids more responsive. The revolutionary potential of smart grid technologies exceeds basic operating needs, promising change that could be the equal of the revolutionary impact the Internet has had on everyday life. On this issue, achieving a sustainable compromise between security and operational performance is not a desire; it is required for constructing resilience and ensuring the provision of critical service continuity in today's interconnected world (Heidel et al., 2012) (Figure 1.17).

Figure 1.17. Balancing security with operational efficiency.

Source: https://visualmodo.com/wp-content/uploads/2023/05/How-To-Balance-Security-And-Productivity-For-Business-Growth.png.

1.6. CASE STUDIES OF CYBERSECURITY INCIDENTS

The analysis of cybersecurity incidents reveals key trends that are shaping current practice in industrial automation. Perhaps one of the most significant examples is the cyberattack on the power grid of Ukraine, which illustrated the catastrophic capabilities of coordinated cyber actions against critical infrastructure. By gaining access to substation networks through infected supervisory control and data acquisition (SCADA) systems, perpetrators performed disruptive switching operations across multiple substations. This example illustrates how an expertly coordinated cyber intrusion can result in mass power blackouts and systemic instability. Experts describe that such incidents call for detailed examinations of potential nightmare scenarios, thereby urging asset owners to invest further in securing their cyber defenses. Therefore, knowledge of real breaches informs us about the weaknesses of the industrial base and emphasizes the need for enhanced cybersecurity practices, especially with increased reliance on automated processes (Ginter et al., 2018). Given persistent cybersecurity attacks, the development of effective incident management procedures is most critical. Organizations commonly release elaborate technical briefings to facilitate detection and response to security incidents that involve automation technology, but these releases are typically marred by semantic imprecisions that reduce their applicability. A structured approach based on semantic methods could enhance the comprehensibility and applicability of such procedures so that professionals can better understand and respond to incidents. Through the provision of advanced analysis of security requirements in terms of semantics, business enterprises are able to enhance documentation quality and externalize crucial knowledge on managing incidents. The more clearly this is defined, the easier it is to train workers and develop response plans so that future breaches of cybersecurity are reduced. These advances are particularly opportune in light of growing industrial system sophistication as they evolve to counteract emerging threats (Borrego Díaz et al., 2016). Furthermore, the study of real-world cybersecurity incidents offers valuable lessons into adaptive responses that organizations can implement in an effort to minimize risk. As industrial automation continues to permeate, attackers also grown more creative in their methods, often targeting weakest links in cybersecurity infrastructure. Based on the analysis of several incidents like the Ukraine blackout and others, it is evident that a reactive stance towards cybersecurity is not enough. Instead, organizations should adopt proactive methods, such as regular assessments of their cyber stance and investing in high-tech surveillance technologies. A system consisting of risk assessments, real-time anomaly detection, and targeted training can foster resilience against threats. As the companies learn from past incidents, they will enhance their security models and develop a culture of vigilance that espouses both real-time response and strategic security planning (Borrego Díaz et al., 2016; Ginter et al., 2018).

1.6.1. Notable Cyber Attacks on Industrial Systems

Industrial automation has delivered significant productivity and efficiency improvements; however, it has also exposed systems to cyber-attacks. Historic industrial system cyber-attacks such as Stuxnet and WannaCry illustrate the growing convergence of industrial activities and cybersecurity. Stuxnet, which was the first known digital weapon to be recorded, infected the Iranian nuclear infrastructure, exemplifying the way a cyber-attack can disrupt critical infrastructure through smart means. The attack highlighted the extent to which cyber attackers can exploit the special vulnerabilities that are inherent in industrial control systems. With industries becoming increasingly integrated with automation technologies, the potential threat of disruptive attacks increases, necessitating a heightened sense of prudence with regard to cybersecurity in industrial environments. Cybersecurity incidents on industrial systems demonstrate the potential threat for far-reaching effects, not only on a single organization but also on entire economies and societies. The 2021 Colonial Pipeline ransomware attack is a case in point, demonstrating the vulnerability of critical infrastructure to cyberattack.

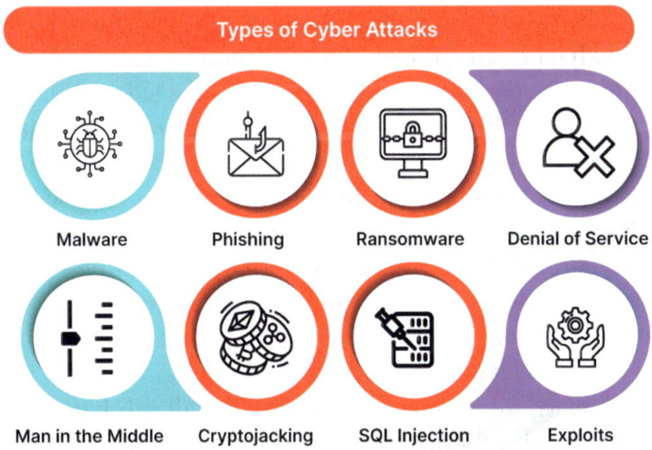

Figure 1.18. Notable cyber-attacks on industrial systems.

Source: https://www.fortinet.com/content/dam/fortinet/images/cyberglossary/types-of-cyberattacks.png.

The attack resulted in fuel supply disruptions across the Eastern United States, a by-product of the interconnectivity of modern industrial processes and the potential for cascading effects between industries. These incidents underscore the critical necessity for end-to-end cybersecurity solutions that are able to address the specific challenges of industrial automation. By learning from such high-level attacks, organizations are better equipped to combat such threats down the line and assist with keeping their operations stable based on their integrity and reliability within a variable cyber world. Managing the threats of high-level cyber-attacks entails an integrated approach of cyber risk assessment, particularly for industrial system IoT projects. The introduction of IoT devices has the potential to enhance operations efficiency but also extends vulnerability to new cyber threats. Research indicates that businesses must undertake self-assessments of their IoT cybersecurity standing in a way that ensures they can effectively anticipate, avoid, and respond to future cyber threats. This reinforces the necessity of adopting a goal-based dependency model, which provides an established framework to measure and enhance an organization's cyber risk management practices. With the IoT market growing, there is a need to be ahead of the threats that accompany these innovations because experts foresee the IoT to expand significantly while posing constant cybersecurity threats (Anderson et al., 2017; Ani et al., 2020) (Figure 1.18).

1.6.2. Impact of Cyber Incidents on Operations

The impact of cyber-attacks on operational frameworks within industrial settings is significant, with increasing reliance on operation technology (OT) increasing the vulnerabilities. As organizations deploy sophisticated automation solutions to automate operations, the level of cybersecurity complexity grows in parallel. Cyber attackers exploit the weaknesses in these tech platforms to create substantial chaos in operations. For instance, an offshore oil drilling platform with high reliance on OT has unique issues as outlined in a recent graphical model for Adversarial Risk Analysis-based cybersecurity risk assessment proposal. This model highlights the requirement of a formal approach to identify and address risks that are specific to the operating environment, identifying the necessity of high-level cybersecurity standards in critical infrastructure industries, particularly those that are quickly digitalized, like the oil and gas industries (Houmb et al., 2014). In addition to short-term operational disruptions, cyber-attacks may have lasting effects on an organization's reputation and financial health. When there is a successful cyberattack, not just production is halted but sensitive data is also leaked, adding to the damage. Companies can suffer from stakeholder confidence loss, loss of competitiveness in the market share, as well as regulatory fines, all which are being caused by bad cybersecurity measures. With changing industrial automation environments, organizations should be concerned with establishing integrated cybersecurity systems that encompass prevention methods and response mechanisms. Various segments of the cybersecurity landscape must be addressed, from technical underpinnings to human factors such as training and awareness, which demand a holistic approach to risk management (Anglano et al., 2018). With the threat environment changing, resilience of industrial operations has grown to the forefront. Organizations need not only spend in cutting-edge cybersecurity technology but also foster security-culture awareness in the workforce. Cybersecurity awareness training initiatives, emphasizing the importance of cybersecurity, can also decrease human errors, which are one of the prime causes of most cyber incidents. Moreover, organizations must have regular risk assessments to ensure that their components of cybersecurity are kept up to date along with evolving threats. Integration of the mechanisms of risk management mechanisms with responsive technology will be imperative in safeguarding operations against cyber-attacks. By bringing operational resilience together with cybersecurity preparedness, industries can ride out the impact of cyber-attacks as well as keep on functioning unproblematically and protecting their vital assets (Anglano et al., 2018).

1.6.3. Lessons Learned from Cybersecurity Breaches

Industrial automation domain cybersecurity incidents are not only as threats, but as valuable learning points for organizations. One of the key learnings from such breaches is the need to have robust security protocols in place within an organization. This means implementing strong cybersecurity frameworks that consider both the technology and the human aspect. The majority of the breaches are a consequence of inadequate training and awareness among employees, proving that employees are usually the weakest link in an organization's security landscape. Thus, developing a security awareness culture in which employees are regularly trained and engaged in performing cybersecurity operations assumes central importance for safeguarding sensitive data and assets within the industrial setting. The need for extensive training is also underscored by recent intelligence reporting that underlines the role of employees in identifying and containing cyber-attacks effectively (Zhang et al., 2020). One of the key lessons derived from cybersecurity incidents also includes the importance of establishing incident response plans. Organizations must realize that even with strong controls in place, it is possible for certain breaches to occur. Therefore, devising incident response plans periodically and updating them helps responders to be ready to contain the damage of such incidents. Effective responders' plans enable organizations to respond immediately as well as help them assess the magnitude of the breach and recover critical data on time. Furthermore, conducting post-incident analysis allows organizations to identify the vulnerabilities that were exploited during the breach and seal such holes in advance. Such a cycle of improvement through real experience with what works and what does not is essential to continuous improvement in cybersecurity policies, especially in areas such as industrial automation where business continuity is paramount (Zhang et al., 2020). Finally, the consequences of cybersecurity lessons from breaches reinforce that there is a necessity for collaboration within organizational silos. Cybersecurity is not an IT function matter alone; it requires collaboration by way of all employee levels in all functions. By promoting cross-functional collaboration, organizations are able to enhance their collective knowledge of cyber threats and build an anticipatory culture for mitigating prospective vulnerabilities. Additionally, sharing threat intelligence with other industry stakeholders and cooperation in collaborative efforts can lead to widespread awareness and preparedness against cyber-attacks. With an increasing number of organizations in the industrial automation sector yet to acquire new cyber threats, such combined efforts can contribute significantly towards creating a viable cybersecurity model (Pidlisna et al., 2024).

SUMMARY

In concluding the review of industrial automation and cybersecurity, it is apparent that the integration of advanced digital technologies into operating systems brings with it immense opportunity accompanied by significant vulnerability. The perpetual expansion of the IoT serves as a foundational catalyst for this dyad, where increased connectivity drives operational efficiency alongside increasing cyber threat. The imperative need for organizations to undertake an exhaustive self-assessment of their IoT cybersecurity position is especially emphasized by changing threats that demand a proactive response to risk. Organizations have no alternative but to accord maximum importance to measures that not just forewarn and forestall cyber-attacks but also increase their ability to respond forcefully to incidents, which is critical in safeguarding their digital assets and operational integrity. Aside from this, the complexity of industrial control systems—such as those underpinned by the IEC 61850 standard—calls for a multi-dimensional analysis of cyber-attacks that go beyond conventional paradigms of cybersecurity. The recognition that these systems were initially created with a focus on operational efficiency and not cybersecurity represents a foundational difference calling for examination and research at present. Experimental testbeds and empirical studies are necessary here to enable the creation of a deep understanding of the potential cyber threats as well as the real-world implications of incidents that may take place. The operation configurations under investigation in various studies indicate the need to replicate these environments in order to test countermeasures and, therefore, to create a necessary knowledge base for organizations to prioritize and use measures against cyber threats. The creation of goal-oriented dependency models offers a potential pathway to enhancing IoT cyber risk estimation. The identification of effective methods for quantifying an enterprise's cyber risk position is most important during the heightened utilization of networked devices. As argued in recent works, these models provide a solid foundation for the evaluation of uncontrollable statuses of risk in advanced IoT systems, enabling organizations to align their cybersecurity approach with their overall business objectives. Shifting toward predictive and preventative strategies for cybersecurity, as revealed by the examination of most risk evaluation models, is a fundamental step in the industrial automation discussion. Ultimately, the synergy between operational resilience and cybersecurity is critical, with recommended strategies laying the groundwork for resilient industrial environments capable of withstanding future cyber threats.

Industrial automation and integration of cybersecurity is a revolutionary shift in manufacturing and operational practices today. Underlying this change is Industry 4.0, the merging of digital technology into manufacturing processes, creating the concept of smart factories. The use of newer technologies such as the IoT, AI, and big data analytics has not only improved the efficiency of the processes but elevated the significance of protecting these networked processes. In the academies, institutions like the Technical University of Catalonia have experienced the imperative of educating engineers with competences in accordance with such technological advancement, befitting initiatives like the I4Tech Lab that bring about active learning in engineering education 4.0 (Prieto et al., 2019). As industrial automation relies increasingly on real-time data exchange and interconnectivity, cybersecurity issues have also increased exponentially. Cyber-Physical Systems and social machines, which make this even more complex, are born out of human interactions with technological platforms that engender new possibilities and new threats. The introduction of IoT devices adds complexity with the security and privacy dimension, and for one to comprehend, one must have a very good idea of how these socio-technical systems function.

The concept of Cyber-Physical Social Machines identifies the intertwined relationship between humans and machines, highlighting that security must evolve in synchronization with these technologies to adequately protect sensitive information and industrial procedures against threats (Creese et al., 2018). Essentially, the convergence of industrial automation and security is revolutionizing manufacturing's landscape, requiring a multi-pronged response through education, technology, and security. Institutions need to keep reforming their curricula to build the required competencies for future engineers, which will enable them to cope successfully with the challenges of this digital revolution. At the same time, it is important to build robust cybersecurity frameworks to safeguard industrial infrastructure and implement promising technologies that drive productivity and efficiency. This integration of education, technological advance, and risk management will be pivotal to driving the sustainable development of industries in the era of Industry 4.0.

REFERENCES

1. Al-Khateeb, H., Burnham, K. J., Epiphaniou, G., Hammoudeh, M., et al. (2019). Preliminaries of orthogonal layered defense using functional and assurance controls in industrial control systems. https://core.ac.uk/download/196185875.pdf (accessed on 13th September 2025).
2. Allodi, L., Cremonini, M., Massacci, F., Shim, W., et al. (2018). The effect of security education and expertise on security assessments: The case of software vulnerabilities. https://core.ac.uk/download/187985486.pdf (accessed on 13th September 2025).
3. Anderson, J., & Rainie, L. (2017). The Internet of Things Connectivity Binge: What are the Implications? https://core.ac.uk/download/89472988.pdf (accessed on 13th September 2025).
4. Anglano, C., Aniello, L., Antinori, A., Armando, A., et al. (2018). The future of cybersecurity in Italy: Strategic focus area. https://core.ac.uk/download/188830442.pdf (accessed on 13th September 2025).
5. Ani, U., Anthi, E., Burnap, P., De Roure, D., et al. (2020). Dynamic real-time risk analytics of uncontrollable states in complex internet of things systems, cyber risk at the edge. https://core.ac.uk/download/491119431.pdf (accessed on 13th September 2025).
6. Bartsch, B., Coka, D. A., Jungbluth, C., Laudien, T., et al. (2020). Learning from Trump and Xi? Globalization and Innovation as Drivers of a New Industrial Policy. Bertelsmann GED Focus 2020. https://core.ac.uk/download/287647968.pdf (accessed on 13th September 2025).
7. Bjerkem, J., Dhéret, C., Giuli, M., Pilati, M., et al. (2019). An Industry Action Plan for A More Competitive, Sustainable and Strategic European Union. European Policy Centre Issue Paper, 7 November 2019. https://core.ac.uk/download/237467861.pdf (accessed on 13th September 2025).
8. Bobák, R., Chromjaková, F., & Hrušecká, D. (2017). Production process stability–Core assumption of Industry 4.0 Concept. https://core.ac.uk/download/97495160.pdf (accessed on 13th September 2025).
9. Borrego Díaz, J., Chávez González, A. M., Matos Arana, V., Pro Martín, M., et al. (2016). Semantics for incident identification and resolution reports. https://core.ac.uk/download/225123898.pdf (accessed on 13th September 2025).
10. Burov, O., Krylova-Grek, Y., Lavrov, E., Lytvynova, S., et al. (2020). Кибербезопасность в образовательных сетях. https://core.ac.uk/download/322373029.pdf (accessed on 13th September 2025).
11. Creese, S., De Roure, D., Hall, W., Madaan, A., et al. (2018). A storm in an

IoT cup: The emergence of cyber-physical social machines. http://arxiv.org/abs/1809.05904 (accessed on 13th September 2025).

12. Delgado Prieto, M., Fernández Sobrino, Á., Fibla Biosca, P., Romeral Martínez, M., et al. (2019). Active learning-based laboratory towards Engineering Education 4.0. https://core.ac.uk/download/294830056.pdf (accessed on 13th September 2025).
13. Díaz González, F., Gallart, R., Girbau Llistuella, F., Heredero Peris, D., et al. (2018). Smart grid architecture for rural distribution networks: Application to a Spanish pilot network. https://core.ac.uk/download/157811722.pdf (accessed on 13th September 2025).
14. Diederichs, F., Gaitanidou, E., Loukea, M., Mathis, P., et al. (2021). Acceptance, a mandatory requirement for the transport of the future. https://core.ac.uk/download/586551478.pdf (accessed on 13th September 2025).
15. Diep, Q. B., Truong, T. C., & Zelinka, I. (2020). Artificial intelligence in the cyber domain: Offense and defense. https://core.ac.uk/download/323112801.pdf (accessed on 13th September 2025).
16. Evangelopoulou, M., Johnson, C. W., & Pavlova, T. (2017). Applying lessons from cyber-attacks on Ukrainian infrastructures to secure gateways onto the industrial Internet of Things. https://core.ac.uk/download/111557942.pdf (accessed on 13th September 2025).
17. Fischer, B., Gruber, T., Křena, B., Magnin, B., et al. (2019). The AQUAS ECSEL project: Aggregated quality assurance for systems–co-engineering inside and across the product life cycle. https://core.ac.uk/download/211121117.pdf (accessed on 13th September 2025).
18. Gattinesi, P. (2018). European Reference Network for Critical Infrastructure Protection: ERNCIP Handbook 2018 Edition. https://core.ac.uk/download/158646578.pdf (accessed on 13th September 2025).
19. Gharehpetian, G. B., Marzband, M., Moradi, J., Nafisi, H., et al. (2019). Attributes of big data analytics for data-driven decision making in cyber-physical power systems. https://core.ac.uk/download/305121394.pdf (accessed on 13th September 2025).
20. Ginter, A., Ten, C. W., Vasilakos, A. V., Yamashita, K., et al. (2018). Impact assessment of hypothesized cyberattacks on interconnected bulk power systems. http://arxiv.org/abs/1801.01048 (accessed on 13th September 2025).
21. Green, B., Hutchison, D., & Krotofil, M. (2016). Achieving ICS resilience and security through granular data flow management. https://core.ac.uk/download/76960566.pdf (accessed on 13th September 2025).

22. Heidel, T. D., Kassakian, J. G., & Schmalensee, R. (2012). Gridlock in 2030? https://core.ac.uk/download/10129205.pdf (accessed on 13th September 2025).
23. Horowitz, M. C. (2018). Artificial intelligence, international competition, and the balance of power (May 2018). https://core.ac.uk/download/211333744.pdf (accessed on 13th September 2025).
24. Houmb, S. H., Insua, D. R., & Vieira, A. C. (2014). A graphical adversarial risk analysis model for oil and gas drilling cybersecurity. http://arxiv.org/abs/1404.1989 (accessed on 13th September 2025).
25. Johnstone, M. N., & Peacock, M. D. (2014). An analysis of security issues in building automation systems. https://core.ac.uk/download/41537512.pdf (accessed on 13th September 2025).
26. Kabir-Querrec, M., Mocanu, S., Savary, E., Thiriet, J. M., et al. (2016). A test bed dedicated to the study of vulnerabilities in IEC 61850 power utility automation networks.
27. Manubolu, G. (2024). A comprehensive security testing framework for PLC-based industrial automation and control systems. https://core.ac.uk/download/619888591.pdf (accessed on 13th September 2025).
28. Tuliainen, T. (2025). Cybersecurity in operational technology networks: NIS2 as a guideline. https://core.ac.uk/download/657106930.pdf (accessed on 13th September 2025).
29. Zhang, M. (2020). National Security Space Launch. https://core.ac.uk/download/305123572.pdf (accessed on 13th September 2025).

CHAPTER 2

NETWORK ARCHITECTURES IN INDUSTRIAL CONTROL SYSTEMS (ICS)

Contents

INTRODUCTION

The landscape of industrial control systems (ICS) has dramatically transformed in recent years, driven largely by technology innovation and the increasing demands for efficiency and reliability. Traditionally, ICS operated on isolated and legacy systems with little networking capability. This is now changing as the adoption of advanced network architectures has emerged as a significant paradigm shift. These structures enable real-time data interaction among devices, making it possible to include necessary factors such as control systems, sensors, and actuators. As a result, integration of these technologies enables improved monitoring, analysis, and automation, which ultimately leads to improved operational efficiency. This shift does not only address today's issues inherent in industrial environments but also facilitates building more solid and adaptable networked systems. The concept of Smart Grids (SGs) is a paradigm example of the transition towards highly interdependent industrial ecosystems. As emphasized by recent scholarship, SGs bring together information technology and energy supply, resulting in bi-directional flows of energy to increase efficiency and performance (Copeland et al., 2019). This transition is also accompanied by the Internet of Things (IoT), which enables spontaneous interworking among many devices installed throughout power generation and transmission systems. IoT sensors and smart meters allow more tracking, monitoring, and management of energy systems. IoT in ICS does not just contribute to enhancing operational efficiency but also allows for proactive maintenance and rapid response to any possible disruption. Hence, information regarding the network architectures driving such systems becomes critical to unlocking their full potential. Another essential characteristic of contemporary industrial control systems is visualizing networked streams of information, which facilitates stakeholders to analyze and interpret complex interactions among such systems. Interactive visual investigation has been found successful in recent studies, particularly when applied to numerous aspects of industrial automation and telecommunication (Holmgren et al., 2008). Through a streamlined process of visualizing data, decision-makers can better comprehend the performance of operations and identify areas for enhancement. Incorporation of visual analytics into ICS guarantees better comprehension and management of network topologies, thereby enhancing overall system security and dependability. Through research on such innovative approaches, this essay aims to explore various dimensions of network architectures in ICS and submit a comprehensive report on their implications for industrial operations in the future.

Definition of Industrial Control Systems (ICS)

ICS are very important frameworks in the context of industrial operations intended for controlling and automating various complex processes. These systems have diverse components, varying from sensors, controls, and human-machine interfaces, all integrated to facilitate real-time control and monitoring in industrial settings. Primarily applied in sectors such as manufacturing, energy, and transportation, ICS facilitate smooth coordination of equipment and operations, thus ensuring efficiency and safety in machine-intensive and data-intensive operations environments. As particularly intriguing is the classification of ICS that further details their respective roles, differentiating between subcategories like Distributed Control Systems (DCS) and Supervisory Control and Data Acquisition (SCADA) systems, each specifically tailor-made to satisfy specific operational requirements. Understanding this definition is critical because it opens up to the study of the intricate network topology underlying such systems. One of the most important properties of ICS is their inherent coupling with cyber-physical systems, particularly under Industry 4.0 that encourages the use of digital technologies in manufacturing. This integration leads to the phenomenon that has been widely referred to as cyber-physical production systems (CPPS), which are ideally constructed for the use of high-end computational processing power along with physical parts. Employing sensors and computation, these systems are not only capable of monitoring automatically but can even shift dynamically based on changing operating conditions through insights drawn from learned data.

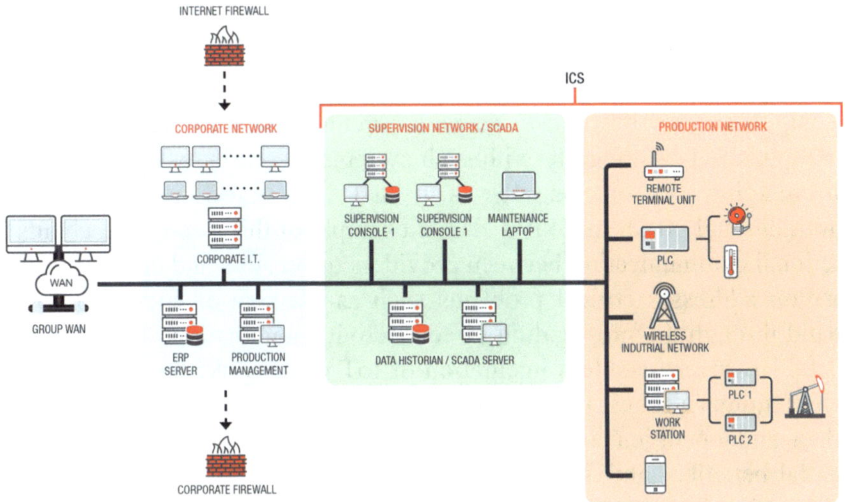

Figure 2.1. Industrial control systems (ICS).

Source: https://documents.trendmicro.com/images/TEx/articles/ICS-System.jpg.

Therefore, the development of standardized reference architectures in this area becomes a necessity in providing interoperability and enhancing the cognitive functions of ICS from different models and vendors, as explained in some recent research work committed to closing gaps in flexibility and generalizability in existing models (Bartz-Beielstein et al., 2019). Additionally, full understanding of ICS must take into consideration the challenges of the day in the form of the threats to cybersecurity, particularly with these systems becoming more interdependent. SCADA systems, a significant class of ICS, are fundamental to the control and monitoring of infrastructure sectors that are essential, such as power distribution. As SCADA systems may be attractive targets for cyber-attacks, an awareness of their architectural sophistication is paramount in order to defend against that which attaches. A systematic architectural model for SCADA systems is critical to stakeholders so that security can be more easily analyzed and system integrity enhanced (Ma et al., 2012). Emphasis on utilizing a layered design allows practitioners to manage the inherent complexity of these systems and protect them from changing threats. Finally, an understanding of the intricacies of ICS definitions and their varying architectures is imperative to addressing the operational and security requirements that industrial environments of the modern world are faced with (Figure 2.1).

Importance of Network Architectures in Industrial Control Systems (ICS)

The significance of robust network architectures in ICS cannot be overemphasized, particularly with such systems increasingly being integrated with emerging technologies such as the IoT. The transition to Smart Grids from traditional power grids is a classic example of this trend, as it entails bi-directional communication between providers of services and consumers. This transition addresses critical problems such as wastage of energy and rising demand through the propagation of connectivity, automation, and monitoring of data. However, seamless integration of IoT devices like sensors and smart meters requires a robust network architecture to enable real-time transmission of data and operation efficiency. Without a solid architecture, Smart Grid potential benefits—and, by extension, those of the ICS infrastructure—could not be realized, and inefficiencies and, potentially, vulnerabilities that could be exploited by malicious players would result. The definition of network architectures is thus key to ICS evolution and safety in modern applications (Copeland et al., 2019). Beyond operation functionality, network architectures in ICS are important in presenting robust security measures against threats that always evolve. As the field of ICS continues to grow, particularly with their integration into Critical National Infrastructure (CNI), the systems present an intriguing target for malicious operations. Consequently, security frameworks instituted constantly receive examination on the issue of effectiveness against sophisticated threats. Traditional layered-defense strategies are found wanting on occasion, which leads to searching for other options, such as the incorporation of orthogonality. Adding orthogonality into network architecture design allows ICS to have confidence and control of operations at every layer, thereby enhancing security without affecting the fundamental real-time transmission of command and control traffic. This two-pronged approach provides a critical remedy for the complex threat landscape, highlighting the imperative of an smart architectural design (Al-Khateeb et al., 2019). The eventual evolution of complete systems network architecture in ICS has long-term ramifications for safety and efficiency. The architectures must provide seamless communication between millions of devices as well as possess new security mechanisms to preventively counter possible attacks. Because of the sophistication of modern automated systems, it is essential that network engineers are sensitive to scalability and adaptability so that they can accommodate future technologies. Most importantly, a strategic approach of architecture design can perhaps increase resilience against disruptions—whether from cyber-attacks or process failures. By prioritizing the implementation of next-generation network architectures, organizations are not only able to enhance their operational efficiency but also enhance their resilience such that critical infrastructure systems maintain their integrities when novel challenges arise. Hence, a network architecture focus is key to the overall success of Industrial Control Systems (Al-Khateeb et al., 2019; Copeland et al., 2019) (Figure 2.2).

Figure 2.2. The significance of robust network architectures in industrial control systems (ICS).

Source: https://www.cisco.com/content/dam/cisco-cdc/site/images/legacy/assets/swa/img/anchor-info/network-architecture-628x353.png.

2.1. TYPES OF INDUSTRIAL CONTROL SYSTEMS (ICS)

The landscape of ICS has evolved significantly, particularly with the way the systems are segmented and combined. In the past, ICS comprised three basic types: Supervisory Control and Data Acquisition (SCADA) systems, Distributed Control Systems (DCS), and Programmable Logic Controllers (PLCs). SCADA systems better suit large operations, managing and supervising geographically dispersed assets. SCADA works primarily through the collection and processing of real-time data, which contributes to major decision-making in power and water utilities industries, among others. DCS systems are best applied to sophisticated processes in manufacturing facilities, facilitating coordinated control of several units of production. PLCs are utilized as adaptive devices that program specific industrial processes, with high response rates and adaptability. As industries trend towards higher operational efficiencies, awareness of the diversified functionality of such control systems is vital. With increased emphasis on connectivity, the earlier insulation of industrial control systems is eroding. Such systems once operated in isolated environments — conventionally explained as being 'air gapped' — minimizing exposure to external attacks. Industrial protocols such as Modbus, which are widespread in PLC usage, were developed to be self-standing as opposed to normal networking paradigms. With more recent innovations, there has been a coming together with enterprise management systems such that data exchange is maximized to enhance efficiency in operation. This new paradigm with the use of the Internet of Things (IoT) brings greater interconnectivity that can enhance manufacturing processes but concurrently risks greater exposure to cybersecurity threats. The industry is at a point where the benefits of networked systems must be weighed thoughtfully against the threats they present, and the need for a proper grasp of these interactions exists. While industrial control systems become adapted to suit contemporary network architectures, the relevance of the connectivity demands a strict examination of security protocols and safeguards. The incorporation of ICS into organizational IT infrastructures makes it complex and involves critical safety and security matters that must be managed. The majority of industrial systems, which were once isolated, are presently exposed to cyber threats due to increased Internet Protocol (IP) functionality and other connectivity enhancements. Incidents of occurrences which have hit the critical infrastructure via cyber-attacks, such as cyber-attacks in Ukrainian systems, raise the potential consequences of inadequate cybersecurity management in ICS systems. Organizations are making more investments in aggressive security solutions to protect their resources from external threats as well as comply with industry standards. The creation of ICS architecture, therefore, is not only a sign of technical advancements but also highlights the importance of integrating cybersecurity as a fundamental aspect of modern industrial processes, particularly those outlined by arising trends within the sector(Evangelopoulou et al., 2017; Johnson et al., 2016) (Figure 2.3).

Figure 2.3. The landscape of industrial control systems (ICS).

Source: https://images.wevolver.com/eyJidWNrZXQiOiJ3ZXZvbHZlci1wc-m9qZWN0LWltYWdlcyIsImtleSI6IjAuODZieHN3YW5jcWtSZWFsVGltZVBy-b2R1Y3Rpb25Nb25pdG9yaW5nU3lzdGVtYXBwbGljYXRpb25pblNtYXJ0S-W5kdXN0cmlhbE1hbnVmYWN0dXJpbmcuanBnIiwiZWRpdHMiOnsicmVza-XplIjp7IndpZHRoIjo0MDAsImhlaWdodCI6MjAwLCJmaXQiOiJjb3ZlciJ9fX0=.

2.1.1. Supervisory Control and Data Acquisition (SCADA)

The integration of SCADA systems into industrial control systems (ICS) is an integral step in operating efficiency and information management. SCADA is a basic structure that allows it to be possible to monitor and control a range of industrial processes in real-time, from manufacturing to power provision. The architecture of SCADA systems typically comprises three main components: the field devices, the SCADA software, and the communication infrastructure. Through effective integration of these components, SCADA systems ensure unencumbered information exchange between operational technology and information technology. Such compatibility is important to maximize decision-making processes, improve operational efficiency, and ensure system reliability in various applications like remote monitoring of infrastructures and resources. With industrial settings trending towards decentralized and automated applications, the application of SCADA systems is increasingly sophisticated. A newer philosophy in SCADA design emphasizes the need to blend latest technologies, including telecommunications and power electronics, particularly for rural distribution systems. Current research presents a modified SCADA system framework that includes decentralized management systems integrated with current structures to realize enhanced resource control and operational flexibility for rural settings. The integration of such advanced methods can go a long way to enhance the capacity of traditional SCADA systems to interact with distributed energy resources, which in turn can make operators enable optimal levels of energy efficiency and increase the viability of the grid. Such adaptability is an essential evolution in the context of Smart Grid initiatives, proving the transformative potential of SCADA in smart industrial networks (Díaz González et al., 2018). Despite the valuable benefits associated with SCADA systems, security concerns present real threats to their operational integrity, particularly in the implementation of the smart grid. The merging of communications technologies with electric infrastructure introduces vulnerabilities that can be exploited by malicious parties, penetrating the confidentiality, integrity, and availability of information. Research has illustrated the necessity for a comprehensive cybersecurity plan to address such vulnerabilities. Standard security solutions have been used to respond to individual components or attack vectors, but in order to prevent the complete architecture from being vulnerable to ever-evolving threats, responding to the entire architecture is needed. Through technical and environmental risk analysis of SCADA installations, subsequent efforts can make countermeasures more effective. This common focus on security will be the central factor in making SCADA systems secure and dependable as they increasingly support vital industrial infrastructure (Ghazi et al., 2018) (Figure 2.4).

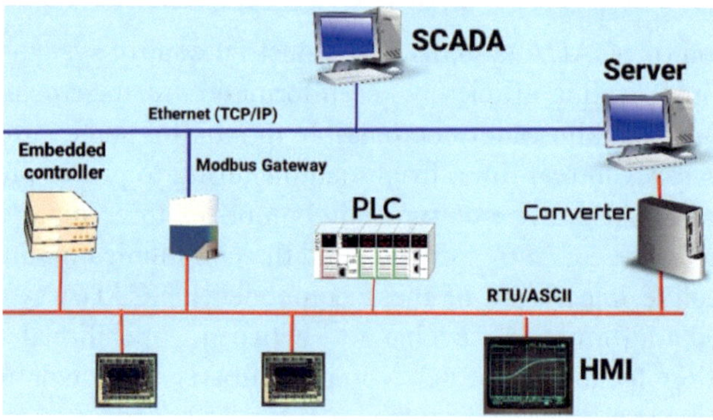

Figure 2.4. The integration of supervisory control and data acquisition (SCADA).

Source: https://cdn.open-pr.com/T/2/T219322798_g.jpg.

2.1.2. Distributed Control Systems (DCS)

DCS has emerged as a pivotal architecture in modern industrial control systems, facilitating decentralized process management across industries. A DCS typically consists of networked control loops with local control of individual units executed in parallel with extended process monitoring. This type increases flexibility and reliability through enhanced response to process changes. Internet Protocol (IP)-based network integration of DCS is a shift from traditional isolation to more integrated architecture with enhanced capabilities and monitoring encouraged. With industries increasingly requiring optimization, DCS is not only a control tool but also an exchange tool for data, ultimately leading to improved efficiency and production outcomes. The evolution of DCS is greatly reliant on the surfacing connectivity trends brought about by the Internet of Things (IoT). Incorporating IoT technologies into DCS provides new opportunities to capture and analyze data in real time and to enable informed decision-making by the operators based on full information.

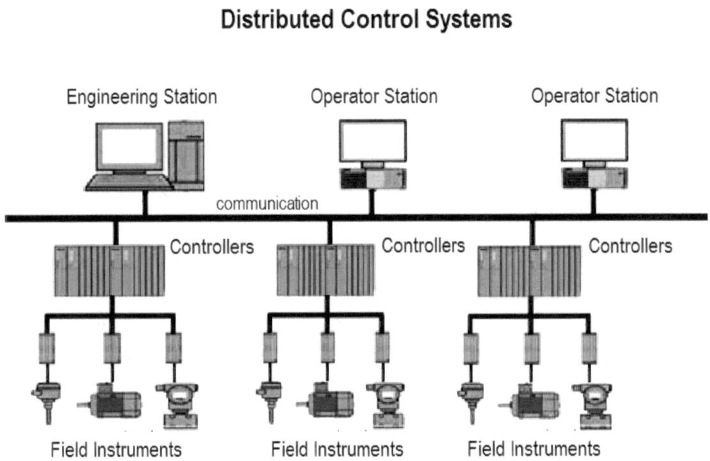

Figure 2.5. Distributed control systems (DCS).

Source: https://www.dcsmodule.com/js/htmledit/kindeditor/attached/20211014/20211014161706_66048.png.

Greater connectivity allows dynamic interactions between system elements, from sensors to actuators, hence optimizing processes and facilitating predictive maintenance. Incorporating DCS into enterprise systems also aids in the optimization of production workflow and resources. This transition to connectedness raises pertinent concerns regarding cyber security and safety. As the reports indicate, heightened connectivity brings on drastic threats that call for instituting sound defense mechanisms that can guard against new risks that arise from vulnerabilities in cyberspace and the disruption of services (Johnson et al., 2016). The role of DCS within the Smart Grids context also addresses the revolutionizing potential of this technology in industrial networks. Smart Grids have the two-way exchange of information between producers and consumers, which ensures the highest energy efficiency and reliability. DCS play a significant role in automation and control of these grids for the purpose of enhanced resource allocation and demand response. The adoption of automated equipment such as smart meters and sensors enhances monitoring and facilitates real-time adjustments that correlate with dynamic patterns of consumption. Overall, the incorporation of DCS within the infrastructure of the Smart Grid not only increases efficiency in operations but also facilitates the greater goals of energy sustainability and reliability in the face of increasing demand and security threats that are byproducts of this complex interlock of technologies (Copeland et al., 2019) (Figure 2.5).

2.1.3. Programmable Logic Controllers (PLC)

PLC are now essential in modern industrial control systems due to the flexibility and scalability that they provide. PLCs were initially designed for discrete control applications, but now they can support advanced automation processes that seamlessly integrate with various network topologies. The evolution is evident as it has shifted from proprietary to open standards wherein the PLC can communicate using commonly accepted protocols such as Ethernet/IP and Modbus TCP. These technologies not only facilitate interoperability between heterogeneous devices but also facilitate better data exchange between an organization's levels, from field devices up to the enterprise management systems. The interconnectivity allows for more productive production optimization, which is increasingly essential in today's competitive industrial setting, as reported by Johnson et al. (2016). For the case of embedded systems, PLCs are employed in order to guarantee the automation solutions continue running and responding suitably. With industries trending towards real-time data processing, the integration of PLCs with embedded platforms has gained greater significance. As these kinds of systems are resource-limited, there is an issue in implementing advanced control solutions. The traditional separation of design and implementation in control engineering may lead to long development times and increased expenses. However, such drawbacks can be overcome by the developments in the area of low-cost reconfigurable platforms emphasized by Bucher et al. (2008). These permit the simultaneous design and implementation of control systems and take advantage not only of open-source software but also of collaborative development approaches, thus shortening the overall process flow. With every newer network design in industrial control networks, PLCs are designed to handle future cybersecurity challenges. The increased interconnectivity and integration of Internet of Things (IoT) technologies create deep vulnerabilities that have the potential to undermine system safety and integrity. In response to these threats, PLCs must be constructed with strong cybersecurity features that align with contemporary industry norms. This emphasis on cybersecurity is central to the safeguarding of sensitive operational data and system reliability. Further research into resilient security protocols and fault-tolerant designs must also be pursued to address the unique challenges posed in industrial control environments. These forward-looking strategies are required to ensure that PLCs are not just capable of maximizing operational efficiency but also secure an industrial environment, further necessitating careful consideration of network design within ICS, as indicated in Johnson et al. (2016) (Figure 2.6).

Figure 2.6. Programmable logic controllers (PLC).

Source: https://www.polycase.com/wp/wp-content/media/plc-maintenance.jpg.

2.2. KEY COMPONENTS OF NETWORK ARCHITECTURES

In the context of the industrial control system (ICS), there is an imperative need to understand the key components of network architectures in a bid to ensure both functionality and security. Central to these architectures are devices such as

Programmable Logic Controllers (PLCs) and intelligent sensors/actuators that feature prominently in real-time monitoring and control of industrial processes. Traditionally, these components were in segregated systems, or "air gapped," where they were separated from other networks. However, with the advent of advanced communication technologies, segregation has weakened. Switches and gateways have facilitated integration to local and wide area networks such that data from safety-critical processes can be accessed by enterprise information systems. This intermingling of networks has opened up new vectors for potential vulnerabilities, which require these elements to be viewed and protected adequately within the larger network infrastructure (Evangelopoulou et al., 2017). Use of Supervisory Control And Data Acquisition (SCADA) systems is another critical aspect of network structure in ICS that is used as the basis for monitoring and responding to industrial processes. Such systems are characterized by their ability to monitor multiple safety-purposed functions using real-time data gathering and analysis. SCADA systems are based on an RTU network and HMIs to collect and present data from a range of sensors and PLCs. This information not only enhances operational performance but also helps in safety controls necessary in the prevention of system failure. However, the differences between traditional IT network and ICS/SCADA system architecture and protocols render application of typical cybersecurity methods challenging. Hence, one needs to enjoy the way the unique designs and operational requirements force network architecture (Evangelopoulou et al., 2016). Moreover, as industrial settings continue to get interconnected using the Industrial Internet of Things (IIoT), network architecture complexity skyrockets. The convergence of IIoT devices facilitates coordination of remote assets with a central management entity, which encourages better operational flexibility. The interconnectedness, however, raises cybersecurity issues with a growing number of access points. While moving from standalone systems to networked arrangements, the attack surface is broadened and calls for better risk assessment and mitigation methodology. Forensic analysis of the latest cyber-attacks targeting critical infrastructures signals the necessity for thorough evaluations of vulnerabilities in the network. By mapping the key components and their interactions within the network architecture, stakeholders are better positioned to anticipate and mitigate upcoming threats, bringing consistency and security to industrial control systems for a wide range of applications (Evangelopoulou et al., 2017).

2.2.1. Communication Protocols

The evolution of communication protocols in industrial control systems (ICS) mirrors the fundamental shift from independent, proprietary communications to increasingly networked and interconnected architectures. ICS, in the past, operated within an air-gapped environment where there were bespoke protocols such as Modbus which existed without interoperability such as present-day networking models in the guise of TCP/IP. Isolation was paramount to provide protection against unauthorized intrusion since it required physical access to break the system. But as increasing demand for sharing of real-time data and enhancing performance compelled organizations to adopt more integrated architectures, automatically cybersecurity problems come along with these advances in communication protocols (Johnson et al., 2016). While organizations use connectivity to make procedures seamless and boost productivity, they also face vulnerabilities of integrating these formerly isolated systems.

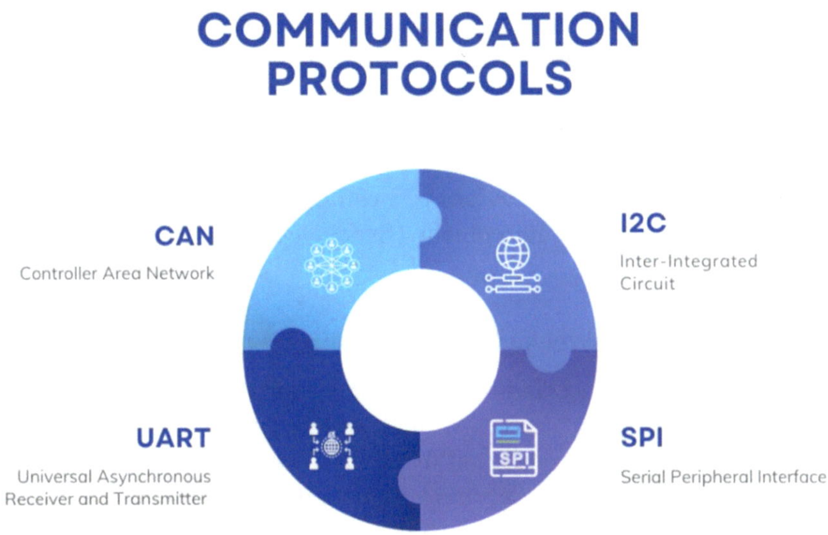

Figure 2.7. The evolution of communication protocols.

Source: https://media.licdn.com/dms/image/v2/D4D12AQEiWnPsQMhGag/article-cover_image-shrink_720_1280/article-cover_image-shrink_720_1280/0/1721976800980?e=2147483647&v=beta&t=lYr1WNTJead41mECCeiFXJHpPZC5mVx6c4ZnkYGUFUU.

Furthermore, the shift to more open communication protocols significantly improves the operational functionality of ICS. With TCP/IP, not only will communications among devices and control systems become simpler, but it is also possible to include safety-related operational technology of high importance into larger enterprise information systems. This revolution has allowed management to leverage information gathered from process control equipment, such as Programmable Logic Controllers (PLCs) and intelligent sensors, towards decision-making (Evangelopoulou et al., 2017). Accessibility is low-cost, but it puts ICS at risk of a wide range of potential cyber hazards. Organizations have to remain vigilant of these dangers and have good security controls in place that can effectively defend sensitive operational data without hampering their networks' performance. Due to the ever-increasing complexities associated with modern communications protocols, ongoing examination of security procedures is still important. With the increasing Industrial Internet of Things (IIoT), the geographically dispersed resources' connectivity allows organizations to control and observe operations remotely from the center, increasing operational efficiency (Johnson et al., 2016). Such a high level of connectivity, though, introduces new points of attack that threaten the safety and operation of critical infrastructures. Analysts and engineers must adopt an active approach to mitigating these risks through in-depth vulnerability analysis and the implementation of defense-in-depth policies that encompass both technological and procedural safeguards. What has been learned from histories of previous attacks on critical infrastructures will be important in dictating these protective measures so that modern ICS can be safely operated within a more interconnected network environment (Evangelopoulou et al., 2017) (Figure 2.7).

2.2.2. Network Devices (Routers, Switches, etc.)

The effective functioning of industrial control systems (ICS) depends significantly on fundamental network devices such as routers and switches. These devices within the network have significant functions to undertake in order to facilitate communication among diverse parts within an ICS infrastructure, for instance, sensors, actuators, and controllers. Routers primarily have the role of relaying data packets from one network to another in order to deliver information to the destination without delay. Switches, on the other hand, are on-the-local-level devices that have the function of uniting devices within a network to facilitate data transfer through direct connections for data packets.

Their implementation and configuration are significant in that they must accommodate the unique requirements of industrial environments, i.e., real-time data processing and greater reliability under potentially harsh operating environments. Their natively inflexible architecture is one of the difficulties of deploying routers and switches within ICS, typically the result of traditional network architecture that defaults to supporting rigid deployment models. For instance, the traditional emphasis on decentralized operations has led to network resource management issues. Organizations are currently increasingly exploring programmable network solutions to address these concerns. Programmable switches and routers integrate to provide more flexibility and responsiveness through dynamic reconfiguration of the network to adapt to changing industrial needs. This capability is particularly important when industrial environments evolve, demanding responsive and flexible network architecture to support improved applications, including real-time analytics and predictive maintenance, thereby enhancing overall operational efficiency.

Further, the intersection of network devices with programmability in ICS is highlighted even more by developments in middleware and cross-industry collaboration. With growing partnership between academia and industry, programmable network management projects are gaining momentum. The continued push in this direction is toward streamlining programmable network architecture integration within ICS to pave the way for future-proof systems that can easily evolve along with new technology. Such innovations are crucial in transcending the constraints of network ossification and inefficiency. Research collaboration, for instance, between BTexact Technologies and the Department of Computing Science at University College London, represents examples of cutting-edge approaches in addressing such concerns and shaping the future of network device manufacturing in ICS (Ahad et al., 2014; Hughes et al., 2002) (Figure 2.8).

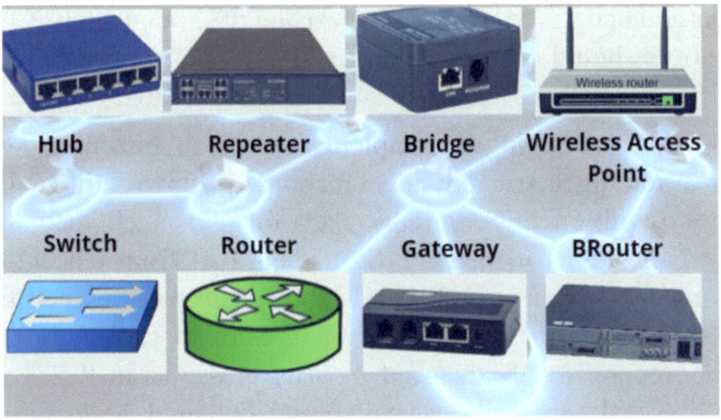

Figure 2.8. Network devices.

Source: https://blogger.googleusercontent.com/img/b/R29vZ2xl/AVvXsEjhfaAX-PTJMH2T84ZFwDqlUsRel4SxA8pSYos_Am9AMRkFaczgXtQLVEuFBeMIox-mWyK04ApCEFZvCCsEgBkaSzEQZ0O8HwpTtMYI2syInIV-PEutJiNGwzcl-Fu4pxz2m9FML3tmX4hsSwyyUKxtKJrXfuRZFcomEcIkKll8oZtDIO7zWLB-d3ZduTwuSw/w640-h362/nt%20devices.PNG.

2.2.3. Security Mechanisms

As industrial control systems (ICS) become increasingly integrated, strong security mechanisms become a point of growing importance. Traditionally, approaches relied on geographical separation, synonymously referred to as air gapping, to protect ICS from external threats. This approach made items inaccessible but failed to meet the growing needs for data sharing and real-time monitoring. The requirement for enterprise management systems with higher levels of sophistication to be able to communicate with process control devices such as Programmable Logic Controllers (PLCs) and intelligent actuators has led to the integration of these systems into wider network infrastructures. Hence, the importance of implementing advanced security features, including firewalls, intrusion detection systems, and encryption methods, has been established due to the fact that they provide critical levels of protection against threats likely to occur in the cyber space and unauthorized user access to confidential information in ICS settings (Johnson et al., 2016). Furthermore, the transformation from isolated systems to those integrated with the Industrial Internet of Things (IIoT) offers a colossal number of new vulnerabilities. As many devices that are linked by protocols like Modbus and Profibus enable communication across platforms, attack surface increases quite significantly. Part of the essential actions involved in enhancing security controls is through regular scans and updates to address such arising risks adequately. Conventional security controls should not only be enacted by organizations but also at least partially align with the changing risk landscape of the IIoT.

This entails establishing a complete security architecture composed of ongoing threat intelligence and incident response plans to actively fight threats. In combination with effective training plans for personnel, such architectures can allow organizations to better protect their ICS from current and future threats (Evangelopoulou et al., 2017). Moreover, lessons learned from recent cyberattacks, say the ones conducted on Ukrainian critical infrastructures, highlight the imperative of advanced security measures in ICS. Such attacks have been live wake-up calls, revealing the potential harm of overlooked vulnerabilities in interdependent infrastructure. Studying these attacks enables organizations to identify significant vulnerabilities in their network architectures, enabling them to implement tailored security solutions. For example, by using machine learning and artificial intelligence, anomaly detection capabilities can be improved to respond to potential threats in a faster way. Other steps are to set up strict access controls and a zero-trust model in order to make sure all devices connected to the network are always monitored and authenticated. In an era when operational technology and information technology lines continue to get intertwined, early implementation of security controls is the key to maintaining the integrity and availability of industrial control systems (Johnson et al., 2016) (Figure 2.9).

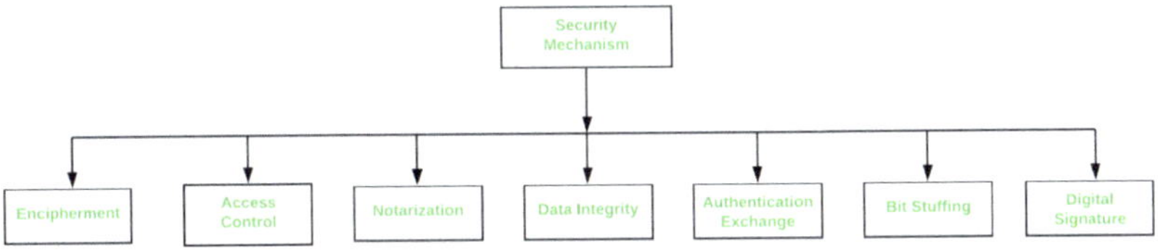

Figure 2.9. Security mechanisms.

Source: https://media.geeksforgeeks.org/wp-content/cdn-uploads/20200910162731/2217.png.

2.3. DESIGN CONSIDERATIONS FOR ICS NETWORKS

Security is the primary focus while developing Industrial Control System (ICS) networks, in keeping with the increased need for robust protection against cyber-attacks. As such systems increasingly rely on network components, vulnerabilities with regards to communication channels become most important. An integrated security solution must cater to various aspects, one of them being providing for secure communications gateways that efficiently control data flow. The research in Marksteiner et al. (2017) addresses the most important security requirements in such systems, which is the protection of lines of communication and data storage. In addition to enhancing the resilience of ICS against cyberattack, this approach also ensures the integrity of sensitive operational information, an element of very critical consideration in maintaining the overall functionality of industrial processes. Also, integrating smart technologies into ICS networks requires a shift to more dynamic architecture. As traditional systems are converted into Smart Grids (SGs), such that bi-directional energy flows and real-time monitoring capabilities are present, design considerations have to advance along with it. Internet of Things (IoT) devices find usage as it is central to leveraging these improved capabilities since they facilitate the automation and connectivity to support efficient grid management. The large-scale survey conducted in Copeland et al. (2019) brings to the fore the necessity of IoT in enabling different network activities through the energy delivery system. This change is not only in line with optimizing operational effectiveness but also with ICS networks' resilience against potential interference, making it important for contemporary design factors. In addition to security and connectivity, scalability is a critical factor in designing ICS networks. Since industrial environments continue to evolve with developments in newer technologies, the infrastructure behind the support network must be capable of sustaining growth and evolution. This implies selecting modules that are of a modular nature such that they can be easily connected or swapped without necessitating significant overhauls of the existing systems. The test is not merely maximizing current resource utilization but also forecasting the demand of the future, demanding careful planning and foresight in the design phase. By making scalability a priority in ICS networks, organizations can ensure long-term stability and adaptability, therefore placing themselves to take advantage of future breakthroughs in technology without jeopardizing the integrity and continuity of their industrial processes.

2.3.1. Scalability and Flexibility

In the area of Industrial Control Systems (ICS), flexibility and scalability are key aspects that have a direct influence on the responsiveness and effectiveness of network designs. Scalability is a matter of how a system can accommodate expansion in such a way that one can increase workload without compromising performance. In ICS, this means that as the demand in production grows, the network should be able to expand simply in terms of having more devices and more data throughput. Scalability becomes relevant through the increased use of Internet of Things (IoT) technology in industry, which generates a vast number of networked devices producing vast amounts of data to be processed efficiently. Scalable systems with their architecture and middleware solutions designed for scalability can therefore address evolving industrial demands better with unbroken, continuous operational effectiveness. Flexibility, though, deals with the ability of a network structure to adapt itself to changing technologies and operations. Flexibility in an industrial system comprises implementing different technologies and the capability to redesign structures whenever needed. Flexibility is important in the face of rapid technological advancements for automation technologies and the movement towards more intelligent systems. The use of Middleware in the Cloud of Things (CoT) is particularly important in this regard because it enables seamless communication between heterogenous devices, thereby enhancing interoperability between ICS networks. Agile architectures are more likely to be capable of supporting innovations such as machine learning and artificial intelligence, enabling industries to react to market volatility and technological disruption while continuing to provide good degrees of performance and service stability (Farahbakhsh et al., 2017). Along with this, scalable and adaptable network structures need to be incorporated for effective data handling in Industrial Control Systems. As ICS is now relying more on data analytics for decision-making and optimizing operations, handling large volumes of data generated by multiple devices becomes the most critical feature. For instance, cloud computing technology, when coupled with IoT, provides the necessary infrastructure to manage big data processing and storage that is generated in the industry. Such massive data volumes, particularly when it comes to aggregation and analysis, can be managed by elastic cloud services that support ease of scaling of resources as demand increases. By employing these technologies, ICS is not only capable of achieving scalability in operation but also flexibility to cater to some analytical demands as well as the evolving standards of the industry (Farahbakhsh et al., 2017) (Figure 2.10).

Figure 2.10. The scalability and flexibility.

Source: https://assets.weareadaptable.com/app/uploads/2024/04/24122832/adapt-able-utilita-2-1.jpg.

2.3.2. Reliability and Redundancy

The importance of reliability in industrial control systems (ICS) cannot be overstated, especially as these systems are increasingly becoming a part of critical infrastructure operations. Reliability ensures that systems will operate correctly in the long term, even when subject to component failure or poor environments. With manufacturing and utilities increasingly being merged with digital technology, the risk of disruption due to system failure makes rugged designs that can survive such adversity more necessary. A number of techniques can be used to enhance reliability, first through redundancy. This concept involves the duplication of principal components or systems to permit ongoing operation when a part is failing. Hence, the design framework of ICS must incorporate mechanisms for reliability that not only have fault tolerance but also enable seamless handover upon component failure so that operations are not suspended. Adding redundancy in network frameworks significantly increases the operational integrity of ICS. Redundant systems can be either hot or cold, based on whether they operate in parallel or are only initialized on the failure of a primary system. For instance, implementation of diverse replicated channels in control systems ensures that when one of the channels fails, another can function without being shut down (Burns et al., 2010). The dual-channel system is an example of a strategic method of improving reliability and addressing potential dependencies among channels. These configurations then must be engineered in a way to evaluate and control risks in terms of aleatory and epistemic uncertainties. Consequently, redundancy not only acts as a fail-safe but also builds a structure that achieves a balance between performance and reliability, thus resulting in higher overall effectiveness of the network architecture in a wide array of industrial applications. Furthermore, emerging developments highlight the need for innovative steps to protect against redundancies' inherent weaknesses, particularly those reliant on external synchronizations such as GPS. While satellite-based systems offer high precision for timing critical in ICS, they are susceptible to interference that may undermine operation (Ahmed faheem Zobaa et al., 2014). It is therefore essential to incorporate an integrated study of these vulnerabilities to ensure the robustness of wide-area monitoring systems. Experts suggest decentralized defense systems utilizing self-organizing sensor networks in order to guarantee synchrony regardless of potential cyber-attacks. Initiatives such as these supplement the ICS reliability framework by tackling potential external attacks. The convergence of reliability, redundancy, and advanced security measures in network design is critical in designing ICS capable of withstanding the advanced nature of modern industrial environments.

2.3.3. Performance and Latency

The performance-latency relationship holds a prominent position in Industrial Control Systems (ICS), where the transmission of data is timely and stable in ensuring efficient operation. Performance in this instance encompasses the throughput of network architecture and computational capacity, and latency is employed to refer to the time delay observed during the data processing and transmission intervals. This relationship becomes especially critical in systems that must support real-time decision-making, say in manufacturing or process control. A high-performance network guarantees that data can be collected and exchanged quickly, reducing latencies that can disrupt automated operations. As industrial systems integrate more and more digital technologies, familiarizing oneself and managing this performance-latency relationship is key to achieving the desired operational outcomes. Latest developments in network architectures, for example, the use of Wormhole-based mesh Networks-on-Chip (wNoC), depict the continued pursuit of optimizing performance without compromising on latency. These architectures enjoy scalability and cost advantages but raise challenges in providing tight time-composable Worst-Case Execution Time (WCET) estimates for safety-critical embedded systems. Bandwidth management is essential in such systems so that WCET computation does not impair average performance significantly. The consequence for ICS is considerable, as industries increasingly rely on precise timing and synchronization of procedures to ensure maximum efficiency. The ability to keep performance intact while accepting reasonable levels of latency is therefore critical to the deployment of complex network architecture in industry environments (Ferrer et al., 2016). Furthermore, the integration of DSP in FPGAs is an actual-world solution to boosting performance with the mitigation of latency. FPGAs inherently offer parallelism to accelerate DSP computations, basically fulfilling the requirement of faster processing of data. With the introduction of dedicated DSP blocks, modern FPGAs have augmented their capacity to process sophisticated algorithms without overstepping diversity. For instance, studies have shown that big convolutional filters are effectively emulated on FPGAs with great frame rates demanded in high-definition video processing. This capability proves to be particularly useful for ICS applications where rapid analysis and response of data are necessary. As sectors continue to integrate high-level digital solutions, the search for utmost performance without delay will remain an underpinning issue in network architecture design in such a scenario (Al-Dujaili et al., 2020).

2.4. COMMON NETWORK TOPOLOGIES IN INDUSTRIAL CONTROL SYSTEMS (ICS)

In an ICS, network topology is important in delivering effective communication among elements. Bus, star, and ring topologies are typical, each with pros and cons. The bus topology, sometimes appreciated due to simplicity and low installation costs, has all devices connected to a common communication line. But with some drawbacks: a failure on the bus can take the entire network down. For comparison, star setup, where everything connects to a central hub, enhances reliability and simplifies troubleshooting. While more immune to failure of individual devices, this central arrangement creates possible points of vulnerability at the hub. As ICS evolves, understanding these topologies helps create more secure and efficient networks, tailor-made for the unique needs of industrial environments. Network design is more than connectivity;

it is central when used for security, particularly in cybersecurity systems for ICS. Communication is the basis of supervisory control and data acquisition (SCADA) systems, which necessitate a thorough comprehension of their architectures. Research highlights that security in SCADA systems is highly reliant on decomposing their structures, which can be complex and multi-layered depending on the application context. For instance, layered architectures not only facilitate the simple segregation of roles within the system but also help establish common ground among the stakeholders, thereby enhancing co-operation in security analysis. By including these security considerations in ICS network design, experts have a greater ability to respond to threats such as cyberattacks on architectural weaknesses, now an even more pertinent concern in today's computer age (Ma et al., 2012). Technology advancements bring into question traditional network topologies within ICS, particularly given the expanding requirement for interoperability with Cyber-Physical Systems (CPS). Modern tools, including the MiniCPS simulation toolkit, provide guidance on ways these topologies can be re-engineered or optimized for enhanced communication and protection from the dangers of existence. Through the simulation of typical CPS components, such as programmable logic controllers, MiniCPS provides a platform for the testing of potential vulnerabilities in network architecture. With the challenges of cybersecurity hitting industries, testing and implementing special defenses through practical avenues such as the use of MiniCPS is achievable in diversified topological environments. Hence, using newer methods alongside traditional topologies is essential to rendering ICS not only functional but also secure against future cyber-attacks, making them usable in today's industrial environment securely (Antonioli et al., 2015).

2.4.1. Star Topology

Star topology is one of the most prevalent network configurations employed in industrial control systems (ICS) due to its inherent advantages in performance and reliability. This is achieved by connecting multiple devices, or nodes, to one central hub or switch. The solitary point acts as a gateway for transmitting data, thus making devices' communication easier. One of the benefits of star topology is fault tolerance; when a node fails, remaining nodes continue to operate without interruptions. This aspect is highly necessary in ICS whereby systems have to be highly available in order to keep running and ensure safety. The centralized nature of star topology also simplifies the management and debugging of the network since any issue can immediately be targeted to individual connections rather than to the entire network. The key to using star topology in ICS is how network structure and Internet of Things (IoT) integration play off each other. As industrial businesses shift to smart grid infrastructure, there becomes a more and more evident requirement for resilient, flexible network architectures (Copeland et al., 2019). Star topology achieves this expansion through the ability to provide a framework where numerous IoT devices, such as sensors and actuators, can easily connect into a single monitoring system. This feature allows data collection, thus enabling real-time analysis and decision-making, which are pivotal in optimizing industrial processes. Star topology also enables simple changes and scalability as it is flexible, thus making it easy for industries to add devices in expanding networks without extensive redesigns. Nevertheless, the use of star topology is not without its limitations, especially regarding the central bottleneck. This is also a problem in applications with intense traffic or numerous attached devices, suggesting the use of good medium access control (MAC) protocols to control data streams properly (Alonso Zárate et al., 2016). Since industrial systems increasingly are becoming interconnected and reliant on the timely transmission of information, overcoming these constraints is required in order to maintain performance standards. Also, as devices increase, the danger of congestion can challenge the reliability of the network, necessitating proactive action. Therefore, constant advancements in the network design and communication protocols are essential to ensure the reliability and effectiveness of star topology in ICS in a bid to meet the increasing demands of modern industrial operations (Figure 2.11).

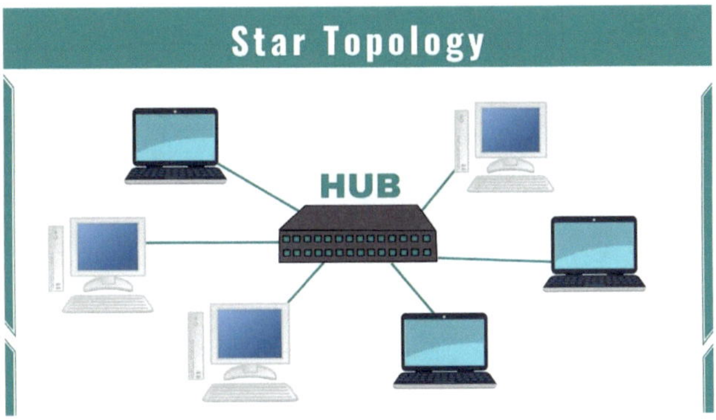

Figure 2.11. Star topology.

Source: https://educatecomputer.com/storage/2024/04/Star-Topology-Diagram.webp.

2.4.2. Ring Topology

Ring topology represents a fundamental network structure with its ring configuration of nodes in such a way that every device has direct links to two devices, forming a loop. This topology provides one-way data exchange that is capable of lowering data collision risks more prevalent in other topologies like bus or star. Ring topology is highly valued in terms of simplicity and simple installation in the scenario of Industrial Control Systems (ICS). It enables efficient data exchange between control systems and automatic machinery, which are crucial to maintaining the precision and reliability required in industrial environments. The systems leverage the guaranteed latency and natural redundancy that can be established in a ring topology, with the guarantee that if one link fails, data can still be transmitted through the other routes. Ring topology use in ICS has not been problem-free.

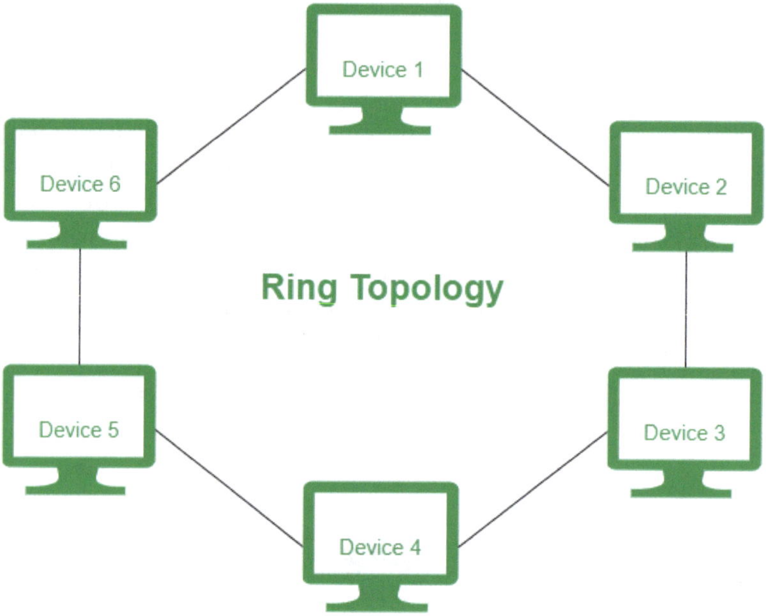

Figure 2.12. Ring topology.

Source: https://media.geeksforgeeks.org/wp-content/uploads/20200526102238/Untitled-Diagram-153-1.png.

For instance, the failure of a single node can cause disruption of communication throughout the network if redundancy is not taken. Using dual-ring topologies will reduce this threat by acting as a fail-safe. Additionally, the performance of the topology's can also be affected by the number of devices interconnected; as more units are interconnected, latency could rise because of the time taken for data to travel the whole loop. In high-traffic settings where data processing must be as rapid and as consistent as possible, network engineers must carefully weigh these trade-offs. The need for high-end network management software to monitor and optimize ring topologies also becomes apparent in highly advanced industrial settings as dynamic situations such as load or equipment status changes can have a significant influence on performance. Further, advanced technologies such as Microelectromechanical Systems (MEMS) and Field-Programmable Gate Array (FPGA) platforms are emerging to advance the performance and flexibility of ring topologies in ICS. Through the use of MEMS, it is possible to form low-cost wireless sensor networks that can be folded into a ring configuration to easily obtain accurate data from various points within the industrial environment. The versatility offered by FPGA technology facilitates real-time tuning of the network's architecture, allowing effective emulation space for the testing of configurations. The breakthrough is vital in the testing of network performance under a variety of conditions as it allows effective data fusion of the participating nodes. These developments highlight the continuous evolution of ring topology in ICS, demonstrating its capability to address the requirements of contemporary industrial uses and improve reliability as well as efficiency in monitoring and control systems (Ardaiz et al., 2025; Braeken et al., 2018) (Figure 2.12).

2.4.3. Mesh Topology

In the context of industrial control systems (ICS), mesh topology represents a strong network architecture developed for reliability and redundancy. In contrast to conventional network topologies, mesh topology interconnects each node with all others, allowing for multiple paths for data transmission. This topology is beneficial in situations where reliability is paramount, as it enables data to travel on alternate routes if a link is down. Therefore, mesh topology's reliability to a great extent minimizes the frequency of system failures, which is especially pertinent in ICS where downtime has the probability of leading to substantial economic loss and safety threats. Given the increasing complexity of industrial applications, scalability and flexibility ensured by mesh topology make it an attractive choice for organizations seeking to ensure maximum operational continuity and productivity. The application of mesh topology in industrial control networks is also usually coordinated with the adoption of more sophisticated networking paradigms such as Software Defined Networking (SDN). This is described in newer research where frameworks like Atomic-SDN specifically capitalize on the changing requirements of low-power wireless networks to optimize communication over mesh topologies. Thanks to its Synchronous Flooding (SF) structure, Atomic-SDN proves the ability to dynamically respond to varied traffic patterns, thereby maximizing data transmission in the mesh network and guaranteeing minimal latency and higher reliability. The optimization is not only based on the topmost requirements of ICS but also prioritizes the synergy of new software approaches and the existing hardware infrastructures. Lastly, the synergistic relationship between mesh topology and SDN can lead to better utilization of resources and enhanced performance in industrial applications, paving the way for future innovation. The need for dependable communication in mesh topologies exceeds conventional networking paradigms, especially in high-risk environments like ICS, where communication failure is unacceptable. Emerging technologies, like nanoscale non-volatile memory, can redefine the dynamic of communication within mesh networks. These advancements are able to address problems surrounding large-fanout connectivity to further enhance the system's ability to maintain large data transfers predominant in industrial processes. By placing importance on the design of focused on-chip routing networks tailored for the needs of mesh architectures, these advancements are capable of significantly optimizing operation efficiency and alleviating latency issues. The implications are profound for ICS, and future network architectures will increasingly rely on cutting-edge memory and routing technologies to support the increasingly complicated requirements of industrial automation and control systems (Figure 2.13).

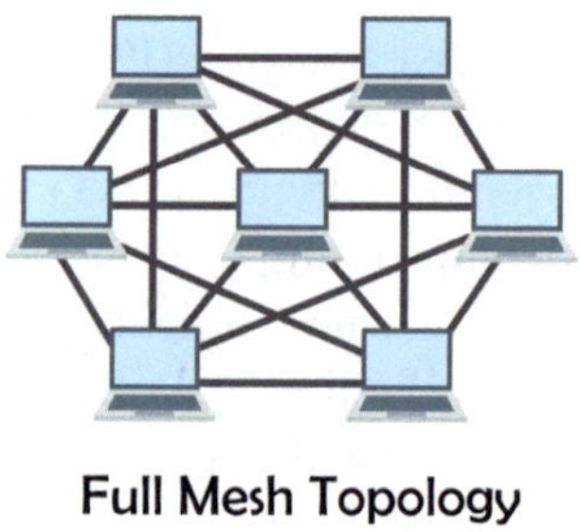

Figure 2.13. Mesh topology.

Source: https://itrelease.com/wp-content/uploads/2021/06/Full-Mesh-Topology.jpg.

2.5. CHALLENGES IN INDUSTRIAL CONTROL SYSTEMS (ICS) NETWORK ARCHITECTURES

ICS's shift from siloed environments to networks has imposed serious challenges. Traditionally, ICS operated on air-gapped networks via proprietary protocols that provided a degree of isolation from the external world. However, with the move towards increased connectivity, weaknesses have been created as such systems interface with enterprise networks to facilitate operating efficiency. In making data exchange possible through TCP/IP communications, organizations can automate production processes, but in the process end up exposing vital control systems to potential cyber-attacks. As explained in Johnson et al. (2016), this transformation from air-gapped systems to networked designs calls for reassessing cyber security defenses, citing the vulnerability inherent in current ICS networks. The state of sophistication of cyber-attacks on the interconnected systems is a harsh reality check of the thin line between operational effectiveness and security. The other pressing concern in ICS network systems is the incorporation of Internet of Things (IoT) technology, adding yet more complexity to security protocols. The Industrial Internet of Things (IIoT) enables a strong web of intercommunicating devices, supporting real-time monitoring and automated decision-making across various industrial processes. Though such progress has potential for improving operations, it also brings in vectors of cyber-attacks. Basing devices on TCP/IP protocols – which were originally developed for non-critical usage – in safety-critical contexts raises the risk of compromise. According to Evangelopoulou et al. (2017), merging the operational technology with safety-critical applications and enterprise information systems has been particularly worrying, as it makes ICS vulnerable to a larger variety of attack surfaces. Consequently, ICS must fight the dual battle of leveraging IoT innovations while having strict security measures in place. Finally, the complexity of ICS network structures has accelerated with organizations struggling to attain digital transformation. The diverse collection of legacy systems, proprietary protocols, and emerging technology infrastructure adds a labyrinth-like environment to security management. Industrial plants are mostly operating on ancient systems not designed to thwart newer-generation cyber threat attacks, thereby adding to the risk. In addition, merger of various communication protocols like Modbus and Profibus with new technologies adds layers of complexity in sustaining common security protocols. This intricate ecosystem presents regulatory and compliance challenges in that regulatory bodies expect strong safeguarding of critical infrastructure. Addressing these challenges calls for a forward-looking approach, drawing from experience in previous intrusions and continually evolving security practices to ensure ICS resilience, as discussed in Evangelopoulou et al. (2017) and Johnson et al. (2016).

2.5.1. Cybersecurity Threats

The increasing convergence of digital technologies into Industrial Control Systems (ICS) has evolved into a two-bladed sword, driving productivity while concurrently introducing such systems to formerly unknown cybersecurity risks. While organizations seek automation and connectivity, the age-old safety-first approach to ICS design can be overlooked, providing vulnerabilities that are exploitable by attackers. This paradigm change has made a paradigm shift in understanding the threat personality imperative. The requirement to enhance cybersecurity architectures has been reported by the Italian Nation research community with a focus on critical infrastructure and controls that are essential for effective cyber defense. By defining areas of concentration for countering threats, stakeholders can create a more secure ICS cyberspace that not only reacts to, but also proactively acts against vulnerability opportunities prior to their exploitability (Anglano et al., 2018).

Figure 2.14. Cybersecurity threats.

Source: https://www.compuquip.com/hubfs/images/blog-images/cybersecurity-threats-2022-min.jpeg.

Particularly, the ICS cyberattack case has accelerated, targeting the central sectors of such critical infrastructure as energy, transport, and water. Keeping in mind that a majority of such systems depend on legacy hardware, if left unsecured against modern threats, the same can be radically compromised. Even more critically, industrial cyber threats are not theoretical; empirical studies such as test bed-based experiments illustrate the real-world effect of cyber-attacks.." For instance, research conducted on IEC 61850 utility automation systems identifies the importance of understanding how attacks happen and influence operating functionality. Such experimental verifications give insight into the effectiveness of various countermeasures along with physical effects of attacks such as false data injection to reveal vulnerabilities of existing architectures (Kabir-Querrec et al., 2016). Finally, the merging of cybersecurity and ICS must be done with a universal approach to design and implementation. It is important that organizations realize that enhancing cybersecurity is not merely a matter of adding the newest technology, but creating a culture of awareness and preparedness among employees. Training and risk management policies must be a component of cybersecurity policy frameworks because human mistakes are usually what cause breaches. Periodic risk analysis will provide direction for iterative improvement of security measures, thereby ensuring that systems can withstand and recover from probable intrusions within a minimal duration. Through a combination of technology innovation and the keen eye of human observation, the industrial environment can strive to attain an assured network design that minimizes vulnerabilities while maximizing functional effectiveness (Anglano et al., 2018) (Figure 2.14).

2.5.2. Integration with Legacy Systems

The integration of legacy systems into contemporary industrial control system (ICS) network infrastructures is difficult as it is ripe with possibility. Historically, these systems operated in isolation, often "air gapped" from entire networks, thereby limiting access and possible points of attack. As industrial environments evolve to increased interconnectedness, reflected by the introduction of Internet of Things (IoT) paradigms, the imperative to connect these legacy systems to present TCP/IP networks reaches a critical juncture. This transformation supports seamless data exchange between various operation technology (OT) platforms, thereby enhancing the production and enabling enterprise-wide decision-making. The integration must be carried out in a cautious approach, taking into account the array of security threats that are associated with increased connectivity between various systems, necessitating a robust system of secure communication and data exchange (Johnson et al., 2016). One specific issue that is brought up due to legacy system integration is the inherent security vulnerabilities present in these legacy technologies. Legacy systems will rely on proprietary protocol and operating systems that are no longer receiving the maintenance and upgrades necessary to protect against new and emerging cybersecurity threats. The progressive deterioration of the air gap that formerly separated these systems has led to the integration of switches and gateways that bridge industrial protocols such as Modbus and Profibus to enterprise information systems. Consequently, while such connectivity makes possible real-time monitoring and optimization of production processes by senior management, it also brings to risk foundational infrastructure. The challenge is to balance operational effectiveness with security and integrity needed to shield safety-critical systems from adversarial intrusions (Evangelopoulou et al., 2017). Additionally, the integration activity extends beyond connecting links; it encompasses detailed considerations of architecture, data flows, and associated risks. Developing a network architecture that incorporates legacy systems requires extensive knowledge of the infrastructure at hand and its limitations. Stakeholders must assess not only the technology infrastructure but also organizational processes that govern system interactions and uses of data. Taking a holistic approach that puts policies on connectivity as well as security policies in place can help organizations sidestep threats while making valuable insights from legacy systems. Lastly, building a safe space for legacy system integration serves as the gateway to operational technology innovation, enabling industries to enjoy the benefits of modern network architecture without compromising safety and reliability (Evangelopoulou et al., 2017; Johnson et al., 2016) (Figure 2.15).

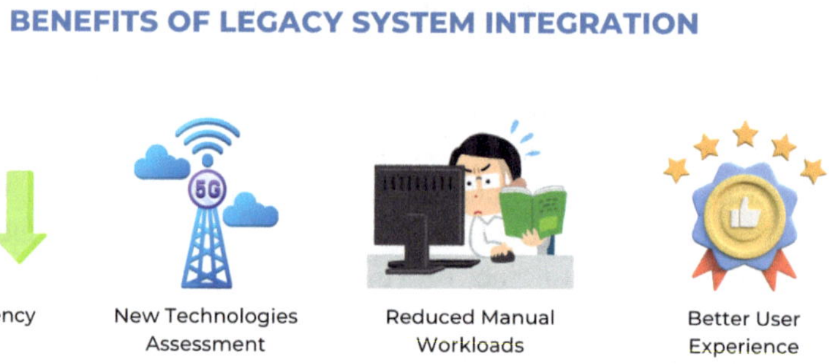

Figure 2.15. The integration of legacy systems.

Source: https://miro.medium.com/v2/resize:fit:1400/1*uN6Tzdx4N_mjkyGxy-ILjGQ.png.

2.5.3. Compliance with Industry Standards

Industry compliance in the ICS global community is not a regulatory requirement; it is among the top requirements for holding on to operational effectiveness and cybersecurity. As more organizations embrace new technologies, especially those in connection with the Internet of Things (IoT), they must upgrade their current infrastructure to accommodate shifting compliance requirements. This involves not just adhering to established standards but also conducting self-evaluations to determine their cybersecurity position. By conducting such evaluations, organizations can determine vulnerabilities that may arise due to integration of IoT devices, allowing them to proactively work on minimizing potential risks. The lack of standard self-assessment methods for quantifying IoT cyber risk emphasizes the necessity to develop new methods in alignment with organizational security goals and operational needs, which results in more resilient ICS networks against outside attacks (Ani et al.,

2020). Moreover, industry standards call for ongoing education and training of personnel performing operating and controlling ICS. With the continuous advancements in technology and the increase in frequency of cyber-attacks, organizations should make appropriate investments in comprehensive training programs that encompass regulatory compliance as well as best practices in cybersecurity. Organizations should make sure they convey to employees the value of compliance with the standards, as their immediate actions contribute to the overall security and reliability of the system. In addition, businesses ought to remain current on any emerging developments or changes in relevant standards so that their policy of compliance is consistent with the rate of progress in the industry. This commitment to ongoing education not only fosters a culture of compliance but also equips employees with the technology required to react effectively to emerging threats to the digital world (Ai et al., 2012). Finally, the convergence of state-of-the-art technologies, such as multi-agent systems, highlights the need for compliance frameworks to keep pace with the advancement of ICS architecture. Such decentralized intelligent systems support a smooth and agile operation environment, which allows for increased interaction among components. Properly leveraging such systems, however, hinges upon adherence to industry standards that govern their design and implementation. Organizations not only need to abide by existing standards but also need to be proactive when addressing the design problems of emerging technologies. Through the establishment of a proactive compliance attitude, organizations can take advantage of the features of modern architectures while having secure and robust networks against possible threats. Such harmonization of technology and compliance is critical to the evolution of secure and efficient industrial processes (Ai et al., 2012).

2.6. FUTURE TRENDS IN INDUSTRIAL CONTROL SYSTEMS (ICS) NETWORK ARCHITECTURES

Future ICS network architecture trends are a revolutionary step towards greater connectivity and interoperability. Traditionally, ICS operated in isolated environments with proprietary protocols like Modbus and private systems, limiting them to being integrated seamlessly with generic IT infrastructures. However, increasing demands for instant access to information and for operational efficiency are driving a paradigm shift toward ever more interconnected systems. Such a shift is in the spirit of the Internet of Things (IoT), where sensor and actuator data from disparate devices can be combined as seamlessly. As industries implement these transitions, there is an imperative need to address the consequent cybersecurity and safety challenges that these interconnections introduce, once more highlighting the requirement for robust solutions as outlined in existing literature (Johnson et al., 2016). Greater advancement in ICS network structures is further mandated by the concept of programmable networking that allows for responsive and adaptive design systems. The erstwhile rigid structure, plagued by the shackles of legacy, is being augmented with more programmable and dynamic frameworks. Techniques such as Software Defined Networking (SDN) and cognitive radio networks are gaining momentum, providing greater flexibility in the manner in which networks respond to varying operational requirements and traffic patterns. Such programmable networks are core to addressing the inefficiencies introduced due to the historical ossification of the Internet architecture. Such programmability not only optimizes resource use but also fortifies security defenses through the ability to dynamically adjust responses against emerging threats. Consequently, industries can attain operational excellence while safeguarding their networks against likely vulnerabilities (Ahad et al., 2014). In addition to programmability and connectivity, the future ICS network architecture also looks to incorporate more resilience through advanced data analytics and artificial intelligence. As control systems become more embedded in enterprise-wide processes, the ability to read large volumes of data in real-time will be crucial. Predictive analytics can help in making better-informed decisions by identifying patterns and potential anomalies well ahead of time before they escalate into serious issues. Additionally, embedding AI-based tools will help in speeding up automated responses, reducing downtime, and maximizing productivity. The coming together of these technologies is a new approach, with an emphasis on the pivotal role played by intelligent and flexible systems within industrial control. As industries continue to mature and incorporate these innovations, future studies will become imperative in exploring new issues and opportunities in creating next-generation ICS network architectures (Ahad et al., 2014; Johnson et al., 2016).

2.6.1. Adoption of Internet of Things (IoT) and Industrial Internet of Things (IIoT)

Implementation of the IoT and the IIoT is a revolutionary paradigm shift in the Industrial Control Systems (ICS) world. Historically, ICS relied on some level of isolation, typically employing an air-gapped architecture to ward off attacks pertaining to digital networking. The approach provided a sense of security since physical proximity was required for accessing control systems. However, with ongoing technology development, such segregation has gradually been diminished, enabling greater interconnectedness between disparate parts of industrial environments. With the convergence of IoT and IIoT technologies, organizations now have the capability to coordinate decentralized resources efficiently without losing operating integrity. Convergence of such networked systems enables real-time analysis of data, enhancing decision-making processes and operating efficiencies that were previously unachievable. Thus, the understanding of the effects of IoT and IIoT integration becomes imperative in order to handle contemporary ICS networks effectively. As organizations tend to embrace IoT and IIoT paradigms more and more, ICS architectural modifications are required due to the complex interdependencies introduced by such technologies. Interestingly enough, the transition to more integrated architectures, as noted in Boyes et al. (2018), has involved moving away from conventional zoned systems that appeared to rely heavily upon firewalls and other security measures to separate operational technology from informational technology systems.

Figure 2.16. The adoption of the internet of things (IoT).

Source: https://www.cleo.com/sites/default/files/internet-of-things-iot.jpg.

This transition calls for a re-evaluation of security architectures to fight probable vulnerability that arises from interconnected systems. The expansion of these dependencies has been accompanied by an increase in security concerns, particularly after recent cyberattacks on infrastructures around the globe. As these networks become more complex, security measures should also keep up at the same rate to be able to counter possible means of attack effectively to allow industrial systems to run securely in a connected world. Moreover, the IoT and IIoT transformation also necessitates better comprehension of their effects on energy efficiency, process improvement, and system robustness. The shift towards centralized decision-making, as discussed in Evangelopoulou et al. (2017), allows industries to more effectively track real-time streams of data, optimizing processes while simultaneously providing space for Industry 4.0 concepts. This not only improves productivity through improved use of resources but also allows industries to respond effectively to forces of the market and operational problems. Additionally, a careful analysis of vulnerabilities inherent in such advanced networks can assist in the formulation of full-fledged strategies for managing risks. As companies keep innovating and expanding their IoT and IIoT capabilities, they require ongoing research and analysis to both harness the benefits of these technologies and safeguard industrial control systems against potential threats (Figure 2.16).

2.6.2. Implementation of Cloud Computing

Implementation of cloud computing for Industrial Control System (ICS) network architectures is a paradigm shift in data management, processing, and use. Traditional ICS were localized systems that were prone to weakness in scalability and data accessibility. With the implementation of cloud computing, organizations can enhance their operational efficiencies and flexibility, enabling real-time data processing as well as analysis of vast networks. This paradigmatic change is necessary, particularly as companies increasingly struggle with increasing energy requirements and complex operational requirements. To have the ability to store and process large volumes of data from the cloud not only improves decision-making but also enables improved predictive maintenance as well as improved operational insights, which are required in modern industrial environments. One of the primary aspects of the deployment of cloud computing's in ICS is its intersection with the Internet of Things (IoT) in creating a robust infrastructure for Smart Grids (SGs). The parallel existence of IoT and cloud computing allows communication to flow bi-directionally between consumers and providers, facilitating a platform where energy management is optimized and made more responsive. IoT devices such as sensors and smart meters are capable of transmitting data directly to cloud platforms, enabling advanced monitoring and control capabilities. Cloud technology application in SGs enhances device connectivity and tracking, making real-time intervention possible. Such integration boosts the creation of innovative applications addressing long-existing issues in traditional power systems, such as reliability, security, and energy wastage, as outlined in Copeland et al. (2019). Moreover, the concept of the Cloud of Things (CoT) refers to the combined ability of cloud computing and IoT in creating intelligent industrial environments. Since the volume of data generated increases, management, aggregation, and storage issues arise. Cloud computing acts as a savior to address these issues by providing virtually infinite space for storage and processing. However, addressing the heterogeneity of devices within ICS remains the biggest issue.

Middleware technologies emerge as a central factor to facilitating smooth interaction among heterogeneous devices and applications. In carrying out their roles as intermediaries, they facilitate interoperability on any platform and architecture, thus enhancing the efficiency of service-oriented ICS as offered by Farahbakhsh et al. (2017). The judicious deployment of these cloud-based solutions has the potential to make substantial contributions to the resilience and efficiency of industrial operations in the modern world.

2.6.3. Advances in Artificial Intelligence (AI) and Machine Learning (ML)

Advances in AI and ML have had a significant influence on many sectors, one of which is industrial control systems (ICS). The application of advanced AI algorithms in ICS enables real-time data analysis and decision-making, giving way to enhanced operational efficiency and system reliability. For instance, machine learning algorithms can analyze vast sensor data and predict equipment failure, thereby minimizing downtime and also maintenance costs. Also, these advanced techniques enable automation of numerous control procedures, which makes it possible for high precision in controlling complex industrial settings. With more reliance by organizations on smart technologies, understanding and adopt the latest advancements in AI and ML is critical to maintaining competitive advantages and ensuring sustainable operation within the manufacturing sector. Interestingly, recent advances in deep learning have fueled enormous amounts of research centered around its applications to robotics, which is poised to affect industrial control systems profoundly. Advances in research concerning deep artificial neural networks hold the potential to enhance operational capability in numerous robotic systems utilized in ICS environments. The applications, benefits, and limitations of deep learning techniques were extensively examined in the literature, highlighting how these technologies can make automation of tasks and response in systems more efficient. The increasing integration of AI-based robots into ICS not only increases efficiency but also revolutionizes traditional practice to make industrial processes adapt to rapid technological advancements (Gashler et al., 2017). These advances are a paradigm shift in industrial control system design and implementation, parallel to the broader impact of AI for industry at large. However, as the context of AI and ML matures further, competition and security considerations come to the forefront, particularly in terms of national and international ramifications. The story around the revolution that AI technologies will produce is significant because those innovations will reshape military power and global geopolitics significantly. The revolutionary nature of AI is similar to other classic examples of technological revolutions, like the discovery of electricity or the internal combustion engine, with the inference that their impact is far greater than for industry. The projections anticipate that nations heavily investing in AI research and its application to the military can potentially achieve great dividends, potentially reshaping the balance of power worldwide (Horowitz et al., 2018). In ICS, this calls for a double focus on leveraging cutting-edge AI solutions as well as safeguarding these systems from emerging cybersecurity risks, creating resilience in an increasingly connected and competitive world.

SUMMARY

In Summary, the evolving world of ICS necessitates in-depth knowledge of their network topologies in order to provide extended functionality as well as security. As set forth in research on SCADA systems, an architecture approach consisting of multiple facets is one essential prerequisite to secure infrastructures like power grids. The layer architectural viewpoint proposed gives a blueprint permitting more comprehensible communication among actors and thereby opening up for effective security analysis. This stratified perspective, stringent in nature, calls for the application of a holistic approach towards the remediation of security vulnerabilities and operational performance within ICS systems. The multi-layered approach opens scope for more detailed dialogue regarding the application of security, which is required to safeguard these vital systems in this day and age (Ma et al., 2012). Furthermore, deployment of emerging technologies such as smart energy solutions makes security dynamics in ICS networks even more complex. Ensuring robust cybersecurity frameworks is crucial, particularly when communicating with systems that are electric storage technology-based. This is because requirements set out for communications gateways highlight the delicate balance of architecture versus security. These suggestions not only enhance security of the communication lines but also address significant points such as data storage and privacy. All these steps are part of the greater process of modernizing and securing the energy grid to minimize threats owing to increased complexity in networked industrial environments (Marksteiner et al., 2017). Ensuring the resilience of such systems ultimately fulfills greater global aims for energy efficiency and security in the context of increasing cyber threats. Ultimately, as industrial ecosystems continue to become increasingly interconnected, the demand for good network architectures in ICS will continue to be increasing in significance. Cyber-attacks accentuate challenges imposed by increasing the demand for a strategic approach that is both security-focused and operationally optimized. Seeking a common architectural framework capable of accommodating diversified stakeholders to obtain system integrity is crucial to future growth. The consensus approach detailed via layered architectures is a feasible means of navigating these intricacies. By collaboration on security architectures and solution designs, future development can become strong ICS networks that not only execute vital functions but also safeguard against continually evolving threats during an age of technological innovations. Strategic coordination of such magnitude is crucial to sustaining the operational effectiveness of industrial sectors, hence facilitating underpinning economic and social infrastructure.

In light of the changing environment of network architecture in ICS, the advent of millimeter-wave (mmW) technology is a critical leap forward. Specifically, the 60 GHz unlicensed band has become a profitable option for enabling gigabit short-range wireless communication. Based on the IEEE standard 802.11ad, or more commonly known as WiGig, the frequency band enables high-data-rate transmission that is necessary for bandwidth-hungry applications common in ICS environments. However, the inherent mmW transmission limitations of path blocking susceptibility and channel propagation loss necessitate the use of several mmW access points (APs) deployments to ensure proper coverage. This consequently implies coordinated management techniques among the APs in an effort to avoid packet collisions and enhance the overall network capacity, thereby optimizing data flow within ICS architectures (Kusano et al., 2015). Moreover, safety of ICS, specifically from a SCADA systems standpoint, is of utmost concern. They are trusted to manage imperative infrastructure that provides a direct contribution to society's operation, like power grids. With SCADA systems being more and more the focus of cyber-attacks, it is imperative to implement a robust architectural base to ensure safety against pending threats. The complexity of SCADA systems renders it very difficult; therefore, through the establishment of a layered architectural model, security analysis and implementation can be simplified. Through the organization of the architecture into four interconnected layers and applying viewpoint concepts, stakeholders are able to address specific subsets of the system with precision, thus fostering an integrated approach to security that is vital for ICS operation integrity (Ma et al., 2012). In general, the integration of next-generation network topologies, including mmW technology and SCADA system topology, is key to enhancing Industrial Control Systems' performance and security. The parallel focus on high-capacity wireless communication and stringent security measures is a reflection of the necessity for ICS to adapt to present challenges. With businesses embracing digitization and automation more and more, emphasizing orchestrated mmW deployment will facilitate smooth and secure data exchange, whereas sound architectural know-how will shield systems from cyber-attacks. Thus, technology enhancement and security infrastructures collectively chart the future state of ICS, where optimization and robustness are assigned topmost priority to guarantee business continuity (Kusano et al., 2015; Ma et al., 2012).

REFERENCES

1. Abella Ferrer, J., Cazorla Almeida, F. J., Hernández, C., Panic, M., et al. (2016). Improving performance guarantees in wormhole mesh NoC designs. https://core.ac.uk/download/81578332.pdf (accessed on 13th September 2025).
2. Ahad, N., Ahmed, N., & Qadir, J. (2014). Building programmable wireless networks: An architectural survey. http://arxiv.org/abs/1310.0251 (accessed on 13th September 2025).
3. Ahmed Faheem, Zobaa, A., Vaccaro, A., Behrendt, K., Celano, T., Dickerson, B., Dion, A., Formato, G., et al. (2014). Vulnerability analysis of satellite-based synchronized smart grids monitoring systems. https://core.ac.uk/download/29204066.pdf (accessed on 13th September 2025).
4. Ai, A., Beck, M., Belkacemi, R., Bellifemine, F., Brown, N., Catterson, V., et al. (2012). Practical applications of multi-agent systems in electric power systems. https://core.ac.uk/download/9033895.pdf (accessed on 13th September 2025).
5. Al-Dujaili, A., Fahmy, S. A., & Ioannou, L. (2020). High throughput spatial convolution filters on FPGAs. https://core.ac.uk/download/323057966.pdf (accessed on 13th September 2025).
6. Al-Khateeb, H., Burnham, K. J., Epiphaniou, G., Hammoudeh, M., et al. (2019). Preliminaries of orthogonal layered defense using functional and assurance controls in industrial control systems. https://core.ac.uk/download/196185875.pdf (accessed on 13th September 2025).
7. Alonso Zárate, J., Alonso Zárate, L. G., Kalalas, C., Laya, A., et al. (2016). Goodbye, ALOHA! https://core.ac.uk/download/81571287.pdf (accessed on 13th September 2025).
8. Anglano, C., Aniello, L., Antinori, A., Armando, A., et al. (2018). The future of cybersecurity in Italy: Strategic focus area. https://core.ac.uk/download/188830442.pdf (accessed on 13th September 2025).
9. Ani, U., Anthi, E., Burnap, P., De Roure, D., et al. (2020). Dynamic real-time risk analytics of uncontrollable states in complex internet of things systems, cyber risk at the edge. https://core.ac.uk/download/491119431.pdf (accessed on 13th September 2025).
10. Antonioli, D., & Tippenhauer, N. O. (2015). MiniCPS: A toolkit for security research on CPS networks. http://arxiv.org/abs/1507.04860 (accessed on 13th September 2025).
11. Ardaiz, O., Chacin, P., Chao, I., Cruellas, S., et al. (2025). Proof-of-Concept Application–Annual Report Year 1. https://core.ac.uk/download/pdf/6496817.pdf (accessed on 13th September 2025).

12. Baddeley, M., Nejabati, R., Oikonomou, G., Raza, S., et al. (2019). Atomic-SDN: Is synchronous flooding the solution to software-defined networking in IoT? https://core.ac.uk/download/218597508.pdf (accessed on 13th September 2025).

13. Bartz-Beielstein, T., Bunte, A., Faeskorn-Woyke, H., Fischbach, S., et al. (2019). Evaluation of cognitive architectures for cyber-physical production systems. http://arxiv.org/abs/1902.08448 (accessed on 13th September 2025).

14. Boyes, H., Cunningham, J., Hallaq, B., Watson, R., et al. (2018). The industrial internet of things (IIoT): An analysis framework. https://core.ac.uk/download/158139219.pdf (accessed on 13th September 2025).

15. Braeken, A., da Silva Gomes, B., Domínguez, F., Touhafi, A., et al. (2018). Exploiting partial reconfiguration through PCIe for a microphone array network emulator. https://core.ac.uk/download/158345320.pdf (accessed on 13th September 2025).

16. Bucher, C., Xia, F., Ma, L., Wittenmark, B., Xia, Y., et al. (2008). Integrated design and implementation of embedded control systems with Scilab. https://core.ac.uk/download/18972242.pdf (accessed on 13th September 2025).

17. Burns, A., & Littlewood, B. (2010). Reasoning about the reliability of multi-version, diverse real-time systems. https://core.ac.uk/download/9559398.pdf (accessed on 13th September 2025).

18. Copeland, R., Crespi, N., Rehmani, M. H., Saleem, Y., et al. (2019). Internet of Things-aided smart grid: Technologies, architectures, applications, prototypes, and future research directions. http://arxiv.org/abs/1704.08977 (accessed on 13th September 2025).

19. Díaz González, F., Gallart, R., Girbau Llistuella, F., Heredero Peris, D., et al. (2018). Smart grid architecture for rural distribution networks: Application to a Spanish pilot network. https://core.ac.uk/download/157811722.pdf (accessed on 13th September 2025).

20. Evangelopoulou, M., Harkness, R., & Johnson, C. W. (2016). Forensic attacks analysis and the cyber security of safety-critical industrial control systems. https://core.ac.uk/download/46664183.pdf (accessed on 13th September 2025).

21. Evangelopoulou, M., Johnson, C. W., & Pavlova, T. (2017). Applying lessons from cyber-attacks on Ukrainian infrastructures to secure gateways onto the industrial Internet of Things. https://core.ac.uk/download/111557942.pdf (accessed on 13th September 2025).

22. Farahbakhsh, R., Farahzadia, A., Rezazadeh, J., Shams, A., et al. (2017). Middleware technologies for Cloud of Things–A survey. https://core.ac.uk/download/333623071.pdf (accessed on 13th September 2025).

23. Gashler, M. S., & Pierson, H. A. (2017). Deep learning in robotics: A review of recent research. http://arxiv.org/abs/1707.07217 (accessed on 13th September 2025).

24. Ghazi, H. E., Ghazi, H. E., Kaabouch, N., Mrabet, Z., et al. (2018). Cyber-security in smart grid: Survey and challenges. http://arxiv.org/abs/1809.02609 (accessed on 13th September 2025).

25. Holmgren, F., & Janson, S. (2008). Interactive visual analysis of networked systems: Workflows for two industrial domains. https://core.ac.uk/download/11434682.pdf (accessed on 13th September 2025).

26. Horowitz, M. C. (2018). Artificial intelligence, international competition, and the balance of power (May 2018). https://core.ac.uk/download/211333744.pdf (accessed on 13th September 2025).

27. Hughes, A. (2002). Middleware for managing a large, heterogeneous programmable network. https://core.ac.uk/download/1669445.pdf (accessed on 13th September 2025).

28. Johnson, C. (2016). Securing the participation of safety-critical SCADA systems in the industrial Internet of Things. https://core.ac.uk/download/46664188.pdf (accessed on 13th September 2025).

29. Kabir-Querrec, M., Mocanu, S., Savary, E., Thiriet, J. M., et al. (2016). A test bed dedicated to the study of vulnerabilities in IEC 61850 power utility automation networks.

30. Kusano, H., Miyamoto, S., Mizukami, M., Mohamed, A., et al. (2015). Millimeter-wave wireless LAN and its extension toward 5G heterogeneous networks. http://arxiv.org/abs/1507.04518 (accessed on 13th September 2025).

31. Ma, Z., Skopik, F., & Smith, P. (2012). Towards a layered architectural view for security analysis in SCADA systems. http://arxiv.org/abs/1211.3908 (accessed on 13th September 2025).

32. Manohar, R., & Moradi, S. (2018). The impact of on-chip communication on memory technologies for neuromorphic systems. http://arxiv.org/abs/1809.06016 (accessed on 13th September 2025).

33. Marksteiner, S., & Vallant, H. (2017). Towards a secure smart grid storage communications gateway. http://arxiv.org/abs/1710.03017 (accessed on 13th September 2025).

CHAPTER 3

COMMON THREATS AND VULNERABILITIES IN INDUSTRIAL SYSTEMS

Contents

INTRODUCTION

The introduction to any conversation about industrial systems necessitates the examination of the vulnerabilities and threats inherent in the systems in a more interconnected world. As industries are becoming more reliant on computerized and automatic technologies, they become more susceptible to cyber-attacks. The incorporation of sophisticated networks into industrial control systems, while making them more efficient, also opens up a universe of security issues, especially since these systems are designed with little resilience to meet the threats posed by malicious attacks. It is vital to grasp the topography of such vulnerabilities in creating strong risk management strategies and cyber defense. Besides, effective security measures must be informed by a comprehensive appreciation of technological and human weaknesses that can penetrate industrial networks, evolving into a fluctuating dynamic that is continuously testing the limits. Despite the broadness of the threat landscape, there are some inherent vulnerabilities that always seek common denominators across all industries. Among the most significant is the reliance on legacy infrastructure that, by the nature of software and hardware being antiquated, present soft targets for cyberattacks. As mentioned, the patches and protections put in place on these systems always lag technology advancements, rendering them systemically exposed. Secondly, increased automation, as beneficial, has rendered critical infrastructure vulnerable to large-scale coordinated attacks that result in widespread disruptions. In order to address these weaknesses, industries must give highest priority to assessment and enhancing their cybersecurity mechanisms with protocols that keep evolving with the threats. With open-mindedness towards continuous analysis and improvement, corporations can build stronger defenses against the advanced integration of existing cyber threats. Education is an essential countermeasure for the weaknesses that pervade industrial systems. Knowledge integration across disciplines is vital, with the evolving world demanding employees who possess a strong understanding of the interdependence of the critical infrastructure systems. Emphasized by studies that feature critical infrastructure platforms, building an interdisciplinary attitude prepares future professionals to handle the complexity of the issues of our time. Ongoing need for cyber security and risk management training indicates the imperative need for institutions of learning to be forward-thinking. Building capability and awareness within the workforce results in preparedness for threats affecting not just operational integrity but functionality of society. It is finally a well-prepared workforce with the knowledge and skills required to create innovative solutions that will defend industrial systems from emerging vulnerabilities and threats, setting the role of education in the critical industry.

Definition of Industrial Systems

Industrial systems refer to a set of interdependent constituents, such as machinery, technology, human factors, and processes that act in unison to deliver products or services. The systems exist within intricate networks, where every component plays an important role in the overall performance and effectiveness. Knowing the definition of industrial systems is necessary for determining their inherent weaknesses and common vulnerabilities. As more businesses rely on automation and cutting-edge technology, the systems become increasingly interdependent and susceptible to disruption. Moreover, the integration of sustainability practices into industrial systems presents opportunities and challenges that necessitate a redefinition of conventional operational definitions in terms of evolving technological environments and regulatory systems. In a broader context, industrial systems are not merely mechanical but socio-technical systems characterized by the interaction of culture, technology, and organizational structure. All elements of this system determine its aggregate productivity and resilience, as evident from the ongoing debate regarding business continuity management (BCM) processes. Emergent definitions of BCM reflect the complexity of managing these systems against potential disruptions.

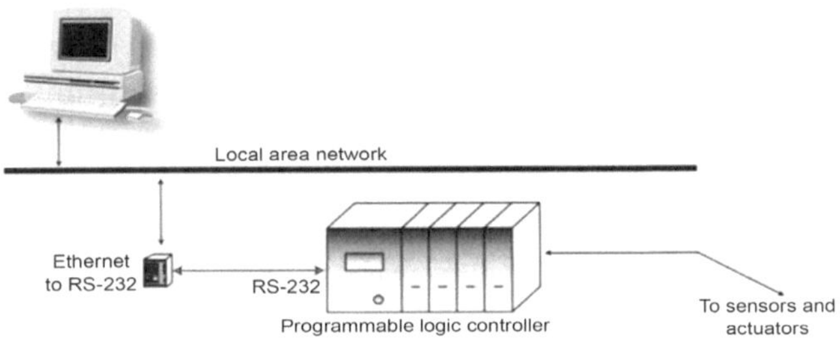

Figure 3.1. The term industrial systems.

Source: https://ars.els-cdn.com/content/image/3-s2.0-B978075065735850028X-f02-06-9780750657358.jpg.

As recognized by recent literature, the connections between academic research and practice-oriented approaches to BCM reveal critical findings to the operational vulnerabilities of industrial systems. The resolution to such sophistications require an understanding of how each component functions under duress, potentially affecting performance and sustainability (Fischbacher-Smith et al., 2017). The recognition of the common threats to industrial systems is crucial in devising effective measures to counter the dangers. For instance, cyber vulnerabilities present unique challenges with manufacturers relying more on connected technologies. Evidence shows that such attacks as data breaches and system intrusions can result in critical business disruptions, where strict security measures are a must. Furthermore, forensic investigation of security breaches reveals systemic vulnerabilities that can threaten organizational integrity and performance. By analyzing real case studies of industrial system violations, researchers can offer constructive comments on the adaptive actions involved with maintaining the resilience of industrial systems. Ultimately, enhancing definitions and practices within industrial systems will enable organizations to enhance their readiness against the evolving threat and vulnerability environment (Maple et al., 2010) (Figure 3.1).

Importance of Security in Industrial Systems

System security is crucial for industrial systems in safeguarding not just business operational integrity but also public safety and national security. The more industrial systems depend on smart technologies and automation, the more they expose themselves to potential cyber-attacks such as hacking, data breaches, and ransomware attacks. These vulnerabilities have potentially disastrous consequences, incapacitating business operations, breaching confidential information, and even threatening the safety of workers and surrounding communities. Therefore, addressing these security challenges is critical in maintaining continuous business operations and instilling stakeholder confidence. Sound BCM plays a key role here since it provides a structured method of minimizing threats and industrial operations' ability to withstand unforeseen challenges (Fischbacher-Smith et al., 2017). This requires organizations to invest in effective security systems and crisis management. Moreover, interdependence between performance operations and security in industrial systems suggests that there is a need for a comprehensive approach to managing risks. While organizations pursue greater efficiency and productivity, they are also expected to struggle concurrently with the complexity of safeguarding their infrastructures against known threats. Inconsistent strategies that fail to balance operational effectiveness with security considerations can lead to nightmare outcomes, particularly with complex socio-technical systems growing increasingly interdependent (Fischbacher-Smith et al., 2017). With high-level security frameworks, risks are not only paid for but also overall organizational performance is enhanced by fostering a culture of safety and resilience. This is not merely compliance, then, but a natural necessity for sustainable development and long-term success for any industrial activity. Furthermore, the advent of smart grid technologies presents a prime example of the multi-faceted security threats that face modern industrial systems. Intelligent grids, through their integration of communication technologies and traditional power infrastructure, must ensure an hierarchical security architecture that neutralizes threats at various levels of operation. For example, effective routing protocols for networks and secure management information systems must exist to preserve data integrity and service continuity. Game theory offers valuable insight into how to frame interactions among system components and probable attackers and highlight the dynamic security issues in such environments. With such advanced analytical techniques, organizations are able to enhance their ability to predict probable threats and design strategic measures for building resilience. Lastly, security in industrial systems cannot be overemphasized because excluding these issues may have far-reaching and negative impacts across different domains (Figure 3.2).

Figure 3.2. The security of industrial systems.

Source: https://cdn.prod.website-files.com/65e795a2f83946f9da1f1bcc/679c68cb78031f4022a44e17_Belfry_Imagery4.png.

Overview of Common Threats and Vulnerabilities

The intricate design of industrial systems, particularly in Supervisory Control and Data Acquisition (SCADA) systems, is especially risky for being exposed to cyber-attacks. These systems, essential for operating infrastructures like power systems and water treatment facilities, are inherently complicated and usually operate on outdated technologies with poor security. Given their central position, a successful cyber-attack would have disastrous consequences in the physical world. Therefore, awareness of potential threats is a requirement in creating relevant security measures. A multi-layered architectural style, suggested in Ma et al. (2012), offers stakeholders with a framework to conceptualize such systems and create targeted security analysis so that proactive protection against potential vulnerabilities that may be exploited by attackers is enabled. Vulnerabilities related to processes are one of the biggest threats to SCADA systems in which unauthorized users have access and manipulate operations, traditionally legitimate in nature. It highlights the challenge of distinguishing normal and malicious behavior in such systems. For this reason, new techniques such as semi-automated log processing have been developed to identify abnormal behavior, as presented in Bolzoni et al. (2012). By focusing on deviation from the normal process of operations, industrial systems are enhanced at preventing interruptions that can occur due to breached access. This proactive detection method is important, as it allows operators to respond promptly against malicious incidents and cut potential threats short before they turn into full-scale operational accidents. In addition to cyber-attacks, industrial systems are also exposed to their physical environment. These include natural disasters such as floods and earthquakes through to human factors such as insider attacks and procedure errors. The use of Internet of Things (IoT) technology in industrial systems also complicates the task of managing risks since, with IoT devices, there can be additional entry points for cyber-attacks. Cyber and physical considerations must therefore be incorporated into security measures to address the multi-directional threats industrial systems experience. As technology evolves, vigilance and quick response to the emergence of new vulnerabilities will become essential to maintaining the integrity and usability of these essential systems.

3.1. TYPES OF INDUSTRIAL SYSTEMS

Classification of industrial systems plays a vital role in identifying the various roles they serve in the contemporary economy and the nature of vulnerabilities they present. The Supervisory Control and Data Acquisition (SCADA) system is one of the prevailing types of industrial systems with a fundamental role to play in running operations across the critical infrastructures, for instance, the energy grid, the water treatment plants, and manufacturing processes. SCADA systems manage monitoring of real-time data and remote control of industrial processes. However, the complexity and widespread interconnectedness of these systems expose them to risks of cyber-attacks. A good comprehension of SCADA architecture is necessary for the efficient deployment of security solutions. The construction of a layer architectural model explains the roles and the interactions within SCADA systems so that all stakeholders can work together to collaborate on security procedures and assessment, and in turn, improve the resilience of these vital systems against intrusions (Ma et al., 2012). Another principal category of the industrial system spectrum is the Distributed Control System (DCS), and it is widely used in manufacturing processes to monitor and control manufacturing procedures. DCS provides a structured approach to process control through distributed networks with localized control while ensuring global system performance. Similar to SCADA, DCS has vulnerabilities as well, primarily because of unauthorized access to its network. An intruder gaining user access privilege can carry out activities that interfere with the control process without being noticed. It is therefore important that there is a good security mechanism that includes real-time monitoring and event logging. Such measures could be in addition to abnormal behavior detection that is typical of potential threats, such that there is integrity in industrial operation (Bolzoni et al., 2012). Awareness of the distinctive features of DCS also contributes to the general assessment of industrial system vulnerabilities. Last but not least, embedding Internet of Things (IoT) technology in industrial systems provides a new dimension of challenges and opportunities. IoT-enabled devices collect vast volumes of information from industrial environments of all kinds, enhancing decision-making and operational efficiency. However, their adoption by a broad range of industrial systems introduces data privacy and cybersecurity threats. Industrial IoT devices, being as vulnerable as their conventional counterparts, can be prime targets for cyberhackers keen on disrupting production or extracting sensitive information. Networked devices need to be guarded with a complete system overview and implementing real-time threat detection mechanisms. Since organizations are increasingly relying on IoT technologies, awareness of vulnerabilities that are native to these systems is important in order to maintain operational continuity along with preventing potential risks of deployment (Ma et al., 2012).

3.1.1. Manufacturing Systems

Manufacturing systems have been significantly defined by the increase in interconnectedness of industrial control environments and broader enterprise networks. Historically, these networks operated in isolation from the external world by virtue of their air-gapped topologies and specialized protocols such as Modbus. However, with the emphasis on enhancing connectivity, previously non-existent weaknesses have been discovered. As manufacturing systems borrowed concepts from the Internet of Things, cyber threats have mushroomed exponentially. Sophisticated enterprise management systems now require uninterrupted data flow that has the ability to simplify the production process, thus complicating the security environment. Legacy systems and emerging technologies come with serious issues when brought together, indicating that efficient cybersecurity solutions need to be adopted so that attacks capable of attacking these newly integrated systems are avoided (Johnson et al., 2016). New technologies in software management of manufacturing systems have changed the manner in which organizations control their control devices, notably Programmable Logic Controllers (PLCs) and sensors.

Figure 3.3. The evolution of manufacturing systems.

Source: https://flowlens.com/wp-content/uploads/2023/06/mech-mind-sTzb90A1Abs-unsplash-1024x683-1.webp.

The potential for firmware upgrade in order to neutralize vulnerabilities enhances the operation efficiency of such devices. While updates enable quick release of security patches, they do so at significant risks, notably if updates are disseminated by means of internet connectivity. Modern events, such as Ukraine, highlight the susceptibility of malware to exploit firmware update streams to target safety-critical programs in industrial environments. Locking devices into continuous loops of firmware upload, attackers can compromise the integrity of critical systems, illustrating the precarious balance between operations updates and cybersecurity risks (Barker et al., 2017). With manufacturing systems increasingly sophisticated, defense against such malicious attacks is paramount. To adequately counter the complexities of modern manufacturing systems, a multi-faceted approach to cybersecurity is essential. Companies must not only integrate technical controls, such as firmware hashing and system update tracking on a regular basis, but organizational measures that further strengthen the supply chain. That includes the establishment of policy for screening and keeping up-to-date third-party components that can interface with critical industrial systems. By solidifying cooperation among manufacturers, cybersecurity experts, and regulatory bodies, firms can establish comprehensive strategies to fight threats. Performing continuous risk analysis and establishing an organizational culture based on security and safety can significantly reduce the incidence of cyber-attacks. Lastly, as the environment surrounding manufacturing systems keeps evolving, the importance of robust cybersecurity remains a need that continues to be part of ensuring their safety and integrity (Johnson et al., 2016) (Figure 3.3).

3.1.2. Energy and Utility Systems

The integration of sophisticated communication technologies into energy and utility systems has revolutionized how they function, advancing significantly towards the achievement of smart grids. These networks employ intricate webs of interdependent equipment that facilitate the exchange of data in real time to improve efficiency and reliability. The increased complexity with these networks come risks that must be addressed by innovative security solutions. The smart grid revolution has underscored the need for the adoption of layered security architectures, whose principle is safeguarding various operational layers from potential attacks. Such protections are necessary not only to ensure the integrity of services but also to ensure consumer confidence and safety. With increased automation of energy networks and interconnection, neglect of their safety can expose them to cyber threats, and current procedures should be reviewed to counteract the threats from harmful sources in this new threat environment (Hauser et al., 2018). Industrial control systems' evolution from isolated systems to networks illustrates the growing pattern of higher interconnectedness in energy and utility frameworks. Previously, the systems were isolated, using proprietary protocols with minimal communication with large-scale information technologies. However, with growing demands for operation optimization, industries are increasingly adopting architectures reminiscent of the Internet of Things (IoT). This integration allows sensors and actuators to be included in the network to facilitate data-driven decision-making and resource management. This connectivity, however, raises a significant concern from the point of cyber security because exposing previously isolated systems to the external network adds new vulnerabilities. Such threats must be addressed to ensure the benefits of increased connectivity are not gained at the cost of safety and reliability, underlining the necessity of robust protective measures (Johnson et al., 2016). In order to respond to the challenge of new energy and utility infrastructure, an overarching approach to security is paramount. Future research is necessary to engineer end-to-end solutions that not only protect against technology-based vulnerabilities but also the human elements involved in security breaches. This integrated knowledge is necessary to design strong systems that can resist both cyber and physical attacks. Game-theoretic models can be useful tools in anticipating and avoiding risks by modeling interactions between possible attackers and system administrators. By these measures, industries can develop responsive security structures that are equally adaptive to the continually changing threat landscape. Finally, progressive development of energy and utility systems calls for a commitment to innovation in security measures so as not to compromise on operational integrity and safety (Hauser et al., 2018) (Figure 3.4).

Figure 3.4. Energy and utility systems.

Source: https://bxenergysystems.com/wp-content/uploads/2024/07/BX-Solar-Battery-Energy-Storage-Systems-109.jpg.

3.1.3. Transportation and Logistics Systems

Transportation and logistics infrastructure are the lifeblood of global commerce, shipping merchandise across long distances. In an increasingly integrated world economy, these networks are the pipelines of economic activity and critical to supply chain integrity. However with growing complexity, they are more susceptible to a host of threats and vulnerabilities. Disruptions may have various causes, including natural disasters, cyberattacks, and geopolitical tensions, all of which pose major risks to operations. A case in point is cyber threats, which have become particularly pervasive, since logistics networks rely on digital platforms to monitor and control inventories. It is essential to have robust cybersecurity in the systems in order to minimize disruptions that ripple across entire supply chains. Besides, the physical transportation and logistics facilities, such as roads, railway lines, and ports, are mostly susceptible through aging infrastructure and maintenance failure. These issues result in inefficiencies and extra operating expenses, which further enhance the impacts of external threats. Climate change is another emergent threat, where extreme weather conditions might destabilize transport links and devastate crucial infrastructure. The combined effect of deteriorating infrastructure and environmental needs necessitates a proactive approach towards vulnerability assessment and risk management. Since the Global Risks Report stresses, new global risks—like climate change—have significant implications for logistics, providing shocks which necessitate adaptive action to build resilience in these vital systems (N/A, 2016). Further, increased reliance on technology in transport and logistics networks presents a new level of vulnerability. Automation, while improving efficiency and reducing labor costs, also presents points of failure that can be exploited by malicious actors. With organizations increasingly adopting advanced technologies, such as Internet of Things (IoT) devices and artificial intelligence (AI), further increasing the attack surface, robust security practices are needed. In this context, one should value the fact that logistical networks are not immune from the broader extent of universal threats, whether pandemics or financial depression. Counteracting these weaknesses involves a commitment not only to enhancing physical and cyber security measures, but also to ensuring cooperation among stakeholders in order to create an even more robust logistics environment capable of withstanding opportunistic and systemic threats (Castillo-Chavez et al., 2002) (Figure 3.5).

Figure 3.5. Transportation and logistics systems.

Source: https://www.referenceforbusiness.com/photos/logistics-and-transportation-390.jpg.

3.2. CYBERSECURITY THREATS

The cybersecurity threat landscape of industrial systems is increasingly advanced with the convergence of information technology (IT) and operational technology (OT). Such convergence introduces vulnerabilities that can be exploited by attackers and lead to potential disruptions of critical infrastructure. Industrial control systems (ICS) are targeted by cybercriminals due to their critical mandate of operating fundamental utilities such as electricity, water, and manufacturing processes. A successful attack also includes not only economic expenses but severe effects on national security and public security. These attacks are highly advanced and involve techniques such as ransomware, phishing, and advanced persistent threats (APTs), which can devastate operations and access confidential information. With more advanced industrial infrastructures, it is vital to understand these threats in order to create effective defenses that ensure operational continuity and resilience to cyber-attacks. Further, the impact of cybersecurity threat extends beyond the immediate operational risk to encompass broader consequences for organizational integrity and customer confidence. Industrial sector organizations are compelled to balance the dual needs of regulatory compliance while simultaneously enhancing their cybersecurity posture.

Cybersecurity standards, such as the NIST Cybersecurity Framework, provide a standard for assessing vulnerabilities and minimizing effective risk management. The importance of a common policy to cybersecurity is emphasized by studies demonstrating a direct correlation between steps taken in successful cybersecurity and the viability of businesses. The passage of time illustrates the importance of continuous training and awareness programs for employees, which becomes a cornerstone of an organization's defense plan. Incurring expenses on comprehensive risk management practice can reduce the risk associated with cyberattacks, thereby ensuring a secure operational environment. Besides, the inherent elements of industrial system cyber security highlight a forward-thinking strategy with provisions for advanced threat detection and response functionality. Based on more recent studies, the use of machine learning and artificial intelligence algorithms can vary significantly improve the capacity to detect threats, allowing real-time examination and rapid action towards breaches. The evolving nature of cyber threats demands the shift from reactive to proactive security measures to ensure that potential vulnerabilities are identified and removed before they are exploited. Industry partnerships with cybersecurity experts can facilitate knowledge sharing, leading to the development of new technologies and strategies. Also, as Anglano et al. (2018) suggests, it is vital to establish the culture of cyber hygiene in organizations so that awareness and best practices become a part of the daily business. This general strategy not only safeguards industrial systems, but also sustains trust in their reliability and security (Figure 3.6).

Figure 3.6. The cybersecurity threats.

Source: https://media.geeksforgeeks.org/wp-content/uploads/20240229165546/Cybersecurity-Threats-copy.webp.

3.2.1. Malware and Ransomware Attacks

The dissemination of malware and ransomware assaults poses threats to industrial networks, undermining their operational integrity and security. Malware is a type of malicious software intended to interfere with, damage, or exploit computer systems without permission. In industrial settings, such attacks can compromise critical infrastructure, leading to massive financial losses and operating downtime. More recently, however, the threats have evolved, using sophisticated techniques that exploit vulnerabilities in the system architecture, and it is crucial to possess robust cybersecurity. Increased digital awareness among policymakers who are responsible for the cyber safety dynamics will be crucial in the fight against such threats, as identified by the EUCPN Secretariat. Understanding the nature of threats is imperative in designing effective malware and ransomware protection, again highlighting the need for strategic initiatives towards enhancing cybersecurity resilience in industrial environments (Vanhee et al., 2017). Ransomware, one of the most malicious forms of malware, encrypts vital information until a ransom is paid, typically freezing operations and services in industrial plants. The potential consequences of such attacks extend beyond immediate costs, as firms may also experience reputational damage and loss of customer trust. With the Internet of Things (IoT) expanding further, linking many industrial devices, such vulnerabilities are likely to become even more acute, offering yet more channels of attack for cybercriminals. The predictions of experts indicate a rise in connected devices over the next few years, which can help efficiency as much as be dangerous if cybersecurity is not made a priority (Anderson et al., 2017). Hence, proactive, holistic cybersecurity measures are necessary to engage all stakeholders in industrial networks so as to collectively fortify defenses against ransomware attacks. Mitigating malware and ransomware threats requires a multi-faceted response involving risk assessment, education, and policy development. Organizations must invest in continuous monitoring and state-of-the-art software solutions that can respond to emerging threats rapidly. Moreover, a culture of cybersecurity awareness is required to equip employees with the intelligence about how to identify potential threats and respond accordingly. The integration of legislative measures and best practices from sources by the EUCPN can serve as a model for organizations to develop specific strategies depending on their specific vulnerabilities and operational needs (Vanhee et al., 2017). By prioritizing cybersecurity education and budgeting, industries can implement a robust defense against malware and ransomware attacks, thereby safeguarding not just their operations but the wider economic ecosystem they are part of (Figure 3.7).

Figure 3.7. Malware and ransomware attacks.

Source: https://www.jonesday.com/-/media/images/publications/2020/10/fincen-issues-guidance-on-ransomware-attacks/imageoverridefortwitter/fincen_issues_guidance_alert_social.jpeg?rev=771db04b362a42ba9ec12e250e8cdd37&hash=FF3A8952A6F8D2454C22D1E1E4E7A0C0.

3.2.2. Phishing and Social Engineering Tactics

The world of cybersecurity is constantly threatened by phishing and social engineering tactics exploiting human vulnerabilities within organizations. These attacks are characterized by the employment of manipulation and trickery, such as manipulating individuals into divulging confidential information or granting unauthorized system access. In industrial setups, where trade secrets and mission-critical infrastructure are paramount, such attacks present a significant threat. Phishing attacks may lead to unauthorized data breaches or ransomware incidents and mostly occur through authentic-looking fraudulent email attacks. As has been observed in recent research, the tactics are changing at a very fast rate, with cybercriminals continually coming up with new ways of evading conventional security protocols, demonstrating the importance of increased vigilance and readiness among industrial system employees. In order to ward off the expanding threat of phishing and social engineering, various strategies have started being implemented within organizations to increase employee resistance against these attacks. Employee awareness and training programs have become the cornerstone in developing a culture of cybersecurity in industrial settings. These programs generally teach employees how to identify suspicious communications as well as the significance of validating information before responding to it.

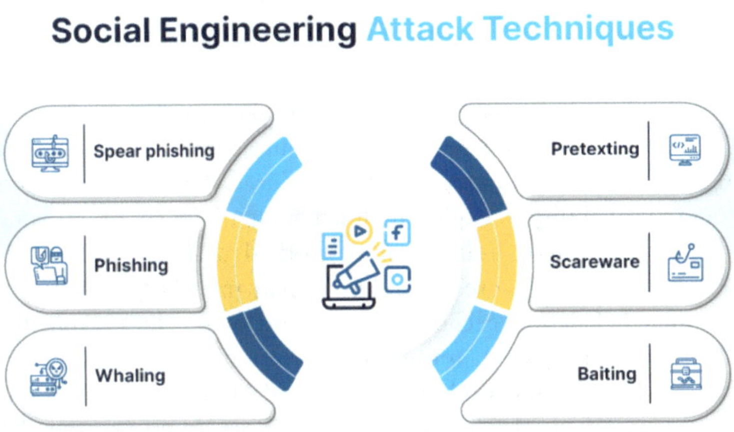

Figure 3.8. Phishing and social engineering tactics.

Source: https://www.extnoc.com/learn/wp-content/uploads/2022/10/Social-Engineering-2.jpg.

The organizations could also carry out periodic security audits to discover vulnerabilities in their defenses. These kinds of analyses can reveal weaknesses not only on the technology front but also on the human side, as evidenced by increased focus on examining the human element in paradigms of security. As highlighted, there has been a need to better penetrate employees using successful penetration testing, where strategies oriented towards humans are employed to measure and enhance their ability to detect potential vulnerabilities (Archibald et al., 2019). The complexity of social engineering and phishing attacks necessitates a multilateral approach to cybersecurity in industrial networks. This entails establishing sound policies that set forth acceptable procedures and practices for handling sensitive information. Organizations are thus encouraged to adopt robust security measures with technical and human elements. For instance, the inclusion of ethical principles within cybersecurity training fosters a responsible workforce that is cognizant of the implications of its actions. Ethical integration in training also aligns with upcoming paradigms like PoinTER, which highly emphasize ethical analysis in human penetration testing. By prioritizing ethics and resilience over other aspects, organizations are able to fortify their defenses against the clever ploys employed by cybercriminals and thus secure their critical infrastructure from the rampant menace of phishing and social engineering attacks (Benedetto et al., 2019) (Figure 3.8).

3.2.3. Insider Threats and Human Error

In the industrial systems area, human error and insider threats are typical vulnerabilities that can pose significant threats to operational integrity. Unlike conventional external threats, which are often easier to identify, insider threats could find their origin in employees or contractors who possess intimate knowledge about internal systems and procedures. Their intimate knowledge gives them the capability to exploit their access maliciously or unintentionally. As industries grow more reliant on networked systems, the cost of such attacks becomes all the more possible. Human mistakes in this regard can lead to unintended data breaches or disruption of services, typically driven by inadequate training or a confused understanding of security protocols, making organizations vulnerable to unexpected breaches. In addition to this, the character of industrial control systems (ICS) adds an additional layer of complexity when it comes to insider threats and human mistakes. These networks are critical to the functioning of critical services, from the provision of energy to the supply of water. Because they are complex, a single small mistake by an insider can lead to catastrophic consequences. Human error is commonly cited as one of the major causes of accidents in these environments. With greater usage of innovative technologies, employees may unknowingly incorporate vulnerabilities through improper configurations or not installing security updates. Thus, organisations must encourage a culture of security that appreciates caution and responsibility of personnel, ensuring that employees are educated about threats and their role in responsibility. To mitigate the threats associated with insider threats and human error, organisations increasingly deploy multilayered security architectures that integrate different technology and people-oriented approaches. By the establishment of robust training initiatives and open communication, organizations are able to heighten their workforces awareness of threat potential and create a culture of resilience. But as indicated by emerging research, including strategies arising out of game-theoretic practice, there needs to be ongoing assessment and reassessment of security processes according to evolving threats. This involves not only strengthening technical defenses, as outlined in the discussion of smart grid security, but incorporating full-spectrum protections that address human vulnerabilities and consequently build greater security for critical infrastructure supporting modern industrial systems (Al-Khateeb et al., 2019; Hauser et al., 2018).

3.3. PHYSICAL SECURITY VULNERABILITIES

The interconnection between physical security vulnerabilities and industrial system operation is critical to understanding the broader risk context. Physical security encompasses controls designed to prevent an organization's assets from being accessed, modified, or destroyed without authorization. In the industrial setting, physical security failures can have catastrophic consequences, such as disruptions to operations, exposure to environmental hazards, or destruction of facilities. For example, with no physical controls, a sabotage actor can enter main control rooms and manipulate the processes to result in total process failure. Physical security is not only a complement to cyber security; both must be combined to formulate a total defense strategy. Securing physical barriers and reinforcing access control can greatly reduce threats that are most likely exploited in industrial environments, such as the trend toward more digital controls being built into traditional physical systems. Designing SCADA architecture is among the main areas where physical security vulnerabilities can be introduced. SCADA systems are generally responsible for monitoring vital infrastructure such as power networks and water treatment facilities and consequently become the prime targets. Malicious intrusion can have disastrous consequences; therefore, right understanding of SCADA architectures is required. An architectural perspective with layers can also make security analysis easier through a better idea of where the vulnerabilities can be in a systems different components. This is in line with stakeholders cooperating to enhance security through proper priority for security in place. Due to the scale and complexity of SCADA systems, an explicitly stated architectural pattern enables organizations to design enhanced physical and cyber security synergies, thereby enhancing overall resilience in the face of multi-faceted threats represented in industrial settings, as highlighted in recent SCADA vulnerability discussion (Ma et al., 2012). Furthermore, the development of technology has added more dynamics to the physical security vulnerabilities of industrial systems, primarily through exploiting the firmware in Programmable Logic Controllers (PLCs) and similar systems. Historically, overhauling the underlying software of industrial devices was a daunting endeavor, with a tendency to produce across-the-board replacements for existing hardware. Contemporary remote firmware updates are accompanied by strengths as well as limitations. While these updates might bring critical security fixes, they come at the expense of other doors being left open to cyber threat vectors, particularly when updates are delivered over internet-connected channels. Recent experience, such as in targeted cyberattacks against critical infrastructure in Ukraine, illustrates malware potential for exploitation during times when firmware is being updated. Due to this, organizations must implement mitigation strategies that integrate technical controls, such as firmware hashing, with organizational processes that enhance supply chain security. This multi-dimensional approach is critical for defense against threats aimed at physical and software industrial system layers (Barker et al., 2017) (Figure 3.9).

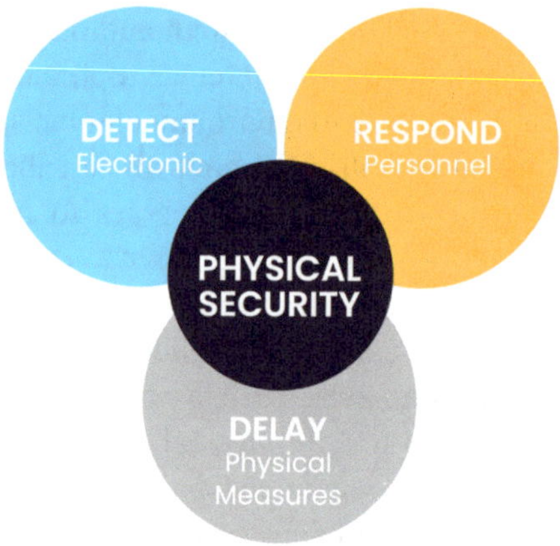

Figure 3.9. The interplay between physical security vulnerabilities and the operation of industrial systems.

Source: https://securithings.com/wp-content/uploads/2025/01/article-physical-security-plan_Img1-1024x967.png.

3.3.1. Unauthorized Access to Facilities

Unauthorized facility access is a significant threat in industrial systems, where physical and cybersecurity threats converge. Intruders' vulnerability to exploit weaknesses in access control systems not only compromises sensitive data but also operational integrity in the process. Usually, physical security in the form of guards and CCTV cameras goes hand-in-hand with digital systems, but such two-layered protection can fail if either system has a flaw. If hackers were to seize control of industrial installations, then they can hijack key systems, leading to disruption and even catastrophe. It is for this reason that physical infrastructures ought to be protected using comprehensive methods that identify untoward access and its implications, calling for heightened vigilance within both security sectors. The impacts of unauthorized access into premises go beyond short-term disruption; they can also produce longer-term weaknesses in industrial systems. For example, physical premise violations can provide a loophole to the intruders for plugging in rogue devices that will subvert system performance or facilitate data theft. For this purpose, it is important to be aware of the exact vulnerability areas in a facility. As security recommendations addressing smart grid infrastructures have observed, an amalgam of robust physical security mechanisms and electronic communication technologies is of paramount significance in blocking unauthorized entry. Organizations must regularly review their physical barriers in addition to incorporating more robust cybersecurity practices in order to minimize the risk of unauthorized access and exploitation of vulnerabilities that lead to operational failures (Qian et al., 2012). Furthermore, unauthorized access must be addressed through consideration of the people aspects of maintaining security measures. The workers are most likely the first line of defense and potential exposure through negligence or poor training. Effective education initiatives can create a security-conscious culture with workers being aware of the paramount significance of procedures governing access to sensitive areas. Further, inequalities in security training can create opportunistic weaknesses in that unauthorized personnel may exploit knowledge loopholes or procedural laxity. Employing an across-the-board strategy integrating personnel training, physical security upgrades, and cyber defenses is capable of making it more difficult for unauthorized persons to gain access into facilities. Such a multi-faceted approach not only protects infrastructures but further boosts the overall resilience of industrial systems against a wide range of emerging threats (Marcelo et al., 2008).

3.3.2. Lack of Surveillance and Monitoring

In industrial systems, there is a lack of surveillance and monitoring mechanisms that provide grave vulnerabilities that can be exploited by unauthorized parties. Industries presently rely greatly on the most modern advanced technologies, including networked devices and embedded systems, to achieve more efficiency in their operations. These technologies are exposed to misuse and cyber-attacks if there are not proper monitoring systems in place. Since the various components are brought together with open networks and wireless media, factory systems unwittingly leave the door ajar for cyberattacks. The sophistication of such systems and the lack of proper security measures necessitate that industries adopt a multi-layered surveillance measure that isn't merely physical security but also cybersecurity considerations. Enhanced surveillance mechanisms, therefore, are important in minimizing risks of unauthorized intrusion and system failure. Further, the effects of ineffectiveness of monitoring go beyond the simple consideration of compromise of cybersecurity; they may result in catastrophic operation interruptions.

Figure 3.10. The lack of adequate surveillance and monitoring.

Source: https://static.independent.co.uk/2024/01/31/12/Severance_Photo_010101.jpg.

For instance, if surveillance systems fail to identify timely intrusions or anomalies, process industrial processes may be vulnerable, leading to potential outages as well as decreased dependability of services. In railway and other infrastructure-heavy sectors, such outages can result in considerable economic loss and harm public safety. The integrated perspective that is being advocated by paradigms such as SHIELD will demonstrate the significance of enhancing surveillance systems' security, privacy, and reliability for industrial settings. With the implementation of an effective surveillance approach that will address such vulnerabilities, industries can better safeguard their infrastructure against interruptions while also increasing operational robustness in the process. Lastly, the ever-increasing IoT focuses on the necessity of filling surveillance gaps in an industrial environment. More devices that communicate and interact with one another expand the attack surface while forcing industries to prioritize fully integrated surveillance strategies. Failure to monitor could also lead to privacy violation, as sensitive data passed between communicating machinery is left open to vulnerability without proper control mechanisms. Although experts are optimistic about the successful implantation of IoT technologies, the potential for cyber risk pushes industries to implement tight surveillance controls to negate retaliation. Capacity to ensure adequate monitoring can provide industries with a competitive advantage, enhancing security networks and business continuity in an increasingly digitized age (Anderson et al., 2017; PRISCOLI et al., 2017) (Figure 3.10).

3.3.3. Inadequate Physical Barriers and Controls

Poor physical barriers and controls in industrial systems are a major weakness that can potentially expose facilities to all manner of various different risks such as theft, sabotage, and unauthorized entry. These physical barriers are the initial line of defense against potential threats. Many parts of the industrial environment continue to utilize old or poorly designed access control systems that are not equipped to successfully deter malicious activities. For instance, security through fences, CCTV, and access control can be ineffective depending on how they have been deployed. The lack of consistency will lead to security gaps, therefore the potential of such incidents causing interruptions to business and putting staff at risk, another indication of the requirement for comprehensive security strategy that includes effective physical obstacles (American Petroleum Institute et al., 2014). Moreover, the ease of having enough physical barriers is also complicated by the underappreciated human factor-technology interaction. Human error, as in the instance of not responding to alarm systems or inadequately securing locations, can cascade and invalidate the effectiveness of security measures. The integration of security management systems that embrace the synergy of both human and technological elements can lead to more robust systems. Compliance checking and efficient training are an essential component of an extensive security policy. In addition, a security-aware culture can be able to manage human oversight-related risks to a high degree, thereby allowing the physical barriers to fulfill their desired function (Marcelo et al., 2008). Additionally, it is advisable to maintain physical security mechanisms current by regularly assessing them in order to address evolving threats in industrial systems. The risks associated with weak controls are constantly evolving, often tied to the emergence of new technologies and changing operating conditions. With changing business norms, firms must ensure intense risk assessments to identify vulnerabilities within their physical security frameworks, wherein they can maintain pace with changing threat conditions. Application of processes that combine management processes, security technologies, and proactive human interactions is crucial in not just securing physical layers but developing overall security resilience (American Petroleum Institute et al., 2014). By constant evaluation and fine-tuning, industrial premises can maintain robust defenses that guard their assets and employees from different threats.

3.4. NETWORK SECURITY ISSUES

As industrial systems continue to embrace networked technologies, vulnerabilities that are inherent in such technologies pose tremendous threats to operational integrity and security. Supervisory Control and Data Acquisition (SCADA) systems and Industrial Control Systems (ICS) are specifically at risk of cyber-attacks that will result in disruption of essential services and catastrophic failure. One of the key concerns is how such systems incorporate communication technologies; while permitting remote monitoring and control, they are also potential points of attack by cyber attackers. A reliable security framework must therefore consider the layered nature of such systems, necessary to provide an end-to-end security analysis. Through adopting a more established theory of system architecture, the stakeholders can recognize threats and vulnerabilities more effectively and thereby enhance the resilience of such key infrastructures (Ma et al., 2012). In neutralizing the inherent vulnerability of industrial networks, there needs to be robust security arrangements across various levels of system architecture. Game theory offers a promising solution in that it mathematizes interaction in the system and between prospective attackers and targets. In doing so, it helps to create cross-layer security that will learn and adapt to shifting threats. For instance, secure routing protocols for the data communication layer can ensure secure access while operational commands are safely relayed. In addition, resilience mechanisms should be attack-proof, as much as possible, to ensure their impact on the system overall functionality and security is negligible (Hauser et al., 2018). Proactively, in this way, multi-layer interactions are provided with a comprehensive understanding, enhancing the defense against probable cyber-attacks. Lastly, the challenges posed by network security threats to industrial systems highlight the need for continuous enhancement of security procedures and policy. As industrial systems become more interconnected, the significance of maintaining a safe climate increases. With increasing complexity in SCADA systems, there is a need to develop security solutions that are multi-faceted in nature and address the various layers and components involved. Building a culture of security awareness among all participants, including operators and technical staff, can also improve the success of these measures. Through the prioritization of a multilayered security approach that encompasses organizational and technical controls, industries can effectively harden themselves against malicious attacks, providing their main services as accessible and dependable (Hauser et al., 2018; Ma et al., 2012) (Figure 3.11).

Figure 3.11. Network security issues.

Source: https://softwareg.com.au/cdn/shop/articles/Orlando-Cyber-Security-Services-768x782.png?v=1707754806.

3.4.1. Insecure Network Configurations

Insecure network settings are a very serious industrial system threat, and they are typically gateways for cyber-attacks that affect operational integrity. They happen when the default settings of network hardware, such as routers and switches, are not tweaked to enhance security. Poor implementation of access controls can allow unauthorized entities to exploit the network, leading to possible data loss or system downtime. The development of cloud computing, through platforms like OpenStack, is the basis for the necessity for secure settings in a de-centralized network. The increasing reliance on cloud infrastructure requires higher levels of adherence to the security attributes of network configurations to deter cyber-attacks and protect sensitive information against malicious forces, as indicated in existing research pointing to the integration of security components in cloud-based installations (Al-Shaidy et al., 2017). Secondly, industrial control systems (ICS) and Supervisory Control and Data Acquisition (SCADA) systems have their network configuration security issues too. ICS and SCADA networks employ proprietary equipment and other communications protocols that are not necessarily conducive to normal cybersecurity practices. It is the complexity built into these systems that can lead to misconfigured networks and thus expose them to cyber-attacks. For instance, lack of monitoring during configuration of firewalls or failure to segment networks can facilitate an attacker to move through the network with ease, and by doing so, possibly breach safety-related functions.

Figure 3.12. Insecure network configurations pose a significant risk within industrial systems.

Source: https://dptechgroup.com/wp-content/uploads/2023/05/Ten-examples-of-how-Insecure-network-configurations-can-lead-to-data-breach-of-ePHI.jpg.

That is a delicate situation; poor configurations not only compromise systems but also risk the production of safety-critical products. ICS/SCADA security vulnerabilities must be addressed by tailored means tailored to their unique architecture and operational requirements, which have been identified by industry experts (Evangelopoulou et al., 2016). Finally, insecure network configurations are best resolved by a holistic method through constant audits, staff training, and implementation of best practices tailored to the industrial sector. Carrying out comprehensive evaluation of network configurations and pinpointing vulnerabilities is key to improving security postures. Organizations need to make strengthening default setting-changing protocols, enforcing robust authentication controls, and regular access control reviews a top priority. Additionally, regular monitoring and testing of network configurations can assist with identifying vulnerabilities prior to exploitation. By developing a plan of continuous improvement premised on risk analysis and incident response plans, industrial systems can fortify themselves against the myriad of cyber-attacks from insecure settings. Not only does this proactive measure secure vital information, but it also ensures the reliability and safety of those critical infrastructure systems that society depends on Al-Shaidy et al. (2017) (Figure 3.12).

3.4.2. Vulnerabilities in Legacy Systems

Legacy systems, which comprise the majority of most industrial environments, suffer from a range of vulnerabilities that leave organizations open to numerous cybersecurity risks. These systems were not constructed with the kind of advanced threats currently riding across the cyberenvironment in mind and are based upon outdated technologies that lack adequate security controls to safeguard them. The lack of timely updating of these systems further exacerbates their weaknesses, and they are extremely desirable targets for malicious attackers. The recent trends have made the promise of firmware updates have certain security benefits, as corporations can provide patches for identified vulnerabilities in Programmable Logic Controllers (PLCs) and other critical devices. However, as highlighted in recent reviews, firmware updates also carry their own risks, primarily when they can turn into malware vectors, interfering with safety-critical operations and compromising system integrity overall (Barker et al., 2017). To top it off, the loss of the "air gap," which used to protect legacy systems somewhat, has presented new exposures for legacy infrastructures to deal with. Previously isolated components such as PLCs and smart sensors are increasingly integrated into enterprise networks, exposing data yet making them vulnerable to remote attacks. Industrial Internet of Things technologies introducing a level of complexity to the security context as information technology converges with operational technology. Convergence could bring a lot of operational efficiency; however, it also increases the potential attack surface. The learned experience from recent cyberattacks on infrastructures in Ukraine serves to underscore the dire ramifications of such vulnerabilities, causing reassessment of security measures of legacy systems and operational resilience against emerging threats (Evangelopoulou et al., 2017). In order for organizations to be able to secure legacy systems, they must adopt a comprehensive security policy incorporating both technical and organizational countermeasures. Mitigation of vulnerabilities is not merely dealing with real-time cyber threats but reinforcing supply chain security throughout the industrial ecosystem. Using firmware hashing is one such way in which the integrity of updates can be guaranteed and malware distribution can be de-risked. Additionally, instilling a security-conscious culture through training and awareness programs can enable employees to detect and act against threats more effectively. By reacting early to enhance security and prioritize legacy system modernization, organizations can protect their critical infrastructure against emerging cyber threats, thereby ensuring their efficiency in operation in a high-risk and evolving digital environment (Figure 3.13).

Figure 3.13. Vulnerabilities in legacy systems.

Source: https://www.blackduck.com/blog/how-to-deal-with-legacy-vulnerabilities/_jcr_content/root/synopsyscontainer/synopsyscontainer/column_copy/colRight/image_copy_copy_copy.coreimg.jpeg/1721185004235/legacy-vulnerabilities-easy-to-exploit.jpeg.

3.4.3. Lack of Segmentation in Network Architecture

Lack of segmentation in network structure has serious industrial system threats and has a direct impact on security and resilience. Segmentation is necessary in traditional networks because it creates a barrier between different parts of an entity, limiting sideways movement of malicious players. Without segmentation walls, the moment an intruder penetrates, they can roam the entire network with ease. This vulnerability is most pronounced in SCADA systems, as such systems have evolved from single-system setups to networked systems for improved performance and dependability. Modernization, as much as it is beneficial, exposes SCADA systems to worldwide cyber-attacks, and thus having good segmentation as part of an excellent security stance is a critical step. With proper segmentation, industrial systems can successfully reduce the attack surface, in addition to improving the overall security architecture. Besides, the rapid expansion of the Internet of Things (IoT) has further aggravated the problem of network segmentation. IoT devices, with their ubiquitous connectivity and often ad-hoc integration into infrastructures, are oftentimes not equipped with intrinsic security features. This disparity makes them more susceptible to exploitation, especially for networks lacking good network segmentation. Each connected device presents a potential entry point for cyber-attacks, complicating security for industrial systems. The built-in vulnerabilities of IoT devices, combined with their diverse operational environments, necessitate a re-examination of network architecture fundamentals. Closing these loopholes by way of strategic segregation can not only enhance security but ensure data integrity and availability, thus ensuring the best operational performance of business-critical industrial operations. Accomplished segmentation may be a significant risk-reducer of risk associated with an interconnected network architecture, but requires intimate knowledge of contemporary vulnerabilities and threats. Limitations of segmentation not only expose industrial networks to unapproved access but can also extend the period and magnitude of security compromise. For instance, a compromise in an under-segmented network can lead to leakage of sensitive operating data, resulting in cascading consequences on safety and financial performance. With industrial activities depending more on integrated networks, notably SCADA systems within national infrastructure, the ramifications of these vulnerabilities extend much wider. As is well documented in existing literature, quantifying these risks from the attacker's point of view is necessary to define weaknesses before exploitation is possible, and hence highlight the importance of strict segmentation protocols in the design of industrial networks (Almalawi et al., 2017; Mendez et al., 2017).

3.5. SUPPLY CHAIN RISKS

The intricacy of modern supply chains maximizes the risk potential of various risks that have a severe disruptive effect on industrial systems. Organisations adopting Industry 4.0 technology expose themselves to multiple economic, social, and technical risks that have the potential to jeopardize operational effectiveness. For instance, although the use of cyber-physical systems and the Internet of Things enables efficiency, it risks leaving the door open to data security and system integration exposures. A poorly managed technology investment can potentially generate great economic losses, while job displacement-induced unemployment remains a major social issue. It is critical that firms identify and assess such supply chain risks ahead of time and thereby establish a proper understanding necessary to attain the complexities involved in Logistics 4.0 and render the threats ineffective (Kavka et al., 2020). Apart from economic and social risks, supply chains also face growing legal and political challenges that can adversely affect industry stakeholders. The legal framework for data ownership and control of sophisticated high-tech logistics systems remains ambiguous, leaving the organizations attempting to innovate here in an extremely uncertain state. For example, businesses operating in environments not well-bound by existing regulations could be impacted by unforeseen changes in regulatory environments, which could impact their supply chain.

Such uncertainties require adjustments like robust compliance mechanisms and agile planning to navigate through future hazards. Addressing these challenges not only improves supply chain resilience but also contributes to long-term sustainability of industrial systems, and hence risk assessment is a priority for contemporary business (Kavka et al., 2020). Moreover, the implications of supply chain risks even extend beyond the immediate interests of businesses, significantly impacting national security and economic stability. The ongoing enhancement of programs such as the United States Space Force's National Security Space Launch (NSSL) highlights the value of consistent supply chains in supporting national defense capabilities. Supply chain disruptions that allow for critical infrastructure such as aerospace and defense sectors can lead to delayed critical military operations and achievement of strategic objectives. With the emphasis of the NSSL being on assured access to space and transitioning to domestically located technologies, vulnerabilities in such supply chains must be addressed so as not to jeopardize national security. Through a comprehension of such interdependencies, industry and policy-makers can develop a proactive response toward mitigating the hazards in supply chains that impact industrial systems at different levels.

3.5.1. Third-Party Vendor Vulnerabilities

In the sophisticated world of industrial systems, third-party vendor vulnerabilities constitute a critical compromise to operational security and integrity. With more and more organizations relying on third-party vendors to deliver software and hardware solutions, they unknowingly place themselves within a chain of risks. The third-party vendors harbor their own varying degrees of security protocols that do not necessarily reflect the main organizations standards. Consequently, a breach by third-party vendors in their own system can cascade down to the parent system, leading to unauthorized access and potential loss of information. The interconnected nature of modern industrial processes serves to point out the utmost importance of frequent assessment and monitoring of third-party vendors' security systems to thwart these gaps and ensure that they continue to be compliant with stipulated best practices. Moreover, the impact of vulnerabilities from third-party vendors also has its influence on not just the direct security problems but also destroys entire supply chains and operational workflows. A vulnerable vendor can result in significant downtimes, which contribute to substantial monetary losses as well as consumer trust loss. In most cases, the effects of such vulnerabilities are aggravated by the nondisclosure of a vendor's cybersecurity practices. Organizations can remain oblivious to the vulnerabilities that exist, hence weakening their capacity to counter any threats. To counter these risks, there is a need to have complete vendor risk management processes involving third-party evaluations, ongoing audits, and contingency planning. These steps not only enhance security but also create resilience throughout industrial operations so that organizations can improve their resistance against likely issues. The emergence of digital technology has further complicated the third-party vendor vulnerabilities issue. With the advent of solutions like digital twins, organizations must put in place sufficient assurance that their vendors also have robust cybersecurity systems for protecting the data employed in these sophisticated systems. Digital twins are virtual copies of real assets and require ongoing data exchange between stakeholders, hence more exposure to probable threats is experienced. As businesses work to leverage the power of digital twins and the Internet of Things (IoT), protection of third-party vendor connections becomes even more vital. By integrating advanced cybersecurity standards into their third-party vendor relationship, organizations will be in a position to protect their assets and enhance the resiliency of their industrial system, encompassing weaknesses and the moral risks associated with handling third-party relationships in a virtual environment (Dwivedi et al., 2020; Rasheed et al., 2020) (Figure 3.14).

Figure 3.14. The third-party vendor vulnerabilities.

Source: https://www.kiteworks.com/wp-content/uploads/2022/01/Third-party-risk-management-glossary.webp.

3.5.2. Counterfeit Components and Materials

The proliferation of counterfeit materials and parts poses a significant threat to industrial systems, and a plethora of vulnerabilities is created. As industries become more reliant on sophisticated supply chains, the risk for inclusion of counterfeit items increases. Counterfeit parts can harm product performance, compromise safety, and cause universal financial loss. Interestingly, sectors such as aerospace, automotive, and healthcare are particularly vulnerable due to the high amount of regulatory standards that are being breached by utilizing fake parts. These systems require a high level of precision and reliability, and hence the utilization of lower quality fake materials carries not only the risk of reduced operational efficiency but end-user safety as well. The implications of such vulnerabilities extend beyond loss of dollars to potential legal consequences and reputational harm that businesses face when counterfeits are discovered in their products. Beyond the technical realities of counterfeits, there is a multifaceted web of regulatory challenges that contribute to the issue. As prescribed in recent estimates, the US import policy and Customs and Border Protection (CBP) role of monitoring incoming materials are key in countering the risks associated with counterfeits. A weak regulatory framework can make it possible for counterfeit components to penetrate supply chains with impunity, leaving industries at the mercy of nefarious actors who exploit these weaknesses for profit. Significantly, the law for customs legislation places emphasis on the requirement for tighter vigilance at frontier controls to prevent penetration by forged goods. Besides, as the threats of counterfeiting evolve into advanced hybrid threats, similar to those in law enforcement environments where military and criminal elements converge, sectors must adopt robust interventions to combat these incursions and protect their operational integrity, such as in serious national security implications over the use of trusted components in national defense systems, such as in Nyagudi et al. (2013). Furthermore, the crisis of fake parts demands a collective effort by regulators, manufacturers, and consumers to get an enhanced industrial system. Such collaboration can be enabled by the creation of traceability technologies, which not only provide the world with a clear perspective of supply chains but also simplify the detection of counterfeit products at every level. While the industry moves towards more sophisticated supply chain management, embedding technologies like blockchain and IoT can become instrumental in developing authenticity in the origin of products. Additionally, promoting greater awareness and education on the dangers of counterfeit materials among stakeholders can help individuals and institutions to make informed decisions. Legislative measures, such as strengthening intellectual property rights and enhancing cargo scanning processes, remain essential to slowing down counterfeiting flows and protecting the integrity of industrial platforms, consistent with prioritized goals outlined in successive CBP reports, which are centered on trade law modernization and trade law enforcement, as addressed in Jones et al. (2013).

3.5.3. Disruptions Due to Geopolitical Factors

Geopolitical events have become major disruptors of industrial systems, impacting global supply chains and financial stability. Political tensions, border conflicts, and trade wars can potentially lead to unexpected shortages in critical resources, leading to inefficiencies in operations and increased costs. For instance, fighting in strategic locations may interdict raw material supply chains required to manufacture a product, prompting companies to turn to secondary sources, which may be less dependable or less economically viable. Interruptions such as these affect not only manufacturing timetables but also force industries to reconsider strategies in attempts to mitigate risks of dependence on specific geopolitical settings.

The need for resilience in industrial systems has thus never been more urgent, as businesses cope with the complexity due to such changing geopolitical environments that call for a robust system of risk management and flexibility. Moreover, communication's role during times of geopolitical crisis cannot be overstated. Effective conveyance of information is paramount in maintaining organizational integrity and functional continuity. As observed in World Economic Forums analysis, the interconnection of risks in the world means that one event can cascade into widespread disruption in various sectors, so that leaders need to adopt proactive communication approaches. Lessons from events such as the Fukushima reveal the importance of having open communication channels and developing strong business models. Organizations must cultivate the ability to react to instant crises as well as to the ensuing anxieties and uncertainties that disruptions in the geopolitical context are likely to evoke among stakeholders.

This is carried out so that there can be a responsive pattern of least disruption to operations and resilience of industrial systems against repeat events. Lastly, the combination of geopolitical risks and long-term sustainability issues forms an aggregated risk to industrial systems. Since the Global Risks report has set out, the interrelationship between economic, environmental, social, and technological drivers means that shocks caused by geopolitical tensions can exacerbate underlying vulnerabilities. An example would be trade sanctions imposed on specific nations, which have more severe environmental implications, as companies have to change their purchasing behaviors, increasing carbon emissions. Emphasizing collective responses is vital, as governments, industry, and civil society can come together and generate an effective framework for addressing such multi-dimensional threats. Addressing the underlying causes of these geopolitical tensions, as well as their effects on industrial systems, will be critical to overcoming future challenges and allowing industries to develop and thrive in an increasingly complex global world (N/A, 2012, 2015).

3.6. REGULATORY AND COMPLIANCE CHALLENGES

In the industrial systems industry, regulatory and compliance concerns are significant issues that have to be addressed by organizations if they are to be able to sustain operational integrity and resilience. There is compliance with various regulations, particularly because evolving threats and vulnerabilities are altering global standards. For instance, the majority of industrial organizations face significant pressure to adhere to stringent safety regulations that involve massive risk analysis and proactive management of disasters. The application of new technologies may facilitate this process, but it introduces complexity because organizations do not marry procedural methods for their operations with regulatory frameworks. This technological-converging regulatory development underscores the necessity to have flexible strategies capable of adapting to evolving regulations without compromising effectiveness in operating systems. Moreover, the challenges of regulatory compliance are not merely technological matters for implementation; they transcend broader organizational and institutional settings. As seen in existing studies, factors such as unplanned urbanization and outdated infrastructure increase the difficulty in meeting regulatory demands in the majority of industrial settings. In most cases, a failure to possess effective regulatory standards can complicate organizations' ability to effectively manage disaster resilience threats. This was best illustrated by empirical evidence in Sri Lankas cities, where weak governance structures, coupled with limited funds and resources, yielded significant obstacles in building resilient systems. It highlights the need for ongoing discussion on the updating of regulation requirements for sufficient adaptation to realities on the ground in various industrial environments. Additionally, regulatory problems are not just confined to the physical and engineering infrastructure of industrial systems but also reach into the realm of cybersecurity, particularly as Medical Cyber-Physical Systems (MCPS) become increasingly used. With these advanced systems combining technology with healthcare provision, organizations become more vulnerable to threats while being required to comply with a myriad of regulatory provisions designed to protect patient data and system integrity. The analysis of potential invasions of such systems presents serious challenges to compliance since they are programmed in a complex way and even deal with tremendous amounts of information. The incorporation of forensic science rules into MCPS design is significant, offering the regulatory framework that not only respects compliance but also enhances an organization's capacity to respond to exposure effectively (Choo et al., 2017). This diversified approach will help organizations to be capable of de-risking against threats as well as navigating through dense landscapes of regulatory issues impinging on them.

3.6.1. Industry-Specific Regulations and Standards

Industry-specific regulations and standards are critical in safeguarding industrial systems against an array of threats and vulnerabilities. These regulations are formulated to address unique challenges pertaining to various industries such as manufacturing, energy, and information technology. For instance, the energy sector encounters certain cybersecurity issues due to its reliance on networked infrastructure and increasing adoption of smart grid technology. Reliance on digital control and communication technologies in the event of smart grids necessitates robust security platforms to prevent potential failures and costly breaches (Qian et al., 2012). Standards such as the National Institute of Standards and Technology (NIST) Cybersecurity Framework are important in shaping organizations to adopt an assertive cybersecurity stance, thereby enhancing the resiliency of critical infrastructure. Industry standards not only make security better but also facilitate interoperability between various technological systems, which is more critical in the current era.

Figure 3.15. The implementation of industry-specific regulations and standards.

Source: https://ksvatc.in/wp-content/uploads/2024/07/regulatory-service-768x462.jpg.

The Internet of Things (IoT) is a classic case in point where every type of device across various industries employs IoT for optimizing efficiency and automation. Yet, the more IoT technologies are employed, the higher the likelihood of attacks. There is a need for industries to enact regulations that counter these very threats and offer protocols for secure communication to keep potential attacks at bay. A smart home, for example, operates under regulations that consider not just user convenience but also security and efficiency in energy consumption, which suggests the importance of all-rounded standards in achieving interdependent systems' security (Romano et al., 2019). In addition to encouraging security, industry standards are also essential in achieving compliance and accountability among organizations. Adherence to established standards promotes a risk culture where companies are obliged to review their cybersecurity frequently and deploy updates as technology evolves. This is also emphasized through audits and assessment that quantify compliance with relevant law. By firmly putting in place a regulatory framework, sectors are able to manage risks of cybersecurity attacks, thus achieving enhanced operational reliability. As industrial systems continue to grow more complex with the addition of additional advanced technologies, the dynamic regulation environment will need continued reevaluation to adequately address new vulnerabilities, with safety and security being paramount in all sectors (Figure 3.15).

3.6.2. Consequences of Non-Compliance

The consequences of non-compliance in industrial systems occur in numerous unfavorable manners, with immediate effects on operational effectiveness and organizational trustworthiness. Non-compliance can have severe legal repercussions, including penalty payments and sanctions that can strain funds. Besides direct financial penalties, organizations stand to incur long-term litigation costs arising from non-compliance with regulations. These financial impacts are augmented by potential operations disruption, whereby non-compliant organizations can be forced to halt operations or services to address compliance gaps. Furthermore, the tainted reputation that arises from regulatory transgression can lead to lost consumer confidence, which ultimately converts to the loss of market share and growth possibilities. As industrial systems increasingly shift towards networked technologies and regimes, business leaders must realize that compliance is not merely a legal requirement but is an integral component of strategic resilience. Besides, the effects of non-compliance extend beyond monetary loss and reputation to endanger the safety of infrastructure and personnel. Non-compliant industrial systems can expose workers and surrounding populations to harm, eventually resulting in workplace accidents or ecological disasters. Such accidents pose not only short-term health issues but also a lengthy recovery period for the damaged ecosystems and populations. For industries like electric power infrastructure, where compromising safety reviews highlight malicious attacks as a priority, non-compliance threats greatly potentiate such attacks. Standard security measures, as proposed by industry research, become key to preventing natural and man-made threats, supporting the requirement for strict compliance with standards to prevent such disastrous outcomes (Marcelo et al., 2008). In addition to the operational and safety problems, non-compliance could inhibit innovation in industrial systems. Organizations are so concerned about legal repercussions of non-compliance that they would become risk-averse, suppressing their inclination to adopt new technology or systems that have the potential to improve operational efficiency. Such risk aversion can lead to stagnation, particularly in industries that require quick responses to changing market needs or emerging threats. Recent violations, the analysis shows, show that non-compliance makes companies vulnerable to vulnerabilities that newer systems seek to bridge (Maple et al., 2010). Neglect of compliance not only jeopardizes business in existence but also diminishes competitive advantage in a fast-changing world. Ultimately, compliance is at the heart of building a culture of continuous improvement and innovation in industrial systems, once more highlighting the importance of strategic congruence with regulatory mandates.

3.6.3. Evolving Nature of Regulatory Requirements

The regulatory landscape governing industrial systems is dynamic, reflecting the very fast pace of technological developments and the evolving nature of risks to which they are subject. With industries adopting smarter technologies and networked systems, regulators are compelled to adapt current frameworks in order to address new vulnerabilities. This creates a recurring cycle of regulatory adaptation, wherein agencies not only respond to current threats but also adapt in anticipation of those still to emerge. Encouragingly, the move towards preventive regulatory measures—those that seek to pre-emptively identify and contain potential risks—is increasingly becoming a hallmark of efficient governance. Through the integration of rules with the intricacy of modern industrial systems, resilience against physical and cyber threats can be attained by the stakeholders in acknowledgement of the fact that traditional methods may not be sufficient in this fast-paced era (Marcelo et al., 2008). Simultaneously with all these, the importance of Business Continuity Management (BCM) has been brought up to a greater level of priority in regulatory requirements. Regulatory bodies are now seriously looking into the inherent connection between BCM and organizational performance, causing reevaluation of how such frameworks are constructed. Recent academic literature boasts the subject of how the need for BCM processes must not only enhance operational resilience but also correlate effectively with diverse regulatory requirements. Incorporating BCM practices into evolving regulatory requirements enables organizations to fit better into uncertainties while staying in compliance. Closing the gaps that presently prevail between theory and practice will further refine the paradigms that direct industrial systems towards making meaningful contributions to planning and response protocols for resilience (Fischbacher-Smith et al., 2017). Moreover, the globalized nature of industrial systems introduces nuances into regulatory regimes whereby cooperation and harmonization across jurisdictions become necessary. Because multinational actors exist geographically, they must comply with varying regulatory standards, which poses compliance and enforcement challenges. The dynamic nature requires regulators to engage in collaborative efforts towards the development and enforcement of uniform guidelines that combat collective vulnerabilities. Not only does such cooperation enhance the effectiveness of one regulatory regime but also consolidates regulation efforts in countering common exposures in industrial systems. As regulation shape shifts, though, its emphasis must remain on developing an adaptive regulatory environment that can manage the slippery nature of technology innovation and threat possibilities (Marcelo et al., 2008).

SUMMARY

In summary, awareness of common threats and vulnerabilities in industrial systems, particularly SCADA systems, gives rise to the critical need for enhanced security practices. Because these systems are at the heart of managing critical infrastructure, any disruption could have catastrophic effects on operational continuity as well as public safety. The evolving nature of threats from the cyber realm demands that stakeholders not only be prepared for probable vulnerabilities but also take proactive steps to neutralize threats. Through identification and characterization of these threats, the installation of robust security architectures is an underlying component in safeguarding industrial processes. A multilayer architectural model for SCADA systems is required in establishing a consensus among the stakeholders and designing concentrated security controls, thereby embracing the complexity that is inherent in these technologies (Ma et al., 2012). The SCADA system complexity intensifies the challenges posed by their security. Such systems tend to have multiple embedded components that might make it challenging to gain an overall idea of their architecture. This complexity needs to be understood, as it has implications for targeted security practices that can keep unauthorized access and attacks at bay.

As emphasized in latest studies, an effective way to find process-oriented threats involves systematic processing of logs to find anomalies that are likely to signal malicious activity by attackers posing as legitimate users (Bolzoni et al., 2012). Such a detection procedure is a critical component of an overarching security strategy that seeks to protect industrial processes from increasing cyber threats, reinforcing the need for adaptation to the changing threat landscape. Lastly, achieving protection of industrial networks like SCADA requires ongoing attentiveness in monitoring, analysis, and adaptation. Constant education and training on best practices in cybersecurity must be the priority for stakeholders in order to adapt to changing tactics employed by threat actors. Apart from implementing cutting-edge technology solutions, fostering a security-culture within organizations can greatly enhance defenses against attacks. By integrating extensive security analysis into the SCADA system lifecycle, organizations can protect not just against current threats but also build resilience to endure future cybersecurity challenges. Thus, the pursuit of security in industrial systems is an ongoing process that needs to be performed with cross-industry cooperation, adherence to best practices, and the deployment of new security innovations to deliver safety and reliability of critical infrastructure.

Industrial systems today are more vulnerable to a variety of threats that compromise their operational integrity and safety. One of the most dangerous problems is cyber-attack on SCADA systems that are used to operate such vital infrastructure as power grid networks and water treatment facilities. Given that the proper working of society as a whole relies on these systems, the impact of an attack can be catastrophic and lead to service disruption or even physical harm to individuals. It is essential, thus, that organizations prioritize the security of SCADA systems, not only by deploying advanced technological tools but by comprehending the system structure as well. An integrated comprehension of the system allows stakeholders to effectively make security assessments, thereby addressing the vulnerabilities in these complex networks (Ma et al., 2012). Another significant threat industrial systems are subjected to includes the evolving nature of global space operations and associated vulnerabilities. The evolution of programs such as the NSSL points to the necessity of assured access to space for military systems.

Although the program aims at lowering launch costs and enhancing national defense, it also poses numerous risks, particularly concerning the uniformity of fresh launch vehicle designs and replacing established technology. Similar to past reliance on foreign technology, such as the rocket engine built by Russia, there are critical vulnerabilities which require scrutiny (Zhang et al., 2020). The assurance of launch integrity and protection from sabotage are top issues with countries competing to establish their strategic assets in more fragile environments. Therefore, awareness of these vulnerabilities is central to the success of industrial space operations. Furthermore, the interconnectedness of modern industrial systems also increases the threat environment. The addition of IoT devices, as it increases operational efficiency, introduces new risks to cyber-attack. Each device is a gateway of potential entry for hackers, who can take advantage of these vulnerabilities as an avenue to advance to larger networks. The lack of uniform security protocols utilized across diverse industries is also an environment with disparities in protection techniques. Such disparity brings forth the importance of a unified risk management approach that not only identifies and slows down specific threats but also promotes collaboration between disparate industries. Closing the gaps in security between systems provides a robust framework that can meet today's and tomorrow's vulnerabilities for industrial environments, ultimately protecting critical infrastructure in the future.

REFERENCES

1. Al-Khateeb, H., Burnham, K. J., Epiphaniou, G., Hammoudeh, M., et al. (2019). Preliminaries of orthogonal layered defense using functional and assurance controls in industrial control systems. https://core.ac.uk/download/196185875.pdf (accessed on 13th September 2025).
2. Almalawi, A., August, T., Baecher, G., Boldea, B., Bouchti, Y., Brand, M., Byres, E., et al. (2017). Assessing and augmenting SCADA cyber security: A survey of techniques. https://core.ac.uk/download/293885085.pdf (accessed on 13th September 2025).
3. Al-Shaidy, A. N. M. K., Kamala, M. A., Kinteh, S., Kiran, A., et al. (2017). Technical report on deploying a highly secured OpenStack cloud infrastructure using BradStack as a case study. http://arxiv.org/abs/1712.09152 (accessed on 13th September 2025).
4. American Petroleum Institute, Arata, C., Dowell, J., Ellis, T., Fennelly, K., Fontaine, J., et al. (2014). Security risk assessment and protection in the chemical and process industry. https://core.ac.uk/download/30731359.pdf (accessed on 13th September 2025).
5. Anderson, J., & Rainie, L. (2017). The Internet of Things Connectivity Binge: What are the Implications? https://core.ac.uk/download/89472988.pdf (accessed on 13th September 2025).
6. Anglano, C., Aniello, L., Antinori, A., Armando, A., et al. (2018). The future of cybersecurity in Italy: Strategic focus area. https://core.ac.uk/download/188830442.pdf (accessed on 13th September 2025).
7. Archibald, J. M., & Renaud, K. (2019). Refining the PoinTER "human firewall" pentesting framework. https://core.ac.uk/download/228178554.pdf (accessed on 13th September 2025).
8. Barker, T., Cook, M., Evangelopoulou, M., Harkness, R., et al. (2017). Defending against firmware cyber-attacks on safety-critical systems. https://core.ac.uk/download/111557940.pdf (accessed on 13th September 2025).
9. Benedetto, L., Bianchi, G., Conti, M., Dargahi, T., et al. (2019). A cyber-kill-chain based taxonomy of crypto-ransomware features. https://core.ac.uk/download/222790025.pdf (accessed on 13th September 2025).
10. Bolzoni, D., Hadziosmanovic, D., & Hartel, P. (2012). A log mining approach for process monitoring in SCADA. https://core.ac.uk/download/pdf/11483058.pdf (accessed on 13th September 2025).
11. Bosher, D., Dainty, A., Dubbeling, M., Georgieva, K., Godschalk, D., et al. (2014). Challenges in creating a disaster resilient built environment.

https://core.ac.uk/download/30731102.pdf (accessed on 13th September 2025).

12. Castillo-Chavez, C., & Roberts, F. S. (2002). Report on DIMACS Working Group Meeting: Mathematical Sciences Methods for The Study of Deliberate Releases of Biological Agents and Their Consequences. http://ecommons.cornell.edu/bitstream/1813/32199/1/BU-1600-M.pdf (accessed on 13th September 2025).
13. Choo, K. K. R., Glisson, W. B., & Grispos, G. (2017). Medical cyber-physical systems development: A forensics-driven approach. http://arxiv.org/abs/1708.05374 (accessed on 13th September 2025).
14. Delli Priscoli, F., Di Giorgio, A., Esposito, M., Fiaschetti, F., et al. (2017). Ensuring cyber-security in smart railway surveillance with SHIELD. https://core.ac.uk/download/80313548.pdf (accessed on 13th September 2025).
15. Dwivedi, Y. K., Ismagilova, E., Hughes, D. L., Carlson, J., Filieri, R., Jacobson, J., Jain, V., et al. (2020). Setting the future of digital and social media marketing research: Perspectives and research propositions. International Journal of Information Management, 59, 102168–102168. https://doi.org/10.1016/j.ijinfomgt.2020.102168 (accessed on 13th September 2025).
16. Evangelopoulou, M., Harkness, R., & Johnson, C. W. (2016). Forensic attacks analysis and the cyber security of safety-critical industrial control systems. https://core.ac.uk/download/46664183.pdf (accessed on 13th September 2025).
17. Evangelopoulou, M., Johnson, C. W., & Pavlova, T. (2017). Applying lessons from cyber-attacks on Ukrainian infrastructures to secure gateways onto the industrial Internet of Things. https://core.ac.uk/download/111557942.pdf (accessed on 13th September 2025).
18. Fischbacher-Smith, D. (2017). When organizational effectiveness fails: Business continuity management and the paradox of performance. https://core.ac.uk/download/74232623.pdf (accessed on 13th September 2025).
19. Gardner, J. F. (2016). Multidisciplinary game-based approach for generating student enthusiasm for addressing critical infrastructure challenges. https://core.ac.uk/download/74506828.pdf (accessed on 13th September 2025).
20. Hauser, C. H., Lorenz, D. H., Fudenberg, D., Santacana, E., Kuipers, F., Wu, F. F., Xue, G., et al. (2018). Multi-layer cyber-physical security and resilience for smart grid. http://arxiv.org/abs/1810.00282 (accessed on 13th September 2025).

21. Johnson, C. (2016). Securing the participation of safety-critical SCADA systems in the industrial Internet of Things. https://core.ac.uk/download/46664188.pdf (accessed on 13th September 2025).
22. Jones, V. C., & Rosenblum, M. R. (2013). U.S. Customs and Border Protection: Trade Facilitation, Enforcement, and Security. https://core.ac.uk/download/19958888.pdf (accessed on 13th September 2025).
23. Kavka, L., Kodym, O., & Kubáč, L. (2020). Risks associated with Logistics 4.0 and their minimization using Blockchain. https://core.ac.uk/download/322825138.pdf (accessed on 13th September 2025).
24. Ma, Z., Skopik, F., & Smith, P. (2012). Towards a layered architectural view for security analysis in SCADA systems. http://arxiv.org/abs/1211.3908 (accessed on 13th September 2025).
25. Maple, C., & Phillips, A. (2010). UK security breach investigations report: An analysis of data compromise cases. https://core.ac.uk/download/29820726.pdf (accessed on 13th September 2025).
26. Masera, M., & Vamanu, B. (2008). Understanding malicious attacks against infrastructures–Overview on the assessment and management of threats and attacks to industrial control systems. https://core.ac.uk/download/38619766.pdf (accessed on 13th September 2025).
27. Mendez, D. M., Papapanagiotou, I., & Yang, B. (2017). Internet of Things: Survey on security and privacy. http://arxiv.org/abs/1707.01879 (accessed on 13th September 2025).
28. N/A. (2012). Global Risks 2012, Seventh Edition. https://core.ac.uk/download/75783274.pdf (accessed on 13th September 2025).
29. N/A. (2015). Global Risks 2015, 10th Edition. https://core.ac.uk/download/75783277.pdf (accessed on 13th September 2025).
30. N/A. (2016). The Global Risks Report 2016, 11th Edition. https://core.ac.uk/download/71338494.pdf (accessed on 13th September 2025).
31. Nyagudi, N. M. (2013). Post-Westgate SWAT: C4ISTAR Architectural Framework for Autonomous Network Integrated Multifaceted Warfighting Solutions Version 1.0: A Peer-Reviewed Monograph. http://arxiv.org/abs/1401.6379 (accessed on 13th September 2025).
32. Qian, Y., Sharif, H., Tipper, D., Yan, Y., et al. (2012). A survey on cyber security for smart grid communications. https://core.ac.uk/download/12211131.pdf (accessed on 13th September 2025).
33. Rasheed, A., San, O., & Kvamsdal, T. (2020). Digital twin: Values, challenges and enablers from a modeling perspective. IEEE Access, 8, 21980–22012. https://doi.org/10.1109/access.2020.2970143 (accessed on 13th September 2025).

34. Romano, N. (2019). Securing our future homes: Smart home security issues and solutions. https://core.ac.uk/download/213463704.pdf (accessed on 13th September 2025).

35. Vanhee, J., & Verleysen, C. (2017). Cyber safety: A theoretical insight. https://core.ac.uk/download/156960827.pdf (accessed on 13th September 2025).

36. Zhang, M. (2020). National Security Space Launch. https://core.ac.uk/download/305123572.pdf (accessed on 13th September 2025).

CHAPTER 4

AUTHENTICATION AND ACCESS CONTROL MECHANISMS

Contents

INTRODUCTION

The fast pace of technological advancements has made it necessary to have robust security protocols, particularly concerning authentication and access control systems. These systems are the basis of cybersecurity, which safeguards confidential data from misuse by unauthorized parties. Authentication involves the verification of the identity of users or devices before granting access to them to resources, and this can be accomplished through various different mechanisms like username and password, biometric, and multi-factor authentication. On the other hand, access control mechanisms decide the privileges and rights provided to legitimate users, defining what can be accessed and to what extent. Since computers make more use of digital platforms for storing and processing critical information, developers and end-users need to know about these mechanisms. Additionally, different applications necessitate specific security solutions designed to meet their own needs, whereby context awareness is essential in the implementation of authentication and access control. One such is the Lightweight Directory Access Protocol (LDAP), providing a standardized method of directory service access but whose effectiveness is dependent on the built-in security features that protect it. These include the utilization of an application-level firewall, a robust authentication mechanism, and a comprehensive access control system, as emphasized in Chadwick et al. (2000). These all combine to form a security barrier against illegal access while permitting authorized users to interact freely with the system. The effectiveness of these security models is paramount, particularly where dynamic balance between accessibility and security rights is necessary. Moreover, the heterogeneity of various application environments, such as the Real-time Transport Protocol (RTP), requires security mechanisms to be treated with appropriate care. Every environment may have solutions unique to it in order to provide mandatory services such as confidentiality, integrity, and source authentication, which are essential to provide trustworthy transmissions of data. The challenge faced by developers is to negotiate through these diverse choices in deciding the most appropriate security approaches that work best for their respective applications, as emphasized by Perkins et al. (2014). The implementation of strong authentication and access controls is not a matter of technology; it is the essence of maintaining an overall security posture that can keep up with the dynamic threats of cyberspace. Thus, knowing these mechanisms in detail is important for ensuring the security and integrity of digital communications and transactions.

Definition of Authentication

Authentication is an important mechanism in information security that determines the identity of users who attempt to access systems or resources. It entails ensuring that users are authenticating to be themselves, utilizing different means such as passwords and biometric scans, security tokens, and multi-factor authentication. As technology environments evolve, so does the complexity level of the authentication mechanisms, which requires more robust solutions in an effort to overcome illegal access. This complexity level is particularly applicable where sensitive information is stored, authentication failure being potentially catastrophic. By creating a credible method of verification, organizations can best reduce the chance of data breaches and unauthorized actions, thus strengthening their overall security mechanisms. The importance of authentication becomes clear when looking at how it combines with access control policies. Access control specifies what actions authenticated users are allowed to perform within a system, frequently utilizing the authentication mechanism in order to validate user identities prior to granting permissions. Traditionally, access controls have been hard-coded in application components and become rigid security policies that are prone to errors (Silva et al., 2019). Interweaving business logic with security concerns complicates the handling of access controls. By decoupling access control policy from the remaining code base, organizations can enhance flexibility to dynamically update permissions in line with changing security requirements.

Figure 4.1. The authentication.

Source: https://www.miniorange.com/blog/assets/2025/what-is-authentication.webp.

This double-pronged approach—offering reliable authentication in combination with adaptable access control—is at the root of effective security management within information systems. Trends in federated access management over the past few years mirror the dynamic character of authentication mechanisms within university campuses. In Britain, for example, universities have been shifting towards Shibboleth technology to enable transparent and secure access to various campus resources (Jiang et al., 2007). This movement highlights the importance of single sign-on capability, which makes the user experience straightforward while maintaining rigorous authentication standards. The implementation of such technology not only supports user access to systems, including virtual learning environments and student records, but represents an entire philosophy of authentication in incorporating ease of use on both the administrator and end-user sides. Lastly, a robust authentication and access control framework is needed to safeguard organizational assets without compromising on ease of use in more complex digital environments (Figure 4.1).

Importance of Access Control

Access control mechanisms form an important cornerstone in the realm of cybersecurity, where only authenticated users are granted access to sensitive assets. This is especially relevant in educational institutions, where younger users, including elementary school students, are coming into contact with technology and the Internet more than ever. One study involving children in grades three through eight found that although children comprehend the general function of passwords—that is, access control and privacy protection—children have difficulty creating strong, secure passwords because of developmental constraints. With an average of just 13% creating very secure passwords, this highlights the need for proper access control mechanisms that not only protect information but also aid people with varying capabilities so that a culture is developed where the importance of secure authentication is duly communicated to the most junior of users (Choong et al., 2019). The high adoption of Internet of Things (IoT) devices further highlights the need for access control. As more frequent use of these devices is being adopted in daily applications, they present unique challenges regarding user authentication and data protection. The diversity of sensitive information that IoT devices collect demands strong access control to prevent unauthorized use and preserve data privacy. For instance, in the absence of strong access control, there are opportunities for data breaches, identity theft, and unauthorized operation of intelligent devices, which rise exponentially. Besides, application of advanced security technologies such as Blockchain in combination with IoT can address different threats, but this introduces its own complexity in terms of computation and bandwidth (Abdolee et al., 2019). Therefore, it is imperative to have robust access control mechanisms in place in order to secure the interconnected world of IoT devices. Moreover, the significance of access control transcends individual user environments and extends to organizational, governmental, and commercial environments, where confidentiality and data integrity are important. Organizations need to implement multi-level access control methods in a bid to guard sensitive information from external threats as well as internal abuse. A good number of these methods most typically embrace role-based access controls, which grant users access to only data important to their respective roles within an organization. As the nature of threats in cyberspace evolves, so must the sophistication of access control mechanisms. Revision and updating of these mechanisms will not only protect data assets but also sustain users,' customers,' and stakeholders' trust, which relies on the organization to preserve their security and privacy (Abdolee et al., 2019). The importance of access control remains the foundation of successful cybersecurity policy and practice.

Overview of Mechanisms

A robust summary of mechanisms for authentication and access control must begin with the ground-level interactions of directory services. The X.500 data model is at the forefront of this debate because it gives a profile for a secure directory service, allowing for a formalized structure of user information. Based on this framework, the LDAP functions by providing access to directory services while simultaneously addressing security concerns. This dual-use feature highlights the necessity for having several managerial tools in cybersecurity. The entire access control mechanism, complemented by an authentication mechanism, acts as a security bar against outside attacks, highlighting the necessity for having a robust security architecture as described in terms of mechanisms mentioned in the literature (Chadwick et al., 2000). Generally, understanding the interaction of such factors forms the basis for a more effective application of authentication and access control techniques in various scenarios. Even further breakdown of the factors of authentication reveals several mechanisms that facilitate secure practice. Authentication mechanisms that have been standardized for LDAPv3, for example, illustrate the evolution of security techniques necessitated by contemporary applications. From basic password checks to advanced multi-factor authentication, all of these have become commonplace in verifying users' identities before granting access to sensitive information. The impacts of employing strict authentication procedures are extensive, as they disband the gap between accessibility and security, thereby guaranteeing data integrity. Moreover, the threats posed by new technologies require ongoing re-evaluation of authentication measures so that they remain in sync with changing risks and vulnerabilities in digital data handling. This interdependent relationship between authenticating parties and defensive measures guides the development of access controls critical to user data protection in modern infrastructures (Chadwick et al., 2000). The access control environment is not without its barriers, particularly in the realm of devising novel security threats with evolving technology. With cognitive radio networks, for instance, spectrum sharing generates inherent threats like licensed user emulation as well as transaction misbehavior in shared control channels. While such situations are catastrophic in typical networks, they are exacerbated in cognitive radio due to the fact that they are founded on dynamic spectrum access technology. Hence, the security measures that are supposed to provide security must transform and enhance to accommodate these growing threats. While looking at these security issues, it is apparent that a multi-pronged method, incorporating strong access control structures with increasing consciousness about new threats, is required to ensure overall integrity and performance of networked environments. Therefore, a clear comprehension of these mechanisms is most important in formulating effective countermeasures against changing cybersecurity threats.

4.1. TYPES OF AUTHENTICATION

Authentication is a crucial layer in access control techniques, encompassing various techniques that guarantee only the permitted persons can access sensitive information or systems. The dominant categories of authentication can broadly be categorized into three kinds: knowledge-based, possession-based, and biometric authentication. Knowledge-based authentication utilizes the user to enter something which they need to know, such as passwords or security answers. Although heavily used, this system is vulnerable to a myriad of attacks that range from phishing and social engineering that exploit human as opposed to technical vulnerabilities. Possession-based authentication, however, relies on the user having an object, such as a smart card or cell phone, that generates time-sensitive authentication codes. This form of authentication enhances security by bringing together something belonging to the user and the identity of the user, hence reducing the risk coming from stolen data or passwords. Biometric authentication also drew more popularity as it offers a unique and secure way of identification because it utilizes unique physical characteristics, for instance, fingerprints, facial recognition, or retinal scanning. This type of authentication defeats the significant weaknesses of traditional methods through the association of an individual's identity with their biological traits, which is much more difficult to counterfeit or steal. Nevertheless, the application of biometric systems is challenging, particularly in the nature of privacy concerns and the threat of data breaches. Further, while very accurate, biometric system reliability can be affected by the environment and aging. With organizations seeking more secure authentication mechanisms, there is a need to scrutinize how biometric technologies can be combined with current access control models with the challenges of meeting ethical issues and user acceptance. For sectors such as education, in which children begin to interact with technology and undergo phases of cognitive development, having effective authentication systems is crucial. What research establishes here is that children will have multiple passwords but will lack the capacity to create strong ones, and this serves as evidence of cognitive and linguistic development constraints. Above all, children are showing that they know that good passwords must be used for protection and privacy but still create passwords that are not secure according to guidelines. Consequently, learning programs must reinforce not only those good habits but also adapt security instruction to suit children. Incorporating learning about the significance of various types of authentication can introduce healthy cybersecurity habits at an early age, thereby preparing the next generation to cope with an ever more sophisticated digital world and address challenges resultant from rapidly changing technology in various industries, for example, in systems-reliant environments like Building Automation Systems (BAS) (Johnstone et al., 2014) (Figure 4.2).

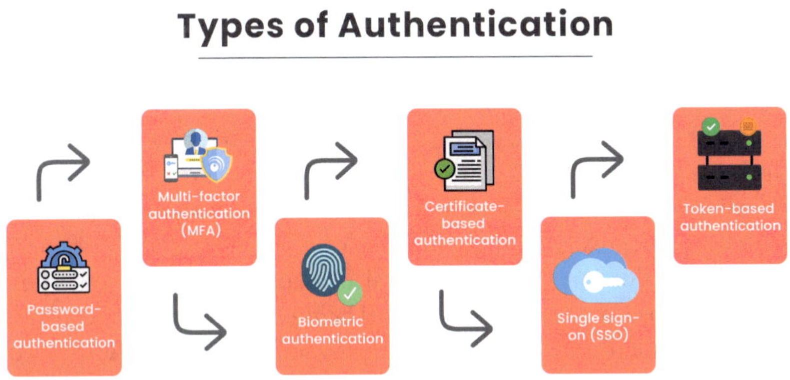

Figure 4.2. Types of authentication.

Source: https://www.infosectrain.com/wp-content/uploads/2023/12/Types-of-Authentication.jpg.

4.1.1. Knowledge-Based Authentication (KBA)

KBA is a technique that depends on the user's possession and memory of certain information, often in the form of passwords, PINs, or security question answers. It is commonly used because of its ease and low cost, making it easily available for organizations to utilize when establishing user identity. Yet, even with how easy it is to implement, KBA is riddled with considerable weaknesses. Users often tend to use weak or easily predicable passwords, and the practice of using the same credentials on numerous sites feeds into security threats. Moreover, the vulnerability of the system to social engineering attacks further raises concerns, as attackers can easily retrieve answers to security questions by looking up information about the users online. Thus, KBA may be seen as a double-edged sword, convenient at first cost but with catastrophic implications on data security and integrity, and with a call for more research on safer alternatives (Erlich et al., 2006). Knowledge-Based Authentication's ease of use is also highly dependent on user activity and consciousness regarding cybersecurity. Users generally are unaware of the importance of using strong passwords or protecting sensitive information and thereby have developed an impression that the weakest link in the security chain is KBA. Convenience, forgetfulness, and lack of knowledge about the consequences of bad security behavior are some of the reasons why users resort to more convenient, less secure methods of authentication. Further, as noted in recent literature, the effectiveness of security controls, such as KBA, can lessen when users perceive they are not engaged towards their cybersecurity, revealing a significant shortfall in delivering security education and training (Sasse et al., 2005). Higher user awareness and education are thus most critical in overcoming the inherent flaws encompassing KBA and developing an environment more secure. Knowledge-Based Authentication alternatives, which consist of possession-based as well as biometric authentication, are more widely popular for companies seeking stronger security measures. Possession-based authentication typically utilizes physical objects, like smart cards or mobile phone tokens, that provide additional verification other than pure knowledge. Biometric authentication utilizes physical characteristics, like fingerprints or facial structure, to ensure that the individual being authenticated is who he or she claims to be. While these alternatives have advantages over KBA, they possess challenges of their own in costs of implementation and privacy concerns. With technology advancing further by the day, the effectiveness and harmony of various methods of authentication will need a high-level understanding of user behavior as well as the quality of security systems. Future research must focus on developing multi-faceted solutions that combine the advantages of KBA with other mechanisms in order to enhance overall security and user experience (Erlich et al., 2006).

4.1.2. Token-Based Authentication

Token-based authentication has emerged as a core mechanism within the larger context of authentication and access control. Token-based authentication provides a safe alternative to traditional password-based systems using tokens as proof of identity. A user is provided with a special token upon successful authentication, which is thereafter utilized for all subsequent interactions without the need for constant reauthentication. This approach increases user convenience by optimizing access to resources on multiple platforms while curtailing the risk of credential loss dramatically. For example, in highly security-oriented settings, token-based systems usually leverage cryptographic techniques to make tokens valid only for a limited time or for a specific transaction, thereby shortening the lifetime of any potential compromised token. Such systems are designed as the demand for intensive security solutions increases with the era of cyber threats. Token-based authentication has advantages and disadvantages that have to be carefully balanced by organizations. A major advantage is that it can support Single Sign-On (SSO) solutions that allow users to access multiple applications using a single set of credentials. This not only increases user experience by minimizing the number of login requests but also enhances security by limiting the number of passwords a user must remember. However, the application of third-party services or centralized token generation can expose vulnerabilities, particularly if these sites are not suitably safeguarded. As highlighted by current controversy over the use of security solutions, the majority of users are still in the dark regarding the specifics of such systems, which may hamper their performance (Sasse et al., 2005).

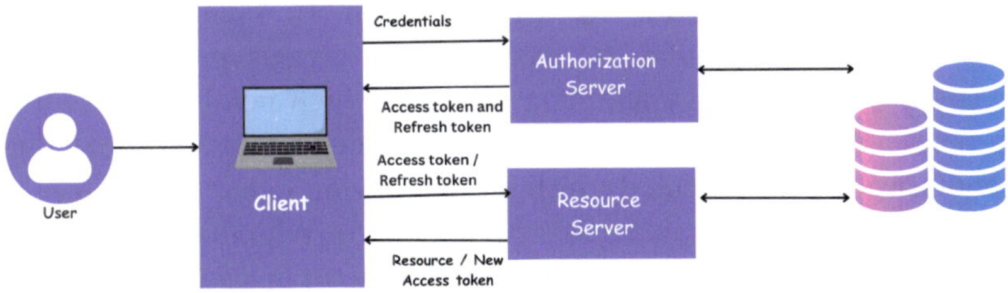

Figure 4.3. Token-based authentication.

Source: https://respected-warmth-1fd8bc18e3.media.strapiapp.com/image3_fd17c0d771.png.

Therefore, ongoing education and successful implementation strategies are essential in order to obtain the optimum level of token-based authentication. Apart from this, token-based authentications role in enhancing secure access control cannot be overstated. By integrating attributes such as attribute-based access control (ABAC) with tokenization, the organization can tailor permissions based on user attributes, roles, and environmental information to enhance security and provide dynamic and adaptive access policies. This adaptability is particularly crucial in hybrid cloud environments where users gain access to resources from various geographies and devices on a regular basis. Research has established that such tools not only increase security but also ease regulatory compliance (Cheverst et al., 2003). Thus, with more companies embracing digital transformation, token-based authentication systems will be an essential requirement for safeguarding confidential data and maintaining secure access across platforms. Therefore, understanding comprehensively both the strengths and weaknesses of token-based authentication is important to organizations that aim to improve their security position in today's dynamic digital world (Figure 4.3).

4.1.3. Biometric Authentication

Biometric authentication has also emerged as a foundational aspect in the area of access control mechanisms in recent times mainly due to its intrinsic ability to utilize unique physical features in the interest of identification. This form of authentication ensures that the system is more secure by employing irreplaceable traits such as fingerprints, facial recognition, and iris structures, hence minimizing the vulnerabilities associated with traditional forms such as passwords. With biometrics, companies can create it more challenging for unauthorized users to access because these biological traits are far harder to replicate or forge than the standard alphanumeric credentials. However, the higher reliance on biometric systems also creates problems from the perspectives of user privacy and data security, which involves implementing strong measures to safeguard these personal characteristics from exploitation or unauthorized use, particularly in light of growing cyber threats that affect individuals and businesses alike (Sasse et al., 2005). With the growing demand for robust security, the use of biometric authentication in decentralized systems is a possible solution to better secure user identity. Present user authentication methods are mostly based on centralized databases, which are a significant point of attack; one attack would expose gigantic amounts of users digital identities (Callahan et al., 2017). While the centralized biometric systems leave them vulnerable to this threat, decentralized ones sidestep this risk by distributing the data on a network, thus not depending on any single database. Apart from increasing security, this new approach also aligns with the new concepts of self-sovereign identity, where the user has control over the biometric data, allowing them to prove who they are without exposing the data to abuse. All of these advancements point to the significance of decentralized biometric systems in writing new rules of digital security. Nevertheless, despite their many advantages, biometric authentication systems have their limitations, particularly in usability as well as user acceptance among different user groups. Studies indicate that success in biometric systems typically hinges on a user-friendly interface that renders interaction an effortless experience, as any kind of complexity will intimidate people away from utilizing such mechanisms (Sasse et al., 2005). Moreover, matters associated with the privacy aspects of biometrics can complicate user interest in using such technology even further. There is an urgent need for the development of biometric systems design that balances security and usability; this balance can greatly enhance public confidence and adoption. By encouraging a culture where biometric verification is safe and convenient, organisations can better secure their assets while shaping a digital culture of literacy that allows users to be in a position to identify and control risks in their online behaviours (Figure 4.4).

Figure 4.4. The biometric authentication.

Source: https://static.wixstatic.com/media/nsplsh_3bbf55a195e64f79bc93dee1626b5b3c~mv2.jpg/v1/fill/w_1042,h_586,al_c,q_85,usm_0.66_1.00_0.01,enc_avif,quality_auto/nsplsh_3bbf55a195e64f79bc93dee1626b5b3c~mv2.jpg.

4.2. ACCESS CONTROL MODELS

Access models are pivotal to the digital platform's security and authentication model. Access models define how the users are authorized, and the model itself determines who can be granted what resources and what they can do. Some of the most used models are discretionary access control (DAC), mandatory access control (MAC), and role-based access control (RBAC). Each model provides unique advantages and limitations, pouring into a master security infrastructure. For instance, DAC provides space for resource owners to be permitted to create permissions, providing a dynamic environment. MAC, however, has rigid policies determined by a central entity, which is crucial in environments that require stringent data integrity and confidentiality. In comparison, RBAC simplifies the management process by providing permissions on a role basis, facilitating simpler scaling with organizations' growth. Therefore, such models need to be scrutinized thoroughly to serve organizational purposes appropriately. With advancements in technology, new access control schemes are being created to improve security and user experience. One such innovative model is Secure Pick Up (SPU), which uses behavioral biometrics to perform implicit user authentication. SPU monitors and learns users' phone pick-up behavior, thereby identifying genuine users from potential attackers without requiring explicit user action for authentication. This model has several advantages, including high accuracy in authentication of 96.3%, which was demonstrated through effective utilization of devices without extensive training data, and capacity for learning across user behavior changes over a long period.

Figure 4.5. Access control models.

Source: https://delinea.com/hs-fs/hubfs/delinea-types-of-physical-access-control-blog.jpg?width=458&name=delinea-types-of-physical-access-control-blog.jpg.

These enhancements in user authentication methods depict the necessity for ongoing development of access control models in a manner that adheres to demands for increasingly complex digital environments that incorporate security and usability harmoniously in modern applications. The wide diffusion of cloud computing renders conventional access control models even more sophisticated, calling for novel methods to secure data storage and sharing. As elaborated by recent research, new methods such as hyperchaotic encryption and new hash functions have been created to ensure the integrity and confidentiality of data stored in clouds. These models not only provide robust security provisions against various cyber-attacks but also incorporate advanced access control provisions to limit illegitimate data interactions. By using cryptographic techniques and authentication protocols, organizations can guarantee secured sensitive information along with facilitating the safe sharing of data among legitimate users. With the virtual realm in a state of continuous flux, it is essential that access control models evolve in tandem, striking a balance of high security with user convenience to protect information in an ever-more connected world (Figure 4.5).

4.2.1. Discretionary Access Control (DAC)

DAC is one of the very first types of access control models in information security, enabling resource owners to manage user permissions for access. This model is characterized by its user-controlled strategy, which allows the owner or creator of an item to control its availability and therefore supply flexibility in sharing resources. In user communities where users frequently interact, such as in organizations, DAC provides a dynamic framework that supports dynamic changes in access levels based on the relationship of users to data. However, this is also a cause of vulnerabilities since unauthorized users are acquiring greater access rights by accident, thereby making sensitive data vulnerable to threats. The effect of DAC extends far and wide into cloud computing environments, particularly when utilizing Database as a Service (DBaaS) clouds. With the flexibility of DAC itself, users can easily disperse information across multiple platforms; but this can also leave database security open. For instance, if there is not proper configuration of access controls, attackers are able to use improperly configured permissions to acquire higher-level privileges or conduct malicious activities. Strong access control techniques would become a necessity for cloud infrastructure, as evidenced by studies pointing towards the risks of working with untrusted service providers. Effective DAC implementation should thus include mechanisms such as role-based access control to minimize any imminent vulnerabilities and enforce security policies that ensure data integrity and confidentiality as documented in Bertino et al. (2016). Besides, advancements in security architectures have led to better methods for the implementation of DAC frameworks for accommodating multi-domain collaborations. The Monterey Security Architecture (MYSEA), for instance, tries to get a balance between requiring secure sharing of information at various classification levels and still allowing users to proceed with familiar activities. This architecture incorporates necessary security protections into cloud services itself, showing the feasibility of expanding DAC outside its traditional scope. It allows for crucial features such as single sign-on and federated controls, thus achieving security and usability needs simultaneously. The robustness of such frameworks ensures that the future of DAC may see more innovations that make it more robust against unauthorized access while ensuring a healthy environment to foster secure collaboration, hence being a great leap in access control mechanisms as elucidated in Gondree et al. (2010) (Figure 4.6).

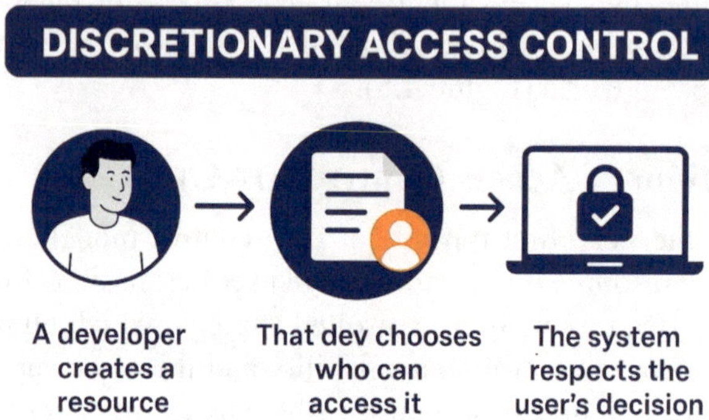

Figure 4.6. Discretionary access control (DAC).

Source: https://frontegg.com/wp-content/webp-express/webp-images/uploads/2025/05/Discretionary-Access-Control.jpg.webp.

4.2.2. Mandatory Access Control (MAC)

MAC is a significant model of access control that enforces rigorous policies dictated by a central authority rather than users discretion. Unlike discretionary access control (DAC) systems, where users control who uses their resources, MAC does not allow users based on preassigned security labels awarded to objects and users. This layered method allows institutions, particularly government and military institutions, to implement more secure mechanisms as it minimizes the risk of unauthorized access by human errors. By establishing strict policies that define how data can be accessed and transmitted, MAC protects confidential data from potential disclosure and threats. The effectiveness of MAC systems is mostly dependent on its usage and adherence to rigid security policies by the entire organization, something which is crucial in maintaining compliance with the regulatory requirements. The role of Mandatory Access Control becomes more solid when applied in particular environments like defense and healthcare, where the sensitiveness of the information should be assured. These environments are the pinnacle of the need for strong security measures, particularly because they handle vast amounts of confidential and classified information. For example, in medicine, a MAC model can ward off unauthorized staff viewing patient data, thereby safeguarding against data breach that can have long-reaching effects on patient confidentiality. In the military, MAC wards off unauthorized individuals from accessing classified documents, further adding to the aura of national security. By enforcing access policies stringently, MAC is a good device for protection and risk management of data, finally leading to increased system trust that handles sensitive data as well as increasing security consciousness culture in organizations. In the current age of cybersecurity, where threats are being faced by systems, using MAC together with other mechanisms for access control can contribute to the overall strength of a system against threats. As businesses shift increasingly toward multi-layered security platforms, an integration of MAC and location-based access control technology can be extremely useful. For instance, location-based control allows users' access to be restricted based on where they happen to be located, complementing MACs strict permission models. With the variety of threats presented by emerging research, including communications and cognitive radio systems threats, the need for advanced security paradigms incorporating these approaches grows ever more urgent (Alhakami et al., 2014; Cleeff et al., 2010). Therefore, embracing new models that converge MAC with emerging access control practices can optimize an organization's ability to protect sensitive data as well as respond adaptively to new cybersecurity threats (Figure 4.7).

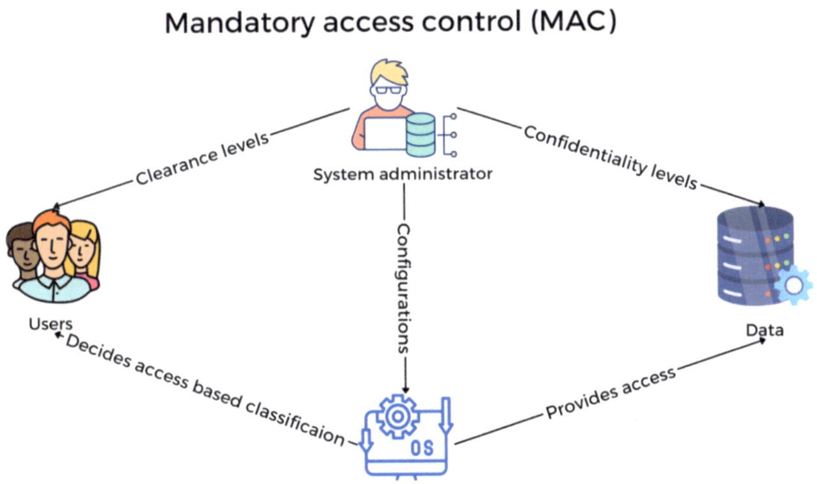

Figure 4.7. Mandatory access control (MAC).

Source: https://aimultiple.com/wp-content/uploads/2024/02/mac.png.

4.2.3. Role-Based Access Control (RBAC)

RBAC) is one of the top models in authentication and access control practices that is defined by its ability to streamline user access rights based on assigned roles in an organization. Such. This approach makes users easier to manage by bundling permissions into roles rather than tying them to users. Subsequently, RBAC not only imposes security through limiting access to confidential information but also imposes efficiency in operation. Through allowing administrators to grant permissions to roles rather than to users directly, organizations will be in a better position to control access rights such that employees possess only access needed for their specific job positions

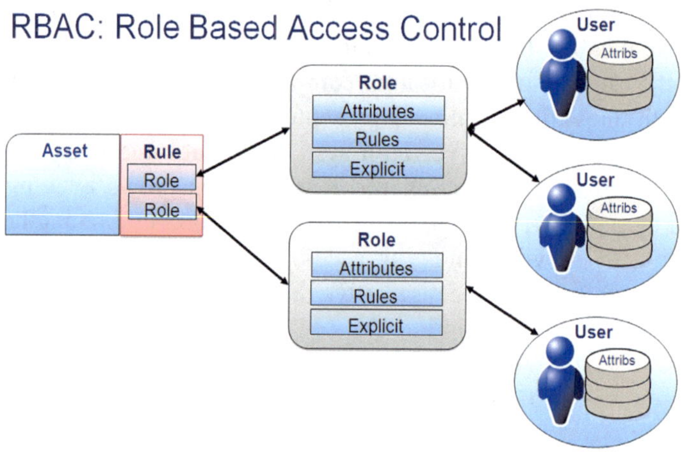

Figure 4.8. Role-based access control (RBAC).

Source: https://idenhaus.com/wp-content/uploads/2017/05/rbac_idenhaus-identi-ty-management-role-based-access.jpg.

This segregation of duties reduces the risk of unauthorized access, thereby enhancing organizational security controls. Secondly, the implementation of RBAC provides enormous advantage in compliance and audit trails, and for this reason, it is extremely desirable for organizations that need to adhere to strict regulatory requirements. With centralized access control within defined roles, implementation of standards such as HIPAA, GDPR, or PCI DSS becomes simpler. The model makes it easy to create an explicit access policy that can be easily inspected and audited, and organizations can demonstrate compliance with regulatory bodies in a structured way. Moreover, because access control decisions can be externally managed and modified without changing system behavior, as noted in integration approaches in various enterprise applications, RBAC allows security postures to be dynamic and adjusted based on shifting security requirements (Silva et al., 2019). Nevertheless, despite its advantages, RBAC also has constraints, particularly concerning scalability and expressiveness in dynamic or high-large environments. When organizations become larger and roles more complicated, the management of role hierarchies and role to permission relationships can get complicated. This not only risks leading to configuration inconsistencies but also inaccessibility in the access control policy. Apart from this, older RBAC systems tend to struggle to adequately accommodate situations requiring increased granular controls or contextual access factors, and thus their utility in intricate operational contexts is compromised (Cleeff et al., 2010). Consequently, even though RBAC is a cornerstone of access control procedures, companies need to weigh its limitations and consider using it together with supporting methodologies to enable holistic security management (Figure 4.8).

4.3. AUTHENTICATION PROTOCOLS

The most core role played by authentication protocols in modern-day cybersecurity cannot be overemphasized. As there has been an upsurge in online interactions on many platforms, it is essential that only authorized users access critical information. Authentication protocols serve as the defense line against unauthorized use, employing various methods to verify user identities. These range from simple password-based ones to advanced multi-factor authentication (MFA) mechanisms using biometrics, hardware tokens, or OTPs (one-time passwords). Layering these security measures enables organizations to readily reduce the threat of breaches. In environments such as wireless mesh networks, where vulnerability is widespread, safe authentication procedures are a lifeline to user privacy and system integrity, as discussed in the comprehensive review of security attacks and countermeasures for wireless mesh networks (Gupta et al., 2013). An authentication protocol's success not only resides in its design but its implementation as well. Secure protocols must adapt to emerging threats and mature with new attack vectors that arise as technologies advance. The flexibility is in using various cryptographic techniques and key management policies to render the authentication process secure. In wireless environments, for example, the identification of protocol layer susceptibilities—e.g., physical, link, or application layers—emphasizes the need for protocols that securely guard such vulnerability points. By comparing existing authentication schemes, the researchers can comprehend the security and weakness of these methodologies. In addition, using trusted third parties during the process of authentication can be more secure at the cost of higher computation and communication overhead (Gupta et al., 2013). As more efficient access control mechanisms are needed, the development and enhancement of authentication protocols is an ever-changing field of cybersecurity. Continued research is necessary to comprehend the nuances of user authentication across varying environments, from single devices to massive corporate networks. Emerging technologies such as blockchain and decentralized identity management are being explored for their scalability and security potential to offer authentication solutions. Keeping time-critical protocols, such as the Network Time Protocol, in sync with the needs of authentication is also an area that requires more research. Making sure that components used for timekeeping are compliant with requirements for authentication is crucial in mitigating spoofing as well as replay attack risks. Therefore, an integrated solution that values the convergence of multiple solutions for authentication will not only enhance user security but also foster trust within cyberspace (Figure 4.9).

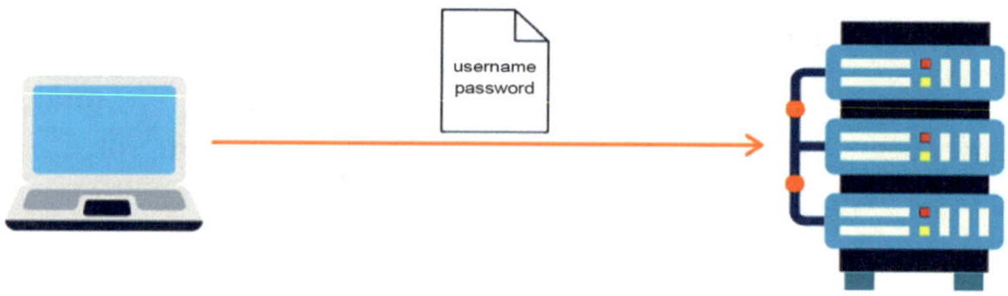

Figure 4.9. The pivotal role of authentication protocols in modern cybersecurity cannot be overstated.

Source: https://cdn.prod.website-files.com/621f84dc15b5ed16dc85a18a/62349fbf0282816e6465acd3_5f87479288363e65e9e5568b_1.png.

4.3.1. Kerberos Protocol

The Kerberos protocol is an architecture component of distributed computing system authentication technology that addresses the fundamental problem of secure identification verification. As a secure method of granting access to networked resources, Kerberos employs a ticketing mechanism that secures user credentials so they can be exchanged safely without transmitting the sensitive information over the network. The protocol adheres to the principle of symmetric key cryptography, where servers and clients both have a common secret key and thereby secure authentication packets with integrity and confidentiality. The robust approach facilitates convenience in streamlined experiences by avoiding the need for repeated password entry on multiple servers or services and enhances usability without sacrificing security. In a global environment where web services are becoming increasingly prevalent, the vulnerabilities of legacy security infrastructures have become an issue, as has been evidenced from the statements regarding cutting-edge access control solutions. Kerberos protocol in particular is the solution to these vulnerabilities by employing a uniform mechanism for credential distribution and access control enforcer among several services. The introduction of Kerberos into modern enterprise settings allows for an improved and more secure access management model, which is required in today's networked systems. In addition, as discussed in studies on deploying security in service-oriented architecture, the inclusion of protocols like Kerberos can plug the existing loopholes in credential management and user access policy, which are critical in maintaining the integrity of web applications (Bemmel et al., 2005). Mobile application development introduces fresh opportunities and challenges for access control systems, with the need for innovative solutions being emphasized. As demonstrated by research into the deployment of next-generation wireless mobile context-aware applications, interspersability between dynamic environments and access protocols is an imperative. Kerberos, being a scalable design, can readily be aligned to accommodate varied wireless technologies while having stringent security protocols. This adaptability is particularly important in most applications that demand tailored user experiences, ranging from location-based services to public internet access. Thus, the power of Kerberos not only fills in the gaps in current technological paradigms but also lays the foundation for future innovations in authentication and safe access control so that user information is protected in more complex digital landscapes (Cheverst et al., 2003).

4.3.2. OAuth 2.0

In contemporary web applications, OAuth 2.0 has emerged as a top framework for authorizing and offering secure access control. Rather than transmitting user credentials straight to numerous applications, OAuth 2.0 makes use of a token-based model under which users can provide limited access to their data without exposing their passwords. This sophisticated technique significantly mitigates security risks concerning sharing of credentials and impacts privacy and user experience. By utilizing an access token mechanism, OAuth 2.0 facilitates ad hoc interactions between users and many services with greater productivity while not compromising security. Such architecture is particularly beneficial to third-party applications, which can acquire a desired user's data without establishing a secure barrier against abuse. Apart from its underlying concepts, the implementation of OAuth 2.0 comprises distinguishing challenges, particularly in terms of security. One of the distinctive features of OAuth 2.0

entails the reliance on redirect URIs and establishing secure communication channels. The protocol must be responsible for protecting token integrity and authenticity throughout the process of authorization. There are many threats like token interception as well as replay attacks that necessitate stringent security measures. For instance, the vulnerability in the underlying web infrastructure poses potential risks for deciding on design focused on protection against unauthorized data retrieval and session hijacking.

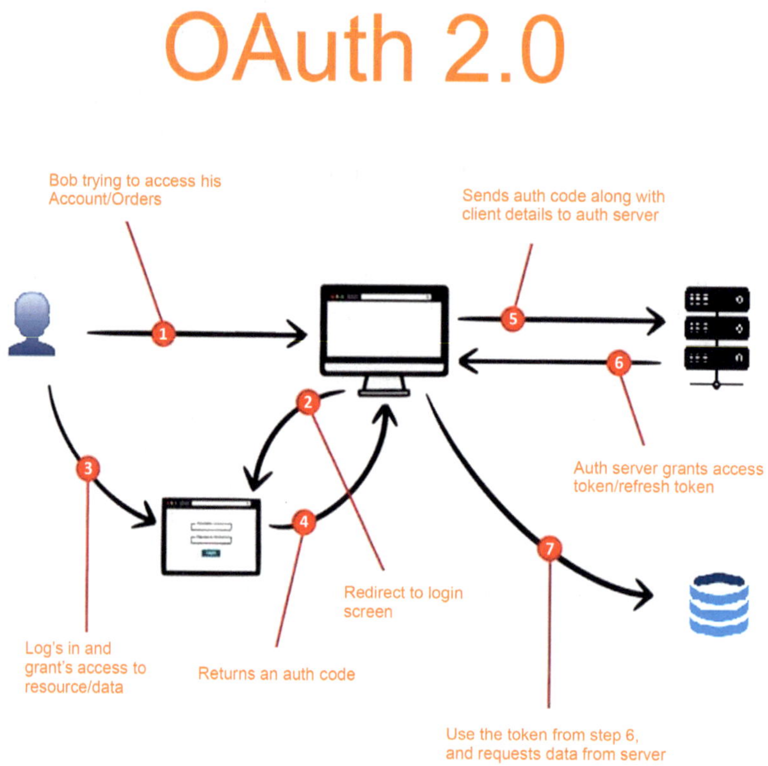

Figure 4.10. OAuth 2.0 has emerged as a predominant framework for managing authorization and ensuring secure access control..

Source: https://miro.medium.com/v2/1*zicBcGvY3wPGvTSKe8-vjQ.png.

Despite some unavoidable weaknesses, the architecture of frameworks enables further enhancement of security mechanisms, thus expanding its strength to protect user information during interaction between various digital ecosystems (Fett et al., 2017). As a response to the widespread need for robust authentication systems, OAuth 2.0 is located in a broader context of API access control and security. The majority of recent research indicates that more than 80% of web APIs support some form of authentication, which underscores the methods centrality in web service usability. OAuth 2.0 not only enables user authentication but also constitutes a critical innovation for web applications that are reliant on diverse streams of data and integrations. Emergence of the semantic web services has further increased the relevance of OAuth 2.0, enabling complete automation and authorization that is tailored to user requirements. As the online world keeps evolving, the OAuth 2.0 principles provide fundamental frameworks that allow developers to secure their applications while ensuring a user-centered experience, effectively closing the security-usability gap (Alvaro et al., 2010) (Figure 4.10).

4.3.3. SAML (Security Assertion Markup Language)

SAML is one of the most important frameworks that offers single sign-on (SSO) and interoperability between heterogeneous security domains within an enterprise environment. With its ability to encapsulate authentication and authorization assertions in XML, SAML enables users to authenticate with a single identity provider and subsequently access multiple service providers without frequent re-input of credentials. All this enhances the usability of systems while reducing administrative burden associated with credential management. Furthermore, based on the assumption of a decentralized model, SAML offers the viability to handle access control policies, again closing the gap between security and user convenience. Such a structure is specifically valuable within contexts where privacy as well as safeguarding sensitive data is paramount since it allows organizations to impose stringent access protocols while easing user access to essential services. SAML presents some challenges, notwithstanding its benefits. Privacy and security concerns would usually mostly pertain to how the authentication data is stored and handed over. Current implementations may lack the stringent controls required to allow users to set and manage access to their data in a complete sense. Out of available research findings, there remains a significant gap in allowing patients or end users to define who gets access to their data and under what scenarios, which is a big deal in sectors such as healthcare. All these problems can be addressed by increased user-friendliness, in which users can actively engage with access control systems. These innovations can ensure that essential security assertions by SAML are not only reliable but also compliant with expectations of user privacy, therefore building user trust in online transactions. The integration of SAML with emerging frameworks, including the obligation of trust (OoT) protocol, ensures it is more practical in managing authorization and privacy management efficiently. The OoT protocol introduces a two-sided model of implementing privacy and authorization requirements, whereby parties may engage in dynamic negotiation of terms. The two-party approach, as contrasted with traditional provider-centric infrastructures, encourages mutual responsibility for specifying access control policies. Through enabling clients as well as service providers to specify their access and privacy preferences, applications are encouraged to foster trust and cooperation within distributed systems. These advances attest to the evolution of SAML from an elementary authentication mechanism to a more comprehensive approach that addresses the sophisticated interaction of privacy, authorization, and user requirements in the digital age, justifying its position in today's discourse of Authentication and Access Control Mechanisms (Figure 4.11).

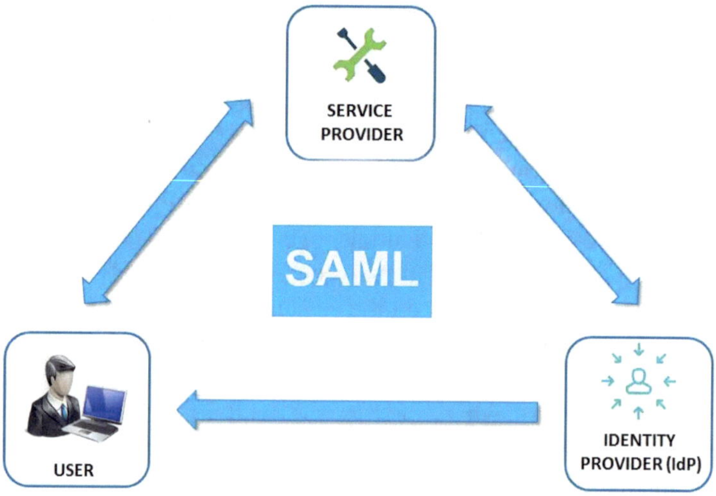

Figure 4.11. Security assertion markup language(SAML).

Source: https://networkencyclopedia.com/wp-content/uploads/2023/05/Security-Assertion-Markup-Language-SAML.webp.

4.4. MULTI-FACTOR AUTHENTICATION (MFA)

In computer security, MFA is an essential process for enhancing user authentication procedures. Unlike traditional single-factor authentication methods, which utilize passwords as the sole form of verification, MFA demands that users enter different forms of verification. These typically consist of something the user knows (e.g., a password), something the user possesses (e.g., a cell phone), and something unique to the user (biometric data such as fingerprints). This multi-factor strategy significantly reduces the threat of unauthorized access, since it is more difficult for attackers to impersonate multiple authentication factors. With distributed applications and systems gaining popularity, the implementation of MFA has become even more important in securing sensitive data and ensuring user privacy in today's fast-paced interconnected world. The significance of MFA not only lies in securing from unauthorized access but also plays a critical role in fulfilling multiple regulatory demands. Businesses in all industries, especially those processing sensitive data, must institute effective security controls, one of which is MFA. By including MFA, companies are able to heighten their security status and show dedication to protecting data that is ever more closely watched under regulations such as the General Data Protection Regulation (GDPR) and the Health Insurance Portability and Accountability Act (HIPAA). By these policies, the implementation of MFA not only helps minimize the threat of security issues but also enhances customer loyalty and confidence. Therefore, organizations must constantly review and change their authentication models to make sure that MFA is a core component in their access control systems.

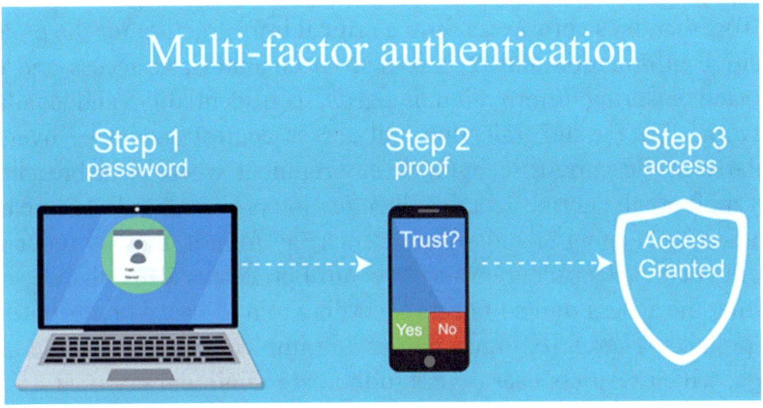

Figure 4.12. Multi-factor authentication (MFA).

Source: https://www.ramsac.com/wp-content/uploads/2021/10/ramsac-MFA-blog.png.

Furthermore, new technologies continue to innovate how MFA is implemented and utilized on various platforms. Current advances have included the integration of artificial intelligence and machine learning to detect user behavioral anomalies, thus enhancing MFA (Mubeen et al., 2024) efficacy. For instance, AI can assess a user's typical login pattern and request additional authentication steps if an anomaly is detected. This adaptive tactic aligns with the approach of Zero Trust Architecture (ZTA), where every entity is assumed to be potentially untrusted and needs to be verified at every point (Ansa et al., 2024). By integrating these technologies seamlessly into existing MFA systems, organizations can not only enhance their security but also maintain their pace with evolving threats within the cyber domain. Finally, one needs to invest in advanced MFA solutions for institutions with an objective of safeguarding their assets and remaining compliant, especially during an era marked by digital transformation (Figure 4.12).

4.4.1. Definition and Importance

In today's rapidly evolving world of cybersecurity, the definition of authentication and access control methods is unequivocally crucial. Authentication is the process of verifying the identity of a machine or user, which forms the basis of secure electronic interactions. Various methods such as passwords, biometrics, and multi-factor authentication possess varying pros and cons. In opposition to that, access control mechanisms specify what can be executed by authenticated entities in a system, thus regulating their actions and rights. Together, these processes form a critical infrastructure for the protection of sensitive information and resources from unauthorized access and cyber-attacks and ensuring information integrity, confidentiality, and availability on networks. Strong authentication and access control cannot be overstated, particularly in the current technology environment where data breaches and identity theft are the norm. Secure authentication processes help ensure critical information by keeping unauthorized users away from it, thereby reducing the threat of cyberattacks. Accurate access control processes also enhance security by limiting the access rights of users according to role-based or attribute-based access policies. Therefore, organizations can implement the principle of least privilege, which restricts user access to the bare minimum necessary for their functioning, thereby minimizing any resulting damage in the event of stolen credentials or an insider attack. This layered security underscores the necessity of a clear understanding of authentication dynamics and access techniques. The importance of this subject lies beyond mere technical definitions; it emphasizes the need for continuing education and sensitization. For instance, it has been proven that children are fast adapting to technology and the need for its authentication, although with huge knowledge gaps in best practice on security (Choong et al., 2019). Therefore, it is imperative to create a culture of cybersecurity awareness that equips the users of all ages with the capabilities to maintain security. The organizations must also be vigilant about maintaining the updated authentication and access control mechanisms in order to keep themselves aligned with current threats and evolving user behaviors. Emphasizing the interaction of such mechanisms is critical to creating a more secure digital environment, emphasizing the reason why this concern necessitates serious scholarly and practical attention (Burstein et al., 2011).

4.4.2. Common Multi-Factor Authentication (MFA) Methods

The authentication environment is rapidly evolving, most particularly with growing utilization of MFA, a prominent mechanism for secure digital interactions. While Single-Factor Authentication (SFA) was sufficient in its early days with only requiring something the user knew, such as a password, it has been found lacking in this era where threats to computers are everywhere. With increasing users' interactions with personal data being conducted online, MFA offers stronger security by requiring multiple factors of authentication, which may involve something the user knows, something they have, and something they are. This multi-layered framework resists the attack vectors of unauthorized access, thereby protecting user data as well as organizational assets. Consequently, the adoption of MFA has accelerated as organizations aim at finding trusted mechanisms for protecting their systems against the sophisticated threats, demonstrating acute sensitization towards the imperative of safe authentication processes (Andreev et al., 2018). Of all the various MFA strategies, Two-Factor Authentication (2FA) is a widely embraced solution that combines a password with another authentication step. The second factor could be a text code, an authentication app, or biometric factor like a fingerprint. Implementation of 2FA extensively minimizes the possibility of successful phishing and credential stealing since access incorporates not just the user's knowledge but also an external verification method. But 2FA is not without its limitations, notably in terms of usability and the possibility of user error, which has the potential to create access barriers if users cannot receive or input the second factor in a timely manner. These issues emphasize the continued necessity to find balance between security controls and user experience so that MFA is both effective and user-friendly (Andreev et al., 2018). Resident technologies are constantly developing in the field of MFA, introducing novel methods such as partial password authentication, which has been specifically in demand from financial institutions. The method includes employing a challenge-response approach where particular characters from passwords are being asked randomly at positions, thus proving challenging for an attacker to intercept the full credential. Even though this method is seen as a good strategy to outsmart keyloggers as well as other forms of malware, their use raises questions regarding uniformity of safety among sampling methods. Moreover, the lack of comprehensive literature pertaining to this method suggests an evident research deficit at the academic level, suggesting further research to study its efficacy and offer guidelines for the safe usage practices (Figure 4.13).

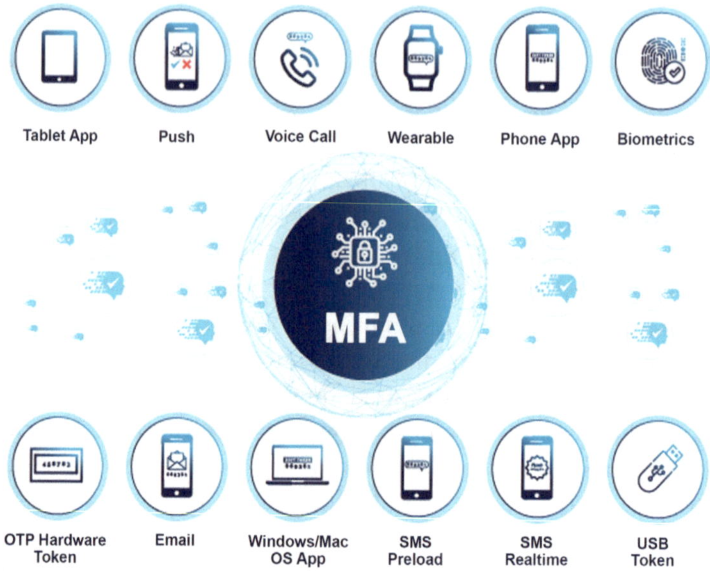

Figure 4.13. Common multi-factor authentication (MFA) methods.

Source: https://securenvoy.com/wp-content/uploads/2022/01/authentication-methods-02.png.

4.4.3. Challenges and Limitations

The authentication and access control mechanism environment is flooded with limitations and challenges that can potentially significantly inhibit their effectiveness. One major challenge arises from a deficiency in consistency among security measures across platforms, resulting in inconsistent access control mechanisms. Deficiency in consistency has the potential to leave one with vulnerabilities that are exploited by cybercriminals. For instance, in the case of the grid computing community, deployment of a centralized certificate authority is prone to scalability issues, as can be seen through the UK e-Science community survey. Therefore, the current practices are not only cumbersome but also hinder the safe and effective sharing of resources, thereby undermining the desired function of such systems and the need for a more robust framework that can accommodate evolving technological needs

(Sinnott et al., 2005). A further critical constraint is related to user experience and ease of use, particularly among individuals with diverse needs. Legacy authentication products, primarily focused on password security, do not meet the needs of all users, especially those with age-related impairments. Pointed out, the challenges confronting older users in accessing their devices and accounts indicate the limitations of conventional username and password-based authentication, which is largely assumed to represent best practice. Such limitations can provide barriers that exclude vulnerable groups from engaging wholly in digital environments, thus recognizing a significant need for inclusive design practices that consider the varying capacities of users. The solutions of these issues not only enhance accessibility but also improve overall system security by forcing user engagement and commitment to security practices (Renaud et al., 2018). Further, the rapid progression of technological platforms presents its own set of challenges that make it difficult for authentication and access control mechanisms to perform. Continual innovation in new software and hardware components makes security ineffective, therefore, increasing the risk of breaks. The grid community is the best example of this challenge because it is faced with finding instruction materials that do not become obsolete in the presence of constantly changing technology standards. Failure to keep pace with innovations in these fields may leave systems open to attack since new technologies always have unforeseen security vulnerabilities that traditional methods of authentication cannot effectively cope with. Such a scenario demands ongoing research and development to render access control mechanisms adaptive and resilient to the new threats while maintaining user confidence and technology integrity (Sinnott et al., 2005).

4.5. SECURITY RISKS AND VULNERABILITIES

Security risks and vulnerabilities are a significant issue in the case of authentication and access control measures on various platforms. Since companies greatly rely on digital infrastructure, the threat level rises, so the malicious users can utilize weaknesses in these systems. For instance, some vulnerabilities are poor password policies that create intruder entry points without permission. Further, weak implementation of authentication technology can result in impersonation attacks or data breaches, jeopardizing the integrity of confidential data. These flaws must be remedied through an interdisciplinary effort, such as the implementation of robust password policies, multi-factor authentication, and regular security audits to ensure compliance and effectiveness. Continuous surveillance of systems for aberrant behavior is also essential because it enables companies to identify potential violations when they are at their most preliminary stages, thereby enhancing the overall security stance of their access control and authentication systems. Cognitive radio networks are unique with regard to security threats because they are dynamic and rely on spectrum-sharing mechanisms. The security concerns typical of traditional wireless networks are exacerbated in the context of cognitive radio by the possibility of unlicensed devices masquerading as authorized users or disrupting access to common control channels. These attacks arise at the cost of significantly degrading overall network performance and jeopardizing data integrity as well as user security. Securing the common control channel is thus a focus in securing cognitive radio networks. There needs to be thorough security solutions that can counter such attacks, like effective spectrum sensing techniques and trusted authentication methods designed particularly for cognitive networks. With the enhancement of security of these communication channels through innovative techniques, organizations will be able to better safeguard themselves against advanced persistent threats and ensure continuous data transmission in a highly connected world (Alhakami et al., 2014). Security of the database, particularly in databases such as MongoDB, becomes increasingly crucial as the dynamics of cyber-attacks continually evolve. With businesses leveraging the agility and expansibility provided by NoSQL databases, authentication practices among users are essential to provide assurance that the databases are secure from unauthorized access. Such exposure to vulnerabilities as injection attacks or privilege escalation has the potential to significantly lead to data breaches, subjecting sensitive data to criminals. Therefore, employing stringent authentication techniques such as SCRAM, x.509 certificates, or LDAP integration is crucial to ward off these attacks. Having proper knowledge of the benefits and limitations of various types of user authentication enables organizations to strategically choose the most appropriate one for their specific environment. By addressing these security problems comprehensively, database administrators can improve the confidentiality and integrity of data kept in their databases and thereby make them more resistant to likely vulnerabilities (Dubey, 2023).

4.5.1. Phishing Attacks

Phishing attacks are a predominant threat in the realm of cybersecurity, which exploits flaws in user authentication and access control systems. Such attacks typically include misleading tactics to deceive individuals into divulging secret information, such as username, passwords, or identification credentials. This fraud often occurs via imitated emails or sites that closely resemble real ones. As a result, users unknowingly leave attackers with their credentials, thereby compromising organizational security. The increasing cleverness of phishing attacks poses a very real challenge to individuals and organizations, calling for robust measures and policies to strengthen defenses against phishing attacks. It is imperative that organizations recognize the battle against phishing cannot be viewed solely as a technical issue but also as a behavioral one, with user education and awareness being essential components of mitigating risks associated with unauthorized access and identity theft (Sedaghat et al., 2017). Also, the greater number of phishing attacks underscores the pressing need for end-to-end security solutions capable of fending off such threats. The economic impact of phishing attacks is devastating, with one successful attack inflicting extensive financial loss and damage to reputation.

Figure 4.14. Phishing attacks.

Source: https://www.usd.de/wp-content/uploads/usd-ag-news-phishing-angriffe.jpg.

The prevailing economic climate also plays a part in further aggravating this issue, as cyber criminals continue to exploit greater vulnerabilities on the part of both enterprise entities and individual users. Efforts in research that have aimed at studying phishing and user behavior habits have demonstrated the necessity of always being watchful and thinking on one's feet within the IT space. The integration of multi-faceted security measures, including strengthened authentication mechanisms and dynamic threat detection mechanisms, can enhance an organization's response to phishing probes. Awareness education courses to instruct users on phishing tactics are also required to strengthen authentication and access control procedures (TABUSCA et al., 2025). Finally, the battle against phishing needs to be tackled with a transdisciplinary action that incorporates technological instruments with awareness on the part of users. As phishing tactics become more sophisticated, organizations must continue adjusting their security processes and embed cybersecurity culture into employees. This dual-pronged strategy not only enhances authentication controls but also allows users to detect and report potential phishing attacks.

Through promoting an informed user base alongside implementing robust access controls, organizations can become more resistant to this pervasive threat. Lastly, encouraging an environment of shared responsibility in cybersecurity is essential in protecting sensitive information during a time when phishing attacks are on the rise and becoming increasingly sophisticated (Sedaghat et al., 2017; TABUSCA et al., 2025) (Figure 4.14).

4.5.2. Credential Stuffing

Credential stuffing is a prevalent cyber-attack where attackers exploit the widespread practice of using the same password to access multiple online platforms. This technique is so dangerous in that it's based on user behavior rather than sophisticated technical vulnerabilities. Once a data breach occurs on one service, the compromised credentials may be used on numerous unconnected services, increasing the rate of successful attacks. The efficiency and ease of credential stuffing attacks can overwhelm an organization, particularly when it lacks an adequate authentication and access control system. These recurrent attacks often result in unauthorized access, data leakage, and justified threats to personal and financial information, highlighting the need for stronger preventive measures in authentication methods. Hence, credential stuffing is not only an organizational concern but also a foundation of user trust and integrity of digital environments. To be able to counter successfully credential stuffing, organizations must put in place layered defense systems, including multi-factor authentication (MFA) and robust password policies. MFA adds an additional verification step beyond a mere password, making it much more difficult for unauthorized access. Organizations must also encourage users to employ unique, complicated, and varied passwords for each service that they use. This practice makes credential stuffing ineffective since the attacker will have to crack multiple password stores rather than one. Moreover, having automated systems monitor unusual login behavior will enable detection and nullification of credential stuffing attempts when they are still small in scale. These pre-emptive measures not only enhance the security posture of an organization but also instill a cyber security culture among end users, thereby minimizing the risks associated with the reuse of credentials. The mounting exposure of credential stuffing defies current digital security dynamics, marrying technical weaknesses with end user behaviors. With the increasing usage of cloud services, as evident from all the reported instances of data breaches attributed to these services, the scope for credential stuffing attacks also increases. Shared technology, multi-tenancy, and third-party dependencies in the cloud stack can create security vulnerabilities of a sophisticated nature, which are typically addressed by attackers. Since cloud services bring multiple components of information together, they become valuable targets, which makes a robust, multi-faceted approach to authentication and access control imperative. Resolution of these challenges demands solid frameworks with legal ramifications for digital forensics and cybersecurity measures to determine indicators of compromise and react accordingly to incidents, with focus on the intertwined relationship between technical solutions and legal aspects in countering credential stuffing (Ali et al., 2025; Ganiyu et al., 2025) (Figure 4.15).

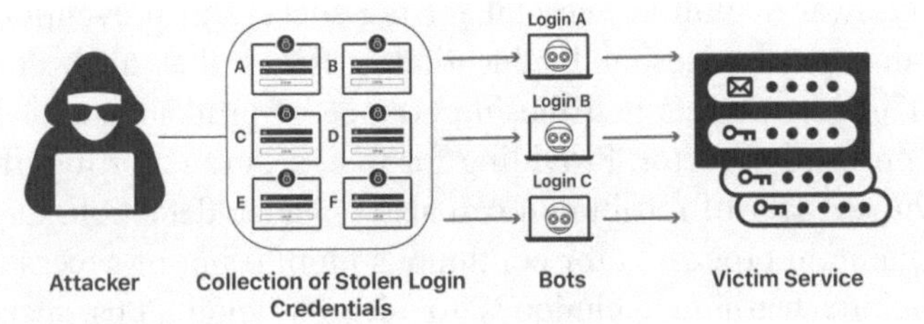

Figure 4.15. Credential stuffing.

Source: https://cf-assets.www.cloudflare.com/slt3lc6tev37/7063QaY9i4PTXesgUmpoYk/cd8e741be8731bca5494bc14d03a134c/credential-stuffing.png.

4.5.3. Insider Threats

Insider threats are among the biggest problems in the cybersecurity domain, particularly in regards to authentication and access control systems. Although outside threats usually prompt certain defense measures, insider threats involve an individual or individuals inside the organization who exploit their permissions. These individuals could be employees, contractors, or even business partners with legitimate access to sensitive information. These behaviors are driven by anything from finances to individual agendas, and they exhibit a broad threat horizon. Human fallibility in behavior renders preventive approaches with complexities of human behavior challenging, such that organizations have to introduce multi-faceted approaches that integrate technology with insider threat psychological drivers learning (Bishop et al., 2010). Prevention and reaction strategies need to therefore incorporate robust access control, effective monitoring, and active incident response in an attempt to mitigate the insider risks. Mechanism design and implementation for authentication and access control are crucial in mitigating insider threats. Organizations must apply strict access controls so that employees possess only minimum information required to perform their jobs—also referred to as the principle of least privilege. Organizations can enhance security by employing multifactor authentication, which adds additional layers of authentication, utilized to discourage probable insider attacks. Periodic access privilege and usage pattern audits can also help to detect anomalous activity associated with insider threats. As has been researched in literature on security in wireless networks, the identification of vulnerabilities in access control protocols is critical for the development of efficient countermeasures (Gupta et al., 2013). Mitigating these vulnerabilities not only enhances the security environment but also helps in the formation of an accountability culture within the organization. Moreover, development of a security-aware culture is crucial for the successful prevention of insider threats. Training exercises and education sessions that alert employees to the risks of insider threats and the importance of prudent practices can be a significant prevention factor. Providing for free, secure reporting of suspicious behavior without fear of retribution can also enhance detection. Organizations must also maintain processes for continuous monitoring of access control and include machine learning techniques to identify anomalous access patterns that can indicate insider malfeasance. Together, all of these are part of an all-encompassing security measure that unifies technology defenses and people-centric measures to make an organization less vulnerable to insider threats. As underscored by current research on insider threats, the combination of technology and people-focused measures is the secret to successful security management (Bishop et al., 2010).

4.6. BEST PRACTICES FOR IMPLEMENTATION

In the pursuit of the right access control and authentication mechanism, best practices in implementation are the building blocks. The practice of doing so involves the transparent blending of multiple authentication technologies without inflicting harm on user experience, particularly in mobile and wireless networks. Where users change from one or more access points or networking technologies, there is a need for transparent end-user authentication. If the solutions that facilitate such transitions are deployed, such as the one obtained from the common secret model between the Universal Subscriber Identity Module (USIM) and the home network, then not only is the security enhanced but also the users' trust in the system is enhanced. Blinding the users to the complexity involved in such transitions emphasizes the need for operational efficiency and reliability in the mechanisms of authentication (Chen et al., 2003). Lastly, the final goal of such best practices is to offer an environment where user mobility occurs without sacrifice in security. Another best practice for deployment is the selection of wise use of effective authentication mechanisms to the organization's specific environment. IEEE 802.1X extensibility allows system implementers to choose among a variety of Extensible Authentication Protocol (EAP) methods with varying strength and limitation. This selection process is guided by factors such as network topology, user requirement, and potential security threats. Through the comparative evaluation of various EAP authentication schemes and their corresponding characteristics, organizations can study and select the most appropriate course of action for their networks. Such systematic analysis not only assists in bolstering access control but also allows organizations to remain agile in dealing with evolving security landscapes (Ali et al., 2007). The implementation of such tailored strategies lays major emphasis on an organizational framework for the implementation of robust, efficient authentication and access controls. Also, continuous evaluation and refinement of authentication practices form a critical component of best practices in deployment. With the progress in technology and the changing threats over the internet, organizations need to have mechanisms for periodically assessing how effective their access control measures are. Through such forward thinking, the authentication system can be updated and enhanced in a timely manner, closing down loopholes that could result from antiquated systems. Analytics and monitoring tools can be critical in the detection of security loopholes or suspicious activity which requires adjustments. Furthermore, involving users through education and feedback can help develop a culture of security awareness, thereby making authentication policies more effective. With constant review and refinement given utmost priority, organizations can ensure that their practices are effective and pertinent to combating emerging threats in the complex digital environment.

4.6.1. Regular Security Audits

Regular security audits are crucial to maintaining effective authentication and access control arrangements within an organization. These audits involve systematic evaluations of security measures, procedures, and policies in a bid to identify vulnerabilities that may undermine the integrity of data and systems. With the fast evolving online business environment, organizations' threats and vulnerabilities change as well. The changing nature of these threats necessitates that organizations take a proactive stand in managing their security with a focus on change and improvement in security controls, Pye et al. (2007). By constantly reviewing security controls, organizations are not just able to improve defenses against emerging threats but also instill customer confidence in their ability to safeguard confidential information. Proactiveness is essential in an era where prospective breaches could lead to enormous financial losses and brand dilutions. Moreover, conducting regular security audits allows organizations to compare their security processes with established standards and templates. This benchmarking task, as documented by Pye et al. (2007), employs well-established IT and information security standards to create and apply relevant benchmarks for online security procedures. The adoption of a formal approach in performing security audits enables organizations to ascertain their adherence to statutory regulation as well as areas where further improvements are needed. Such a process of ongoing review and fine-tuning not only helps to keep the company aware of potential threats at hand but also prepares it to counter potential threats of cyberattacks in the future. Ongoing security audits are thus a foundation for creating secure access control and authentication measures that are strong and dynamic in nature. Aside from the benefits of vulnerability benchmarking and evaluation, there are also regular security audits that contribute to the assurance of a clear understanding of the evolving technology world, including emerging technologies like cloud computing. Since more organizations are adopting cloud-based solutions, there is a need to understand the security features of these environments. The changing environment is accompanied by unique challenges, such as data traffic congestion and server availability, that must be rectified in order to realize successful authentication and access control functions. Monitoring of security audits may enable organizations to establish a good framework, irrespective of the hybrid cloud solutions employed, as in some case studies, such as application in national ID databases. By addressing these problems with periodic audits, organizations can ensure that their access control mechanisms are secure and reliable, enhancing their overall cybersecurity stance (Figure 4.16).

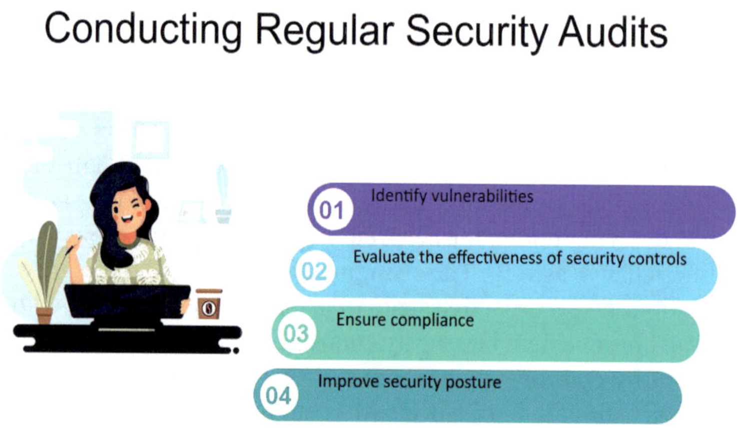

Figure 4.16. Regular security audits.

Source: https://fastercapital.co/i/Cloud-Security--Mitigating-Pilotfishing-Risks-in-the-Cloud--Conducting-Regular-Security-Audits.webp.

4.6.2. User Education and Awareness

User education and awareness are key components in the successful implementation of authentication and access control mechanisms. As threats online grow more sophisticated, users need to be educated about the different security protocols in place and how to best interact with them. Security awareness builds a security culture, which teaches people to identify weaknesses inherent in their behaviors—like poor password habits or unauthorized account sharing. Furthermore, learning about security procedures enhances users' sensitivity to the significance of compliance with these procedures, ultimately leading to increased organizational and individual security. Integrating training activities that improve awareness of the potential threats within organizational policies can make users the first line of defense against cyber threats, significantly decreasing security breaches. User training in cybersecurity policy should not only cover the technical points of authentication protocols but also the social points of cybersecurity. Users are frequently the security chain's weakest component and, as a consequence, directed awareness programs are key to minimizing risk. User training on common tactics employed by attackers has a high potential to enhance users' ability to identify and counter such attacks. Besides, appreciation of the access controls, such as least privilege, enables user compliance with security policy is paramount. As evident from the examination of the consequences of the 1988 outbreak of the worm, the establishment of user accountability and consciousness is essential to maintaining the integrity of the overall networked environment, requiring widespread educational programs in organizations (Denning et al., 2025). The challenges of user education are then exacerbated by the accelerating rate of technological evolution and the assortment of devices in the modern world. With mobile operating systems increasingly a part of the daily routine, as in the recent research described, the users have to continue being mindful of device-specific security protocols and weaknesses (Amen et al., 2016). The need for continuous education is evident as users switch from one application to another and one service to another, each having its own security concerns and authorization techniques. Hence, organizations should not treat user education as a one-time process but as a iterative process that expands along with emerging technologies. This persistent emphasis on user awareness not only optimizes the effectiveness of authentication and access control measures but also promotes an effective security culture among users, ultimately safeguarding valuable information from probable attacks.

4.6.3. Updating and Patching Systems

One cannot overstate the importance of regularly updating and patching systems, particularly in the case of authentication and access control measures.

Since security threats evolve continuously at a rapid rate, malicious users might exploit software vulnerabilities to achieve unauthorized access to sensitive information. Periodic updates not only plug the vulnerabilities but also enhance systems' security posture. For instance, prompt adoption of patches can neutralize dangers associated with known exploits, lessening the incidence of successful intrusions. In an environment dominated by growing cyber threats like malware and ransomware, organizations need to give priority to these updates to ensure the security of their authentication systems and protect sensitive user information. To properly execute an updating and patching strategy, organizations need to develop a proactive security culture with an emphasis on continuous monitoring and assessment. This can be achieved through the use of automated patch management systems, which automatically carry out detecting, testing, and applying patches on various platforms. Besides this, integrating a risk-based approach allows an organization to also prioritize patches by the degree of vulnerability and likelihood to crash key systems. Basing on the foundational principles developed in Al-Khateeb et al. (2019) and Madnick et al. (2015), an organization's ability to respond promptly to emergent threats is important to ensuring healthy access control practices. Through theoretically handling vulnerabilities, organizations are best suited to strengthen their authentication controls so that only valid users can gain access to valuable assets. Ultimately, encouraging an enterprise-wide system update and patching process requires frequent employee training and sensitization to the value of cybersecurity best practices. User behavior is the most significant factor in determining how effective authentication and access controls are. The employee will be more likely to adhere to organizational security procedures if educated on the risks posed by outdated systems or unpatched software. This education should also cover the implications of cyber-attacks on industrial control systems and critical national infrastructure, according to Al-Khateeb et al. (2019) and Madnick et al. (2015). By incorporating user awareness into the technology architecture, organizations can make their defenses more difficult to attack, and thus increase the overall resilience of their access control and authentication systems.

SUMMARY

In finalizing the discussion on access control and authentication mechanisms, it is apparent that the security and integrity of digital environments greatly depend on these mechanisms. The dynamic nature of cyber threats makes strong authentication mechanisms and stringent access control policies mandatory that can evolve in response to emerging challenges. Central to this challenge is the creation of end-to-end protocols that not only confirm user identities but also oversee access privileges in various contexts. An environment rich with devices, particularly with the advent of the Internet of Things, calls for new approaches to address security adequately. These controls constitute the first line of defense, which keeps unauthorized access from leading to breach of critical systems, thus safeguarding sensitive data and business processes. In a review of current authentication systems, it is evident that there are strengths and weaknesses as integrated in current protocols. Passwords have been used traditionally for a long time but are found wanting more often with vulnerability to attacks and human mistakes. Methods like multi-factor authentication and biometric authentication are giant leaps in the direction of enhanced security. These methods, too, are beset by their own list of disadvantages, typically centering on inconveniencing consumers or requiring significant infrastructure investment. Further, consolidation of emerging technologies, like utilizing blockchain solutions, promises to make authentication and access control more secure. This concept is particularly apt in the case of the Internet of Things, where there are a multitude of devices that present new challenges in data security and consumer privacy, hence indicating the need to adopt robust mechanisms in contemporary security platforms (Bellekens et al., 2018). In conclusion, the ultimate decision in this analysis reiterates the necessity for a multidisciplinary approach towards strengthening authentication and access control mechanisms. With the increasing advancement of cyber threats, there will always be a need to evaluate and enhance these mechanisms so that security and confidence in technological systems are guaranteed. Organizations will need to make sure that they not only employ sophisticated methodologies but also foster a culture of security awareness among their users. Appropriate training on usage and potential vulnerabilities will remain key to the effectiveness of such authentication mechanisms. As the environment evolves, harmony between technological innovation and user behavior and awareness will be paramount in tackling the challenges posed by cyber threats (Bishop et al., 1991). The shared objective should be to produce an adaptive and robust security environment that can thrive in an interdependent digital landscape.

The correct knowledge of authentication and access controls is crucial in safeguarding digital property as well as in giving users privacy. In wireless mesh networks (WMNs), possible security threats are unique challenges. A comprehensive overview identifies many vulnerabilities to various levels of the communication protocol stack namely, physical, link, network, transport, and application layers. Each of these layers presents various threats that may compromise the integrity and confidentiality of data. Those vulnerabilities are identified to enable security experts to design tailored defense mechanisms that can prevent probable attacks. This multi-faceted approach not only addresses immediate security requirements but also provides a model for ongoing assessment of the threats that continuously evolve in the cyberworld, as evident in the lengthy considerations outlined in Gupta et al. (2013). Additionally, creation of new technologies and cyber interactions has shifted the threat landscape, giving added significance to robust authentication and access controls. As discussed in the literature today, extensive utilization of information systems has resulted in increased threats associated with physical and electronic assets. Threats now encompass a broader range of cyber threats, from targeted attacks on single data to sophisticated deceptions that take advantage of unsuspecting victims, as cited in Sasse et al. (2005). The reduction of the price to perform various kinds of attacks has further made it necessary for users and organizations to have effective security measures in place. The fact that humans traditionally constitute the weakest link in security systems makes it crucial to incorporate user training into technological countermeasures. Prioritizing the usability of such security measures is also essential to improve the efficacy of authentication and access control systems. Current research indicates that security controls are either not known to be available to most users or cannot be utilized by them to their fullest potential. This loophole prompted researchers to investigate the usability of security controls and identify barriers to optimal efficacy and propose remedial measures. Implementing user-centric security interfaces and accessibility control mechanisms can significantly minimize vulnerabilities, promoting security behavior culture and responsibility among users. Finally, improving usability does not just empower people but also makes the digital world a safer place to be, as evidenced by the work described in Sasse et al. (2005).

REFERENCES

1. Abdolee, R., Maroufi, M., & Tazekand, B. M. (2019). On the convergence of blockchain and Internet of Things (IoT) technologies. http://arxiv.org/abs/1904.01936 (accessed on 13th September 2025).
2. Ahmed, N., & Tarannum, N. (2013). Efficient and reliable hybrid cloud architecture for big data. http://arxiv.org/abs/1405.5200 (accessed on 13th September 2025).
3. Alhakami, W., Mansour, A., & Safdar, G. A. (2014). Spectrum sharing security and attacks in CRNs: A review. https://core.ac.uk/download/29822044.pdf (accessed on 13th September 2025).
4. Ali, K. M., & Owens, T. J. (2007). Selection of EAP-authentication methods in WLANs. https://core.ac.uk/download/334218.pdf (accessed on 13th September 2025).
5. Ali, M. (2025). Information security threats in cloud services: A comprehensive overview. https://core.ac.uk/download/657111775.pdf (accessed on 13th September 2025).
6. Al-Khateeb, H., Burnham, K. J., Epiphaniou, G., Hammoudeh, M., et al. (2019). Preliminaries of orthogonal layered defence using functional and assurance controls in industrial control systems. https://core.ac.uk/download/196185875.pdf (accessed on 13th September 2025).
7. Alvaro, G., Domingue, J., Maleshkova, M., Martinez, S., et al. (2010). Using semantics for automating the authentication of Web APIs. https://core.ac.uk/download/4834.pdf (accessed on 13th September 2025).
8. Amen, B., Mahmood, S., & Nabi, R. M. (2016). Mobile application security platforms survey. https://core.ac.uk/download/30734396.pdf (accessed on 13th September 2025).
9. Anderson, A., Chadwick, D., Cooper, G. S., Park, J. H., & Mbanaso, U. M. (2009). Obligations of trust for privacy and confidentiality in distributed transactions. https://core.ac.uk/download/104938.pdf (accessed on 13th September 2025).
10. Andreev, S., Bezzateev, S., Koucheryavy, Y., Mikkonen, T., et al. (2018). Multi-factor authentication: A survey. https://core.ac.uk/download/146449871.pdf (accessed on 13th September 2025).
11. Ansa, G., & Mpamugo, E. (2024). Enhancing network security in mobile applications with role-based access control. https://core.ac.uk/download/655274678.pdf (accessed on 13th September 2025).
12. Bellekens, X., Ferguson, R. I., & Mackenzie, B. (2018). An assessment of blockchain consensus protocols for the Internet of Things. https://core.

ac.uk/download/228178537.pdf (accessed on 13th September 2025).

13. Bemmel, J. van, Lagerberg, K., & Wegdam, M. (2005). 3PAC: Enforcing access policies for Web services. https://core.ac.uk/download/pdf/11459642.pdf (accessed on 13th September 2025).
14. Bertino, E., Takabi, H., Calero, J. M. A., & Matsumoto, M. (2016). Hierarchical role-based access control with homomorphic encryption for database as a service. http://arxiv.org/abs/1603.00572 (accessed on 13th September 2025).
15. Bishop, M. (1991). A security analysis of version 2 of the Network Time Protocol (NTP): A report to the privacy and security research group. https://core.ac.uk/download/pdf/42812182.pdf (accessed on 13th September 2025).
16. Bishop, M., Coles-Kemp, L., Gollmann, D., Hunker, J., et al. (2010). 10341 Abstracts Collection–Insider Threats: Strategies for Prevention, Mitigation, and Response. https://core.ac.uk/download/62915875.pdf (accessed on 13th September 2025).
17. Burstein, A. J., Mulligan, D. K., & Wang, L. (2011). Privacy in the smart grid: An information flow analysis. https://core.ac.uk/download/288432753.pdf (accessed on 13th September 2025).
18. Callahan, J., & Othman, A. (2017). The Horcrux Protocol: A method for decentralized biometric-based self-sovereign identity. http://arxiv.org/abs/1711.07127 (accessed on 13th September 2025).
19. Chadwick, D. W. (2000). Secure directories. https://core.ac.uk/download/62887.pdf (accessed on 13th September 2025).
20. Chen, H., Plas, D. J., & Zivkovic, M. (2003). Transparent end-user authentication across heterogeneous wireless networks. https://core.ac.uk/download/pdf/11460651.pdf (accessed on 13th September 2025).
21. Cheverst, K., Davies, N., Finney, J., Friday, A., et al. (2003). Network layer access control for context-aware IPv6 applications. https://core.ac.uk/download/377179.pdf (accessed on 13th September 2025).
22. Choong, Y. Y., Prior, S., Renaud, K., Theofanos, M., et al. (2019). "Passwords protect my stuff"–A study of children's password practices. https://core.ac.uk/download/275550949.pdf (accessed on 13th September 2025).
23. Cleeff, A. van, Pieters, W., & Wieringa, R. (2010). Benefits of location-based access control: A literature study. https://core.ac.uk/download/pdf/11478350.pdf (accessed on 13th September 2025).
24. Da Silva, C., Medeiros, W., & Sampaio, S. (2019). PEP4Django–A policy

enforcement points for Python web applications. https://core.ac.uk/download/322489434.pdf (accessed on 13th September 2025).

25. Delgado Mercè, J., & Llorente Viejo, S. (2015). Privacy provision in eHealth using external services. https://core.ac.uk/download/81580819.pdf (accessed on 13th September 2025).
26. Denning, P. J. (2025). The Internet worm. https://core.ac.uk/download/pdf/42823382.pdf (accessed on 13th September 2025).
27. Dubey, A. (2023). Security issues and user authentication in MongoDB. https://core.ac.uk/download/613698547.pdf (accessed on 13th September 2025).
28. Erlich, Z., & Zviran, M. (2006). Identification and authentication: Technology and implementation issues. https://core.ac.uk/download/301376504.pdf (accessed on 13th September 2025).
29. Fett, D., Kuesters, R., & Schmitz, G. (2017). The Web SSO standard OpenID Connect: In-depth formal security analysis and security guidelines. http://arxiv.org/abs/1704.08539 (accessed on 13th September 2025).
30. Ganiyu, A., Oguntoyinbo, O., Oyelakin, O., Yeboah-Ofori, A., et al. (2025). Digital forensics investigations and network security issues in tracking the trails of cybercriminals. https://core.ac.uk/download/653371793.pdf (accessed on 13th September 2025).
31. Gondree, M. A., Irvine, C. E., Khosalim, J., Levin, T., et al. (2010). A cloud-oriented cross-domain security architecture. https://core.ac.uk/download/36700748.pdf (accessed on 13th September 2025).
32. Gupta, A., Mishra, A., Naveed, A., Perrig, A., Shamir, A. A., Franklin, A. D., & Wood, A. D. (2013). Security and privacy issues in wireless mesh networks: A survey. http://arxiv.org/abs/1302.0939 (accessed on 13th September 2025).
33. Jiang, J., Sinnott, R. O., & Watt, J. (2007). The GLASS project: Supporting secure Shibboleth-based single sign-on to campus resources. https://core.ac.uk/download/1397101.pdf (accessed on 13th September 2025).
34. Jin, H., Lee, R. B., Lee, W. H., & Liu, Y. (2017). Secure pick up: Implicit authentication when you start using the smartphone. http://arxiv.org/abs/1708.09366 (accessed on 13th September 2025).
35. Johnstone, M. N., & Peacock, M. D. (2014). An analysis of security issues in building automation systems. https://core.ac.uk/download/41537512.pdf (accessed on 13th September 2025).
36. Komninos, N., Mourouzis, T., & Wojcik, M. (2025). On the security evaluation of partial password implementations. https://core.ac.uk/download/83943036.pdf (accessed on 13th September 2025).

37. Madnick, S. E., & Nourian, A. (2015). A systems theoretic approach to the security threats in cyber physical systems applied to Stuxnet. https://core.ac.uk/download/83235017.pdf (accessed on 13th September 2025).
38. Mubeen, M. (2024). Zero-Trust Architecture for Cloud-Based AI Chat Applications: Encryption, Access Control and Continuous AI-Driven Verification. https://core.ac.uk/download/638725525.pdf (accessed on 13th September 2025).
39. Perkins, C., & Westerlund, M. (2014). Options for securing RTP sessions. https://core.ac.uk/download/42352765.pdf (accessed on 13th September 2025).
40. Pye, G., & Warren, M. (2007). A model and framework for online security benchmarking. Deakin Research Online. http://hdl.handle.net/10536/DRO/DU:30007062 (accessed on 13th September 2025).
41. Renaud, K., Scott-Brown, K. C., & Szymkowiak, A. (2018). Designing authentication with seniors in mind. https://core.ac.uk/download/228178438.pdf (accessed on 13th September 2025).
42. Sasse, M. A. (2005). Usability and trust in information systems. https://core.ac.uk/download/1688826.pdf (accessed on 13th September 2025).
43. Sedaghat, S. (2017). CERT strategy to deal with phishing attacks. http://arxiv.org/abs/1706.02610 (accessed on 13th September 2025).
44. Sinnott, R. O., Stell, A. J., & Watt, J. (2005). Experiences in teaching grid computing to advanced level students. https://core.ac.uk/download/1394009.pdf (accessed on 13th September 2025).
45. Tabusca, A., & Lungu, I. (2025). Optimizing anti-phishing solutions based on user awareness, education and the use of the latest web security solutions. https://core.ac.uk/download/pdf/6612459.pdf (accessed on 13th September 2025).

CHAPTER 5

SECURE COMMUNICATION PROTOCOLS FOR INDUSTRIAL NETWORKS

Contents

INTRODUCTION

More digital evolution in industrial environments has introduced profoundly significant shifts to communication inside and between networks. Traditionally, industrial control systems that handled safety issues operated with an air gap so that items like PLCs were not exposed to external vulnerabilities. This isolation provided a minimum level of protection from cyber-attacks since physical entry would typically be needed to target the system. However, the integration of industrial protocols into enterprise networks has increasingly eroded the effectiveness of this air gap approach. Consequently, industrial networks now face an escalating risk of cyber-attacks that exploit such interconnections, necessitating a re-evaluation of secure communications protocols to insulate critical infrastructure from hostile invasion and data loss (Evangelopoulou et al., 2017). As the Industrial Internet of Things (IIoT) expands, so does industrial network complexity. The connectivity that enables greater efficiencies in operation has introduced new security threats. Organizations are increasingly realizing that they need to implement robust security measures to protect sensitive data that traverses these networks. The move towards integrated communication systems, gateways, and node security configurations ensures secure transfer of sensitive data without compromising the operational functionality essential for operational technology. By completely defining security requirements for such communication gateways, industries are in a position to better mitigate the threat that is associated with connected devices and dependent systems (Marksteiner et al., 2017). The implications of unsafe security in industrial networks are underscored by recent high-profile cyber-attacks, which highlight the imperative need for implementing secure communication protocols. As industries become more reliant on digital platforms to automate and optimize operations, their vulnerabilities in communications infrastructure can no longer be overlooked. The ongoing analysis of past attacks, particularly in critical industries, is employed to guide the development of stronger cybersecurity defenses that are proactive rather than reactive. Such defenses are essential in protecting not only the operational integrity of industrial systems but also the safety and privacy of parties involved in the manufacturing process. By focusing on state-of-the-art secure communication protocols, industries can effectively navigate the complexities of today's networked environments without losing resilience to emerging cyber threats (Evangelopoulou et al., 2017; Marksteiner et al., 2017).

Definition of Secure Communication Protocols

Secure communication protocols for today's electronic systems are required to safeguard data being transmitted over networks from loss of integrity, confidentiality, and availability. Secure communication protocols make use of a variety of cryptographic algorithms to keep unauthorized access and data compromise to a minimum. Secure communication protocols enable data to be encrypted during transmission, effectively making it unreadable for any unauthorized user. Secure channels enable organizations to protect sensitive information from tampering or interception by unauthorized people. Moreover, secure protocols in industrial networks where operations data and control signals must be secure in terms of assuring process reliability and safety are imperative. Hence, it is the comprehension of the nuances and deployments of these protocols that forms the basis of a secure digital infrastructure in industrial environments. A good definition of secure communication protocols includes their roles in authentication, integrity, and confidentiality. The authentication procedures guarantee that the identities of the communicating parties are verified, enabling trust in the network. It is particularly important in situations where there are lots of devices and users communicating, like in industry networks where there are machines and human operators to be communicated to in a secure manner. The protocols also must provide integrity checks to prevent the data from being altered as it travels, thus the reliability of the information being communicated. Confidentiality is also a critical aspect, achieved by applying encryption techniques that hide the content from unintended viewers. With evolving sophisticated cyber threat profiles, industrial sectors are constantly asked to upgrade their secure communications strategies in an effort to shield against potential vulnerabilities, and this demands a good grasp of the definitions and utilities of such protocols. Incorporating the new advances, such as quantum cryptography, into the framework of secure communication protocols is a paradigm shift when it comes to data protection. The traditional cryptography systems are vulnerable to advanced hacking techniques, and therefore, new approaches are needed that leverage quantum mechanics principles to ensure more robust security. Quantum cryptography provides new methodologies for the generation of cryptographic keys that are theoretically invulnerable to interception or duplication. As industries adopt increasingly automated and networked systems, the reliance on similar advanced protocols grows significantly, and the evolving security risk dynamics are handled efficiently as a consequence. This change does not merely illustrate the necessity to safeguard communication protocols but sparks ongoing research into developing standardized models for assessing quality in terms of performance and effectiveness. Actually, secure communication protocols are at the forefront in securing industrial networks because they implement more advanced technologies in their operation (Curty et al., 2014; Hong et al., 2006) (Figure 5.1).

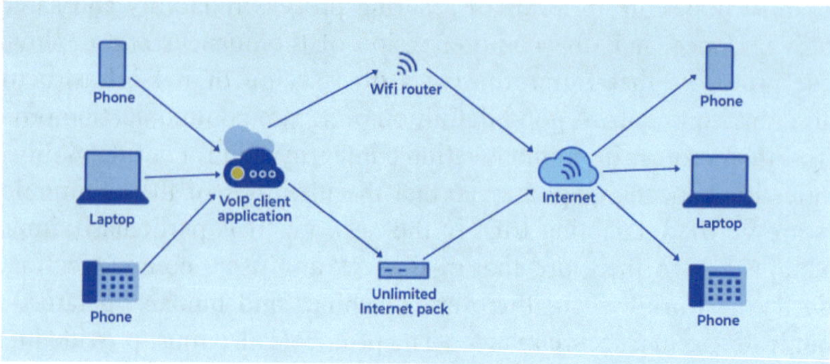

Figure 5.1. The secure communication protocols.

Source: https://assignmenthelp4me.com/assets/img/latest-homepage/protocol3.jpg.

Importance of Security in Industrial Networks

The importance of security in industrial networks cannot be overemphasized, given that today these systems are being integrated more. Although industries rely on technological innovation to improve performance, the likelihood of cyber-attacks has grown. Vulnerabilities in communication networks can lead to uncontrolled access, exposure of data, and even system collapse. Since industrial systems often rely on the constant flow of real-time information for decision-making, the slightest interruption will have a knock-on effect. This danger of business disruption poses the urgent imperative for tight security. Having secure communication mechanisms in place is necessary to keep sensitive information safe and secure the integrity of such operations in such networks and, subsequently, establish trust among stakeholders and assets of an organization. Development of traditional infrastructures into Smart Grids is a show of growing industrial network complexities, which lead to additional security issues. Smart Grids have a large number of devices like smart meters and sensors that provide two-way communication between the consumers and the producers of energy. It enhances efficiency and responsiveness, but at the cost of enlarging the attack surface to be utilized by attackers. The use of Internet of Things (IoT) in industrial systems makes security even more painful, as networked devices lack standard security protocols. Hence, a balanced security model is needed to manage such diverse elements effectively. The measures should include encryption, access controls, and continuous monitoring so that the threats are repressed, and the integrity of the Smart Grid is not lost, thus highlighting the necessity of sophisticated security in industrial networks

(Copeland et al., 2019). In the quest to achieve the goal of making industrial networks more secure from cyber-attacks, elastic security measures need to be prioritized. A multi-layered security solution can effectively counter various channels of attacks likely to be faced by these networks. This is not only in the form of secure communication protocols but also via regular vulnerability scans, penetration testing, and employee training modules on cybersecurity awareness. Moreover, as seen in the HORIZON 2020 project STORY, innovation in secure communications for smart energy systems gives instances of the technological solutions that are being developed to impart strong cybersecurity standards. Through emphasis on privacy matters and safe storage of data, industries can increase the operational resilience of their networks while also promoting innovation. It is thus imperative that industrial networks incorporate holistic security solutions to sustain growth and evolution (Marksteiner et al., 2017).

5.1. OVERVIEW OF INDUSTRIAL NETWORKS

With the evolution of technologies in industries, the role of industrial networks increases leaps and bounds, serving as the backbone to interoperability and performance of industrial control equipment and systems. Industrial networks enable real-time communication and data exchange between sensors, machines, and control systems, thus enabling automation and improving the efficiency of operations. The architecture of industrial networks can vary significantly, enabling numerous different protocols designed to address specific industrial needs. Major network categories include fieldbus systems such as DeviceNet and Profibus, and more modern ones such as Ethernet/IP and Profinet that employ standard Internet protocols to facilitate faster data transportation. These various systems are to be integrated, and integration poses special security issues that have necessitated an urgent need for dependable communication protocols which not only function to enable communication but also are cyber threat-resistant. The emergence of smart grids is a typical instance of the blending of industrial networks and new communication technologies, and thus, there is a need for an emphasis on security mechanisms. As described in current research, smart grids utilize two-way information exchange to maximize energy delivery, incorporating renewable sources of energy and supporting a more environmentally-friendly power usage paradigm. But this increased connectivity provides susceptibility to cyber-attacks that can threaten the integrity and availability of the energy supply. Despite many attack paths being identified via research, most are still fragmented, affecting isolated aspects of the system and not an overall security framework. Developing a safe communication model is therefore paramount, addressing the multi-aspect nature of industrial integrity, confidentiality, and availability requirements, in an effort to provide the foundation for an end-to-end security strategy to safeguard smart grids and other industrial networks (Ghazi et al., 2018). While considering the future of industrial networks, it is essential to pay attention to end-to-end cybersecurity solutions that are tailored to the complexities of such networks. Research currently focuses on communication gateways that are effective in processing mixed protocols and safeguard the data exchange through encryption and authentication mechanisms. In addition, embracing strict countermeasures against potential attacks, such as those targeting data storage handling and issues of privacy, is critical for the adoption of technologies like electric storage on the grid. By means of illustrations in ongoing research projects, the construction of secure smart energy communications not only enhances performance in operations but also secures consumer confidence in these emerging technologies. By addressing both technical and regulatory network security aspects, market participants can defend against emerging threats and ensure the integrity of industrial processes (Marksteiner et al., 2017) (Figure 5.2).

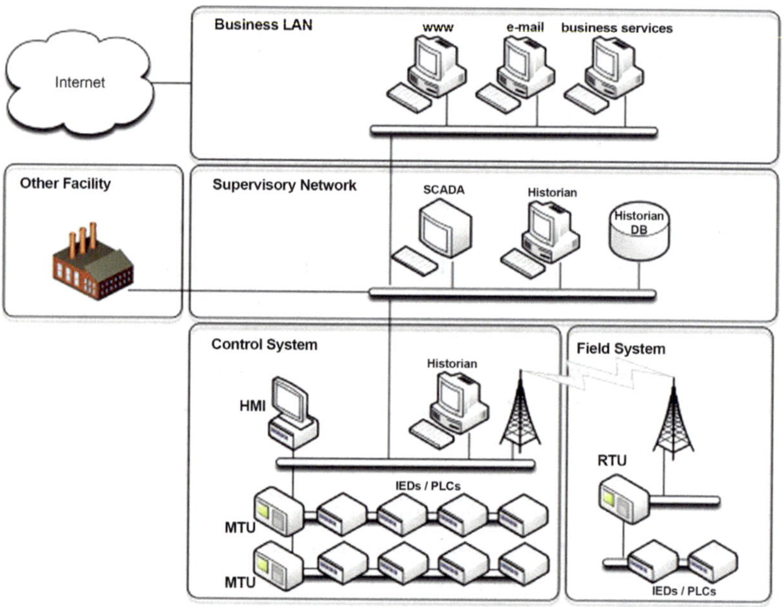

Figure 5.2. The role of industrial networks.

Source: https://ars.els-cdn.com/content/image/3-s2.0-B9781597496452000021-f02-01-9781597496452.jpg.

5.1.1. Types of Industrial Networks

Industrial networks are the backbone of modern manufacturing and production systems, facilitating communication and control between various devices and processes. Such networks can be categorized broadly into various types, i.e., fieldbus networks, Ethernet-based networks, and wireless sensor networks. Fieldbus networks, such as Profibus and Modbus, are designed exclusively for real-time communication in automation applications in industry, enabling devices like sensors and actuators to communicate efficiently. Ethernet-based networks utilize standard Internet protocols to enable support for industrial applications, enabling fast and scalable communication. This transformation, however, has created certain security concerns, as evidenced in the integration of communications capabilities into previously cut-off systems. The movement toward integrated networks highlights the necessity for robust secure communication protocols to accommodate these vulnerabilities and safeguard sensitive operational data within industries. The evolution from traditional "air-gapped" industrial control systems to increasingly coupled and combined setups has radically altered the industrial networking environment. Previously, safety-critical components, for example, Programmable Logic Controllers (PLCs) and sensors, operated independently with limited access to possible attackers. However, growing tendencies to network these components with enterprise information systems using TCP/IP have introduced more vulnerability to cyber-attacks. This is illustrated by the Industrial Internet of Things (IIoT) that facilitates far-end monitoring and control of manufacturing process. Although this innovation offers increased operational flexibility, it also demands a general sense of new threats associated with these networked systems, as noted in recent critical infrastructure analyses by the discipline of cybersecurity. It is essential to mitigate these security challenges in order to enable industrial networks to be resilient to potential cyber-attacks, especially as these systems become integral to business strategies. The complexities of industrial networks continue to improve, underpinned by technological advancements as well as operational efficiency needs. With industrial systems becoming increasingly embedded through the utilization of IoT devices and the use of smart grid technologies, the need for secure communication protocols increases even more. Such protocols must effectively secure against attacks aimed at compromising data integrity, availability, and confidentiality traversing these networks. Specifically, the smart grid leverages advanced information technology to enhance the supply of energy but is simultaneously susceptible to attacks that can compromise the vital infrastructure. Furthermore, existing security is biased toward individual threats, with no strategy in place addressing the different security needs of industrial networks. This highlights the requirement for an essential research gap to be addressed, stressing the necessity of developing end-to-end cybersecurity solutions that comprehensively embrace the vulnerabilities and intricacies of current industrial communication infrastructures (Evangelopoulou et al., 2017; Ghazi et al., 2018) (Figure 5.3).

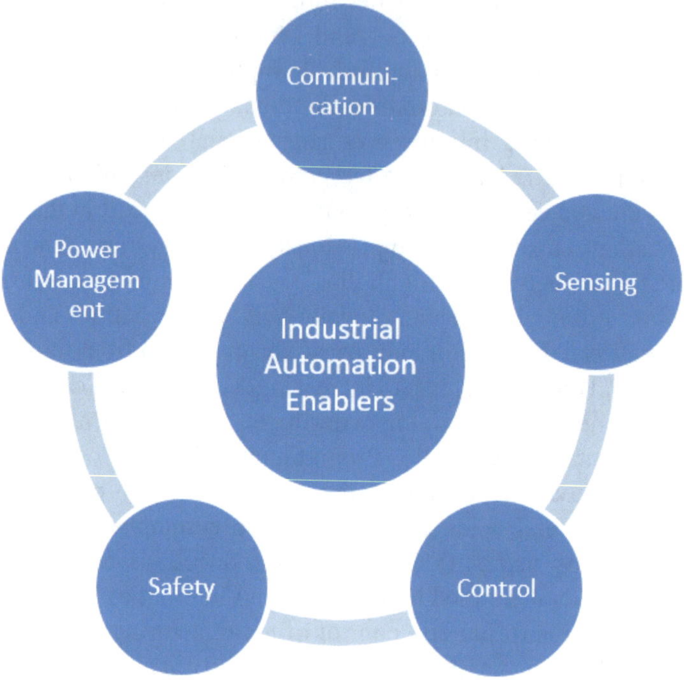

Figure 5.3. Industrial networks form the backbone of modern manufacturing and production systems.

Source: https://www.renesas.com/sites/default/files/media/images/industrial_automation_enablers.png.

5.1.2. Key Components of Industrial Communication Systems

In the modern industrial environments, effective communication systems are the key to the seamless operation of machinery and overall productivity. The basic components of industrial communication systems include sensors, actuators, controllers, and communication protocols which facilitate data transmission from one device to another. All these pieces work in synchronization to achieve automation and real-time monitoring so that processes are maximized for optimum efficiency. Although companies are shifting increasingly towards the integration of such advanced systems, they must address the security threats that come with them. With the advent of technologies such as Industrial Internet of Things (IIoT), connectivity between devices has introduced increased danger from cyberattacks, which demands a robust security model that spans across all layers of the communications stack. One of the individual areas of industrial communication systems is the application of common communication protocols, which are leading in ensuring interoperability among different equipment. Protocols like Modbus and Profibus facilitate data transmission across the network, thereby allowing diverse devices to exchange information effectively, regardless of manufacturer or operational intent. However, as the protocols are at the center of plant operations, so have the natural security loopholes. Past reliance on isolated systems, which are more popularly referred to as air-gapped systems, has decreased significantly and thus become more susceptible to external attacks (Evangelopoulou et al., 2017). As such, an adequate plan to secure such communication protocols against vulnerabilities must be formulated and implemented (Evangelopoulou et al., 2017). Additionally, with the industries moving towards smart grids and more automation, it is important to take layered security measures that secure all dimensions of operational technology. Structural intricacies brought in by interconnected systems require dealing with not just conventional vulnerabilities like confidentiality and integrity but also dimensions like accountability, which has been given a low priority in security reviews so far. As highlighted by recent studies, a generic approach to security is not sufficient; instead, a particular strategy that considers the diverse range of threats must be the area of concern. These problems highlight that measures towards productive cybersecurity mechanisms must not only lower the risks but also raise the bar on accountability and resilience within industrial communication systems (Ghazi et al., 2018). Through the integration of these elements into secure communication protocols' architecture, industries can safeguard their operations as they embrace technological advancements.

5.1.3. Common Applications in Various Industries

With the fast evolving face of technology in the current day, secure communication protocols are more essential in various industries. One very visible application is in the Internet of Things (IoT), where devices must communicate in complete synergy and security without compromising data integrity or privacy. Industries such as manufacturing, health, and transport use IoT technologies to enhance efficiency and reduce costs. However, to effectively manage threats like data leakages and unauthorized access, secure procedures have to be put in place. For example, organizations can make use of blockchain technologies in a bid to create decentralized networks that not only contribute to security but also provide transparency of transaction histories. Such a strategy not only minimizes risks that are accrued in centralized systems but also addresses challenges to compliance yielded by data privacy laws, enabling firms to have faith and accountability in their operations (Abdolee et al., 2019). Another critical application of secure communication protocols is in sectors which have to comply with stringent data governance rules, such as healthcare and finance. These industries rely on sensitive data confidentiality and integrity.

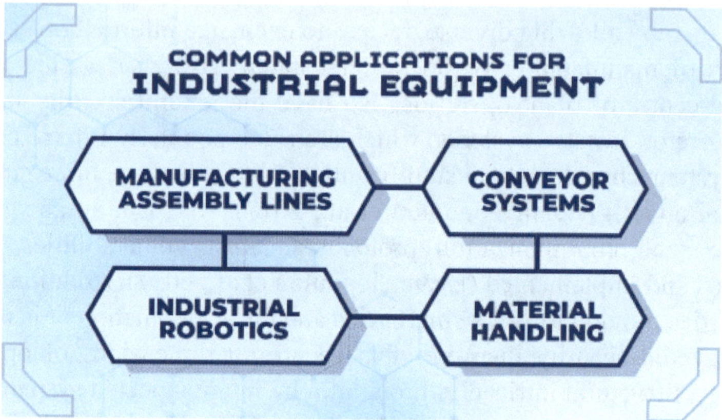

Figure 5.4. Common applications in various industries.

Source: https://hapcoincorporated.com/wp-content/uploads/Industrial-Applications-01-500x281.png.

The innovation in secure multiparty computation (SMC) is changing the paradigm so that privacy-preserving machine learning (PPML) techniques are employed to carry out computations on encrypted data directly without exposing personally identifiable information (PII). These technologies are critical to maintaining compliance with requirements like HIPAA, which mandates best practices for protecting patient data. As can be seen through current trends, secure communication protocols allow data exchange between healthcare organizations and predictive analytics platforms securely, leading to better clinical outcomes in the end. The combination of secure protocols and sophisticated computational approaches thus offers a platform for innovation while upholding the ethical standards of patient care (Cock et al., 2018). Besides, industries have come to recognize the need for strong communication protocols in the process of maximizing industrial automation and control systems. In manufacturing, data transfer protocols that guarantee secure data exchange between devices and monitoring systems help to protect system integrity and prevent potential cyber-attacks. As Industry 4.0 and IT-OT convergence gain ground in factories, security is more critical. For instance, employing end-to-end encryption in communication networks can allow businesses to secure sensitive information about their operations from eavesdropping or interference. Trustworthy protocols in automated systems allow industries not only to enhance operational reliability but also to establish the trust required among stakeholders. Here, the continuous development of secure communication protocols mirrors the growing needs of industries looking to advance technologically while ensuring their critical assets and sensitive information are safeguarded (Figure 5.4).

5.2. THREATS TO INDUSTRIAL NETWORK SECURITY

Increased interconnectedness among industrial networks has amplified the vulnerability of such systems to threats to security. Compared to isolated systems in the past, modern industrial settings such as smart grids and Internet of Things (IoT) applications implement advanced communication technologies and therefore are vulnerable to attack in a broader class. Such development provides avenues for attackers, who can take advantage of inherent weaknesses within network protocols. Different research studies have confirmed that security issues in such environments are normally associated with the lack of standardized defense mechanisms against cyber threats. The majority of existing approaches categorize attacks solely on the basis of parameters such as confidentiality, integrity, and availability, without accounting for other significant issues, such as accountability. Given this climate, it is essential to acknowledge the individuality of these attacks and their potential impact in terms of developing appropriate security measures that protect industrial networks from the sophisticated threats (Ghazi et al., 2018). Among various threats targeting industrial networks, Denial-of-Service (DoS) and data tampering are particularly malicious. These cyber-attacks have the potential to not only disrupt business but also pose serious safety risks in infrastructures. In manufacturing facilities, for example, unauthorized access to data by sensors can result in misconfigurations that compromise manufacturing processes. Moreover, data integrity failures can facilitate further exploits, compromising not only corporate assets but also public safety in places like chemical plants. As there is mounting connectivity of IoT devices, all vulnerabilities in a network are potential attack points. Therefore, industries must adopt a macroscopic security strategy that considers these vulnerabilities in an overall manner along with new technologies that are capable of real-time anomaly detection (Frey et al., 2018). Industrial security needs posed by threats necessitate an active shift towards cybersecurity. Contemporary defenses have been fragmented countermeasures targeting specific weaknesses rather than applying an all-in-one strategy to maintain the entire system secure. In the future, industries need to embrace measures that increase resistance to cyber-attacks in addition to enabling secure data exchange between interrelated devices. The emerging Information Centric Networking (ICN) paradigm has shown promise for safeguarding industrial IoT environments with its content-based model of security, which offers increased resistance to DoS attacks. Utilization of such advanced solutions can significantly bolster the integrity and availability of critical industrial systems, enabling them to maintain operation continuity irrespective of the growing environment of cyber threats (Frey et al., 2018; Ghazi et al., 2018) (Figure 5.5).

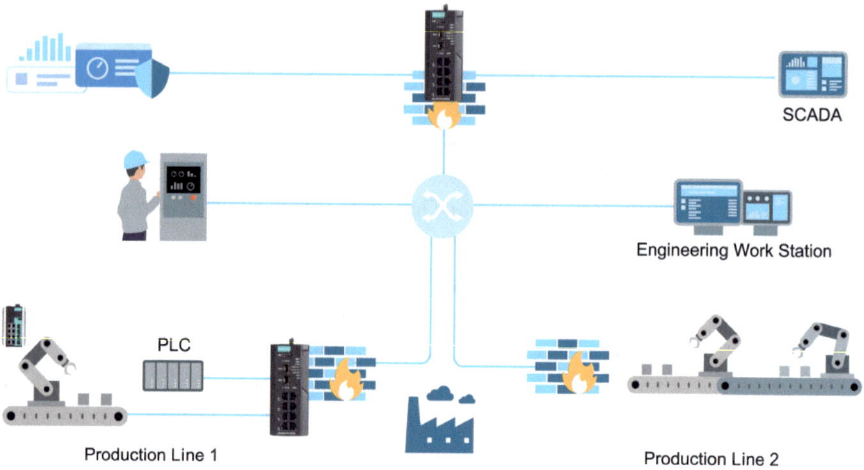

Figure 5.5. Threats to industrial network security.

Source: https://www.moxa.com/en/spotlight/portfolio/industrial-network-security/v23/images/secure-network-connectivity.

5.2.1. Cybersecurity Threats and Vulnerabilities

Cybersecurity threats are seen in numerous forms in industrial networks, exhibiting vulnerabilities that can interfere with critical operations. With industrial automation systems increasingly relying on Building Automation Systems (BAS) that are networked via IP-based communications, the absence of robust security features is problematic. The majority of BAS are developed around legacy protocols that prioritize function over security, and hence, are open to cyber-attacks. Specifically, the use of relying on security by obscurity is insufficient in decreasing unauthorized access and tampering with operational data risks. The importance of an overall evaluation of existing security controls in BAS has been brought out through research. By identifying specific threats and vulnerabilities, industries can better grasp their risk context and implement effective countermeasures, thereby safeguarding their infrastructure from future cyber threats like cyber terrorism and hacktivism (Johnstone et al., 2014). The implications of these weaknesses extend beyond direct operational consequences and can have cascading effects on critical infrastructure. Cyberattacks on industrial networks can have the ability to paralyze critical services and thus have profound economic and societal consequences. One such case is the attack on the Electric Power Transmission System, wherein attackers have the potential to bring down or hijack major components, resulting in massive power blackouts. Such cases are further complicated by the dynamic nature of modern threats, which alter rapidly to attack newly discovered vulnerabilities. News that examine attack vectors against power grids reflect the critical need for proactive security measures, since communication protocol vulnerabilities are so easily exploited by malicious individuals. Stringent security measures are therefore paramount to prevent disabling effects (Marcelo et al., 2011).

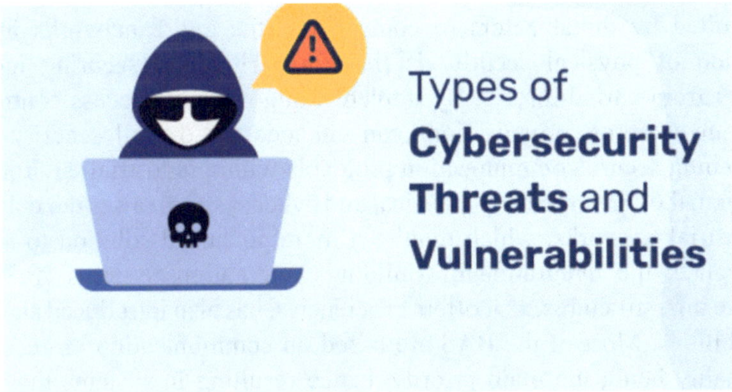

Figure 5.6. Cybersecurity threats and vulnerabilities.

Source: https://cdn-caklk.nitrocdn.com/sJCpruYkVvovUsWvtAEzkQwTuXjD-DQLL/assets/images/optimized/rev-fcf9680/powerdmarc.com/wp-content/uploads/2024/02/Types-of-Cybersecurity-Threats-and-Vulnerabilities.png.

To address such challenges of interdisciplinary and complex in nature, cybersecurity must be addressed comprehensively on all industrial networks. This entails not only the development of safe communication protocols but also ongoing assessment and adjustment of those protocols to thwart new threats. The inclusion of newer technologies, such as machine learning, in intrusion detection and prevention systems can also assist in enhancing the ability to detect and react to potential intrusions in real time. Through the establishment of a security culture with an emphasis on regular updates and ongoing monitoring, organizations can build an immunized infrastructure that can sustain resistance against the ever-changing nature of cyber threats. Ultimately, effective risk management practice, including strong training and awareness initiatives, will assist industrial sectors in fortifying defense against changing communication network vulnerabilities (Johnstone et al., 2014) (Figure 5.6).

5.2.2. Physical Security Risks

Physical security threats within the industrial network realm are a significant concern that demands rigorous examination, particularly as industrial networks grow to integrate digital technologies. Physical security weaknesses are caused by breached access to facilities holding critical infrastructure, such as power generation networks and data centers. These incidents can lead to service disruptions, data exfiltration, or even manipulation of operational technology, indicating the critical interdependent relationship of physical and cybersecurity. Vulnerabilities like inadequate surveillance or inadequate access controls can be exploited by threat actors to compromise the entire networks integrity. Resolution of physical security is therefore critical to securing industrial networks from varied threats. By implementing stringent access controls and monitoring systems, organizations can enhance the overall security system underpinning secure communication protocols within industrial environments. In addition, the convergence of physical and cyber systems raises the risk profile for industrial networks, which require a more advanced solution to security. For instance, the integration of Building Automation Systems (BAS) into corporate infrastructures, as it offers efficiencies, has also introduced significant vulnerabilities. Most of the BAS are based on communication protocols with functionality being the main priority, hence resulting in systems that can be exploited. This is also exacerbated by the increasing practice of remote access for management, albeit convenient, opening such systems to cyber-attacks. It has been noted that researchers have highlighted the imperative of defining threats to BAS in a comprehensive fashion and improving existing detection methods in order to mitigate these threats. Through this, industrial networks can establish a robust structure that synchronizes physical security measures with cyber system integrity to provide a uniform defense against a dynamic threat profile (Johnstone et al., 2014). Furthermore, the use of smart grids in industrial activities has also brought out the connection between physical security and cyber threats. As smart grids enhance operation efficiency and sustainability in the environment, they also offer numerous security risks. The inherent communication weaknesses within such grids expose the system to attacks that may compromise their availability and integrity. There has been an identification by research of a critical lack of sound security architectures that take into account both traditional and emerging threats, especially those which cannot easily fit into traditional buckets like confidentiality or integrity. To effectively respond to such attacks, there must be a global strategy that involves a collection of countermeasures to secure the overall system and not isolated components (Ghazi et al., 2018). Addressing these vulnerabilities with concerted measures is essential in ensuring secure communication processes that are robust enough to resist physical as well as cyber-attacks in industrial networks (Figure 5.7).

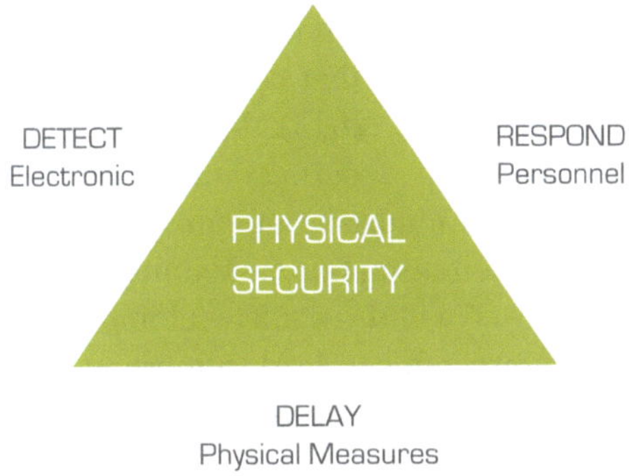

Figure 5.7. Physical security risks.

Source: https://www.charter-global.com/wp-content/uploads/2019/03/Security-trifecta-1024x693.png.

5.2.3. Insider Threats and Human Error

Insider threats and human mistake are the top weaknesses in the context of industrial networks since they threaten the integrity and security of the communication protocols. Insider threats tend to originate from individuals who possess legal access to network resources but present a risk through malicious behavior or carelessness. The complexity of industrial networks that involve combinations of layers of technology and staff heightens such threats. For instance, an employee can intentionally misuse their level of access to the system in order to destabilize operations or inadvertently expose sensitive information due to careless handling practices. Thus, as organizations continue to implement smart grid systems and interconnected infrastructures, they should employ secure communication protocols particularly designed to counter such insider threats and overall improve security levels in industrial environments (Hauser et al., 2018). Human error is intrinsically linked to insider threats, and it is a leading causative factor in security breaches in industrial networks. Human errors by workers can inadvertently lead to unauthorized access or information loss. This is particularly unsettling in the field of emergency management, where computational systems provide the foundation for proper responses and recovery procedures.

Poor training, ambiguous procedures, and the inherent tension associated with emergency circumstances can all increase the potential for human error. While high as the stakes may be for the consequences of such errors - catastrophic system breakdown or breakdowns of emergency response systems, for example - organizations must guarantee ongoing education and good use of safe communications procedures that consider human factors and promote situation awareness (Apikul et al., 2013). Moreover, protection from insider threats and human mistakes in secure communication protocols involves nurturing a security culture in industrial organizations. This involves not just technical controls but also promoting honest communication among employees in revealing potential exposures and risks. Developing procedures that engage employees in the security process can lead to more watchful practice and can significantly minimize risks associated with insider actions, whether intentional or accidental. By adding game-theoretic features to developing such procedures, organizations are able to create a dynamic framework that maximizes resilience against insider threats and human error through prudent decision-making and response mechanisms. This tightening of security in this multi-faceted manner will eventually translate to the robustness of industrial networks and their ability to surmount human and technological challenges (Hauser et al., 2018) (Figure 5.8).

Figure 5.8. The insider threats and human error.

Source: https://lh7-us.googleusercontent.com/RrPom7x4VvSHQl7aeyT4N2XcFciEh_1fb6R-0xkPksEJq2Wu7vZZGbrPQ97cNpcDlBx4vk4P6K5OZTf8wT2CP-T28YMA_n0r9qyLcCtKkUFgO5VRdhlOPwqsBsbRpqN499P0vCGIVTjdO-CYCkaOPdUk0.

5.3. KEY SECURE COMMUNICATION PROTOCOLS

Secure communication protocols in the case of industrial networks are critical to maintaining the integrity and reliability of data transmission. These protocols are the backbone of protecting sensitive information from unauthorized access and other forms of cyber-attacks. Of greatest concern is protocols specifically designed for secure data transfer to ensure confidentiality, integrity, and availability. For instance, protocols such as Transport Layer Security (TLS) and Secure Hypertext Transfer Protocol (HTTPS) are commonly employed in industrial environments to facilitate secure communication. Their multiple levels of encryption not only secure data in transit but also verify the sender and receiver entities. Imposing robust security controls becomes increasingly important as networks become increasingly interconnected, unveiling vulnerabilities that could be used to breach sensitive industrial processes. Furthermore, the proliferation of smart technologies into industrial networks introduces complexity that security solutions must reverse. As identified in other works, i.e., in domains such as smart grids, secure communication protocols must keep up with emerging security risks introduced by such technologies. Smart grids adopt two-way communication to facilitate more efficient energy distribution and user access. This convergence, however, necessarily increases security weaknesses as in recent research that surveyed the spectrum of cyber-attacks on key infrastructures such as smart grids (Ghazi et al., 2018). The shortcoming present in such systems calls for comprehensive protocols that not only cover the traditional security procedures but also transparency and accountability.

This wholesome solution is essential to address the evolving environment of cyber threats that industrial networks face in today's technologically advanced situation. Finally, the adoption of strict security procedures calls for not only accepting existing protocols but also innovating new solutions to further enhance communication security. Research continues to explore the dynamic interaction of various communication technologies and security issues, particularly the integration of electricity storage technologies into the grid (Marksteiner et al., 2017). These technologies need to be driven by optimized configuration guidance and data protection to improve immunity to cyber-attacks. With the growth in business dependance on automation and interconnected systems, these measures need to become more sophisticated to provide complete protection. Given the speedy advancement of technology, ongoing research is necessary to negate the current vulnerabilities and pave the way for future secure industrial communications. By actively working on these protocols, industries are able to significantly mitigate risks and make a positive contribution to overall security in their operational networks (Figure 5.9).

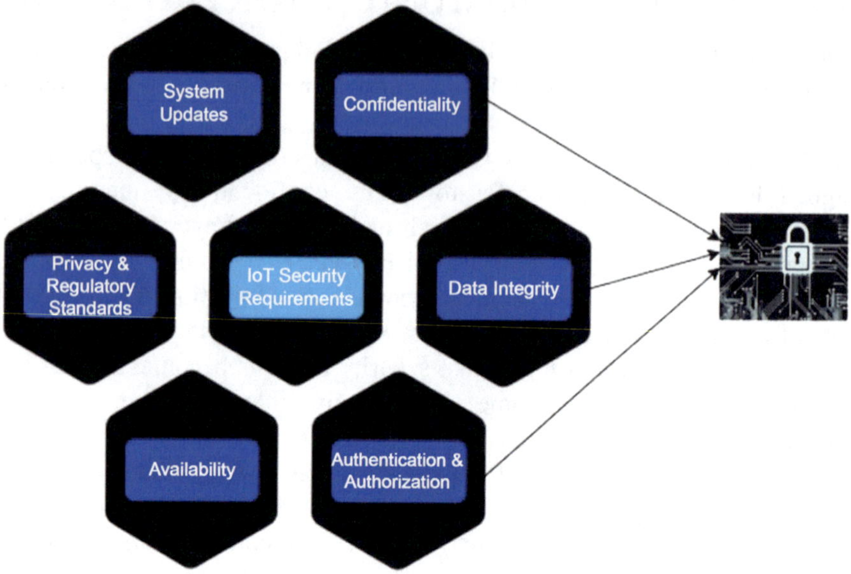

Figure 5.9. The secure communication protocols.

Source: https://media.springernature.com/lw685/springer-static/image/chp%3A10.1007%2F978-3-031-82826-3_6/MediaObjects/637413_1_En_6_Fig5_HTML.png.

5.3.1. Transmission Control Protocol (TCP)

In industrial networks, it is the role of TCP to make communication both efficient and reliable. TCP is naturally designed to provide a connection-oriented service for communication, meaning that TCP establishes a connection between endpoints prior to data transfer. This feature is particularly important in industrial systems, where the accuracy and order of data packets are of great significance. TCP uses a numbering of sequences to organize data packets so that missing packets can be identified and retried. Consequently, those industries relying upon real-time data communication, for example, manufacturing or energy management, gain significant advantages from TCP's inherent delivery guarantee as symbolized in the smart grid case when communications technologies are combined with electric power systems for increased efficiency (Ghazi et al., 2018). In addition, the versatility of TCP has made it possible to integrate many industrial protocols into contemporary enterprise systems through the utilization of the potential of the Internet. With evolving industries, the infusion of digital communications via TCP makes it possible to integrate diverse operational technologies in synchronization, showing how the traditional, independent industrial environments are giving way to converged systems. While the advantages comprise many benefits, the usage of TCP also harbors a variety of vulnerabilities, particularly in terms of cybersecurity. As highlighted in recent research in critical infrastructures, attackers are increasingly able to target such communication channels, and therefore companies must embrace firm cybersecurity strategies tailored to TCP. The transition from isolated systems to interconnected ones via TCP accelerates not only business efficiencies but also exposes them to potential security threats (Evangelopoulou et al., 2017). Fixing security issues of TCP within industrial networks requires an interdisciplinary approach. While TCP's strongest points are its reliability and capacity for complex transmissions, it calls for comprehensive security controls like encryption, intrusion detection, and vigilant monitoring. The movement towards the Industrial Internet of Things (IIoT) also adds to these challenges as devices become autonomous and more networked. It is important that organizations pinpoint and remedy vulnerabilities in their TCP setups so that data integrity and confidentiality during transmission are maintained. Research is continually pointing towards the necessity of developing global security architectures that can keep up with the complexity of modern industrial environments, opening doors for more secure and resilient communication protocols immune to emerging threats (Evangelopoulou et al., 2017; Ghazi et al., 2018) (Figure 5.10).

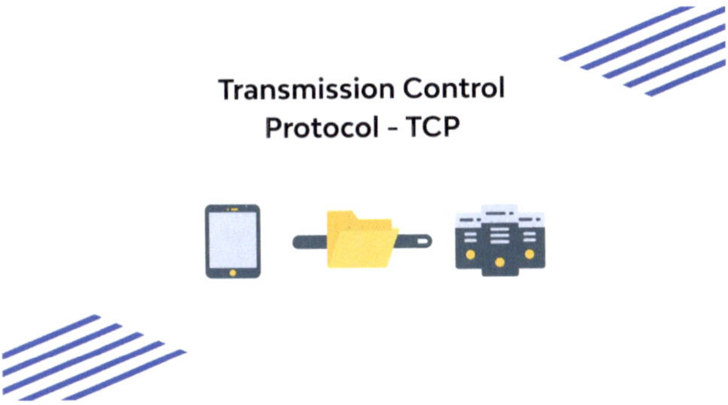

Figure 5.10. The transmission control protocol (TCP).

Source: https://cdn.prod.website-files.com/5ff66329429d880392f6cba2/627cb3f27138814b179c1d86_tcp%20Preview.jpg.

5.3.2. User Datagram Protocol (UDP)

UDP is a fundamental communication protocol with a central position in most industrial network applications, particularly where speed and efficiency are necessary. Unlike its alternative, the Transmission Control Protocol (TCP), UDP is connectionless in that it does not establish an end-to-end connection before transmission of data. This characteristic allows UDP to deliver packets, or datagrams, with less latency, which makes it perfectly appropriate for applications that need real-time delivery of data, e.g., Voice over IP (VoIP) and video conferencing. But at a cost because UDP offers no guarantee of delivery, sequencing, or error correction, which has enormous security implications in industrial applications where data integrity is of the utmost importance. Consequently, the application of UDP must be rigorously tested against the requirement for security and reliability within such networks. UDP security in industrial networks is a critical area of concern in the context of the threat of vulnerabilities inherent within its connectionless structure. Without intrinsic controls for data integrity and confidentiality incorporated within UDP, additional layers of security might be necessary in order to guard against interception or unauthorized access.

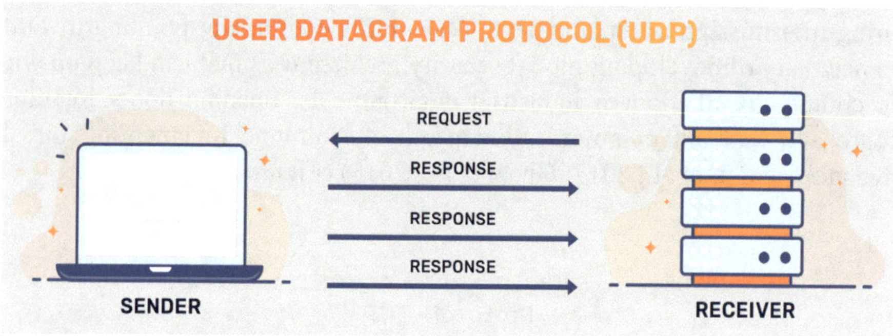

Figure 5.11. The user datagram protocol (UDP).

Source: https://encrypted-tbn0.gstatic.com/images?q=tbn:ANd9GcS-J2o9OaVbgzYtJ7IsP1HfZ9x9huRfHbHlsA&s.

One of these is to integrate Transport Layer Security (TLS) or Datagram Transport Layer Security (DTLS), which can encrypt and authenticate data over UDP. Also, methods such as hole punching have been added to support traversal of Network Address Translation (NAT), as industrial networks use devices behind NATs for security purposes. A survey indicates that approximately 82% of NATs tested provide UDP hole punching, and this demonstrates the feasibility of applying reliable peer-to-peer communications to various deployment scenarios (Ford et al., 2005). Despite the inherent challenges, UDP deployment in industrial networks can be warranted if due care is exercised to ensure secure communication. Its universality enables it to transport a collection of applications beyond ordinary file transfers, making real-time monitoring and data control possible in mission-critical systems. Its recognition of its own limitations has inspired ongoing research to further expand NAT traversal techniques for performance improvement as well as to enhance the reliability of UDP communication in real life (Brunstrom et al., 2016). Moreover, as more sectors implement Internet of Things (IoT) structures, being able to utilize UDP for immediate data distribution without the expense of handling connections renders it all the more appealing. Ultimately, while UDP may not be the most secure off-the-shelf option for every industrial application, its speed and efficiency, along with the appropriate security protocols and methods, can be made to produce a secure communication environment when properly integrated into higher-level system designs (Figure 5.11).

5.3.3. Secure Sockets Layer (SSL)/Transport Layer Security (TLS)

SSL and its successor, TLS, are essentially critical protocols that ensure secure communication on networks, particularly in industrial applications. SSL was originally developed by Netscape for establishing a secure communication channel between two machines spanning across the internet, whereas TLS is a more developed version that includes more robust security features and more efficient cryptographic methods. These protocols prove useful when dealing with establishing encrypted tunnels between a web server and a web browser, effectively protecting sensitive data from eavesdropping and man-in-the-middle attacks. In the case of industrial networks where operational technologies and information technologies converge, the SSL/TLS protocols find extensive usage for securing critical communication between devices to prevent operational and proprietary information from being leaked or compromised, thus denying potential dangers in digital communication based on vulnerabilities. SSL/TLS technologies in industrial networks present many security challenges. New communication systems come with greater exposure to cyber-attacks, which necessitate strong security protocols to be adopted. Industrial sectors such as power and utilities need secure data exchange technologies to facilitate smart grid technologies where information technology and energy systems converge. As per studies, while SSL/TLS effectively addresses risks such as unauthorized access and data tampering, there are problems regarding accountability and integrity of the data. Besides, it should be understood about the specific vulnerabilities to which industrial systems are subject and which in most situations differ from typical IT environments. Facilitating a secure communication platform must account for more than the encryption provided by SSL/TLS but an inclusive solution handling other security parameters, as elaborated in various studies on smart grid technologies and their communication structures

(Ghazi et al., 2018; Marksteiner et al., 2017). Furthermore, the application of SSL/TLS within industrial networks requires an evaluation of advancements and the prospect of cyber security. More sophisticated architectures that are being used more and more integrate Internet of Things (IoT) and machine-to-machine (M2M), where security protocols play center stage due to the level of interconnectivity between devices. The shift from SSL to TLS is a promise of better security standards in a time when cyber-attacks continue to push the limits of security features. As industries implement intelligent technologies, stringent validation and authentication mechanisms become a must, particularly for remote management systems. As such, it is evident that the use of SSL/TLS not only protects sensitive data but also increases the overall robustness of industrial communication infrastructure from ever-evolving cyber threats. This proactive vision towards security will definitely give rise to more innovation in secure communication protocols for industrial purposes (Ghazi et al., 2018; Marksteiner et al., 2017) (Figure 5.12).

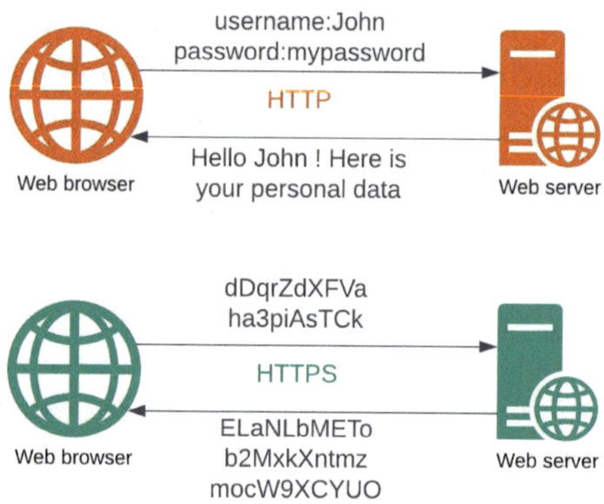

Figure 5.12. The secure sockets layer (SSL) and its successor.

Source: https://miro.medium.com/v2/1*wu2ZW7whIaA6S4wzWIskrw.jpeg.

5.4. ENCRYPTION TECHNIQUES IN INDUSTRIAL NETWORKS

Encryption methods in industrial networks are a necessary security mechanism against unwanted access and cyber intrusions. As industries head towards connected systems, like the Industrial Internet of Things (IIoT), proper encryption methods are more than ever crucial. These encryption methods properly safeguard critical information being transferred among devices that range from programmable logic controllers (PLCs) to other sensors and actuators. Existing encryption methods such as Advanced Encryption Standard (AES) and Rivest–Shamir–Adleman (RSA) can utilize stringent security by encrypting information into unreadable patterns with cipher forms that need corresponding keys to decipher. The lack of a globally standardized encryption process, however, remains a problem, as indicated in the literature, with existing solutions often addressing particular vulnerability in lieu of utilizing pervasive protection against the entirety of the industrial system (Ghazi et al., 2018). Security risks have grown as the traditional air-gapped systems evolved into networked architectures. Industrial control systems, previously, were air-gapped and had inherent security from cyber-attacks since they required physical access for potential attacks. Currently with the integration of network technologies, attackers can remotely exploit vulnerabilities with the old security measures being rendered useless. The development of communication interfaces such as Profibus and Modbus protocol coupling over TCP/IP networks demonstrates the convergence of enterprise information systems and operational technology to improve productivity as well as threats. As depicted in modern research, such as forensic analysis of attacks on infrastructures, the part played by total encryption policies has never been greater (Evangelopoulou et al., 2017). The tale emphasizes the importance of industries to move early towards adopting next-generation encryption standards simultaneously as they prepare to deal with diversified threats from new cyber environments. Moreover, encryption on industrial networks must be followed by robust key management practices in order to realize long-lasting resilience for data integrity. Key management is the linchpin of encryption; it encompasses generating, distributing, storing, and destroying encryption keys, which need to be tightly controlled and regularly updated. Compromised keys can negate even the strongest encryption, exposing networks to intrusions. As the smart technologies continue to advance, industries are prompted to embrace encryption models that not only protect data in transit but also exacerbate current weaknesses through efficient key lifecycle management. The shift toward automated key management platforms, coupled with consistent risk evaluations, will promote a forward-thinking security posture within industrial networks. This multi-layered approach provides a path toward enhanced resilience to cyber-attacks that exploit connectivity in a way aligned with best practices underscored in recent cybersecurity research (Evangelopoulou et al., 2017; Ghazi et al., 2018).

5.4.1. Symmetric Encryption Methods

Symmetric encryption methods are a vital part of secure communication in industrial networks, where data integrity and confidentiality are of top concern. This encryption technique employs a single common key for both the encryption and decryption, thus less complex than the asymmetric encryption algorithms. The efficiency of symmetric encryption in data processing with high-speed demands in environments is best applicable to industrial operations where real-time data communication is critical. With such encryption schemes as Advanced Encryption Standard (AES) becoming popular, organizations can ensure very strong security against unauthorized access. Such means do include secure handling procedures for keys, however, because loss of the shared key would leave enormous security vulnerabilities that could risk compromising the entire communication system.

Figure 5.13. Symmetric encryption methods.

Source: https://www.clickssl.net/wp-content/uploads/2019/12/symmetric-encryption.jpg.

For industrial networks, symmetric cryptographic approaches also need to consider special cybersecurity challenges inherent in interweaving advanced high-tech technologies, such as smart grids. As outlined in various other studies, the application of communicative technologies in these grids has introduced special vulnerabilities to be exploited by hostile actors. For instance, any breach of the communication protocol would detrimentally affect not only the local system but also larger realms of operation which are deeply reliant on secure data exchange. Research has indicated that while traditional security measures have sought to prevent specific kinds of assaults, there is a need to adopt an integrated approach wherein several countermeasures are combined to create an overarching protection against threats. These broad-based strategies must involve symmetric encryption methods in order to enhance the robustness of industrial networks' security (Ghazi et al., 2018). Moreover, the ever-evolving nature of industrial communication systems demands constant upgrades in symmetric encryption techniques to deal with emerging threats. As industries rely more on Internet of Things (IoT) technologies, the challenge lies in how to design encryption schemes that are efficient as well as adaptable to various applications. Recent research, such as the development of secure communication devices that use energy storage technology, demonstrates how to integrate encryption into new architectures. In such instances, privacy and protection from data tampering by means of robust symmetric encryption are key to enabling trust in these infrastructures. These developments not only improve current infrastructure but also provide a basis for future innovation in secure communication protocols. This ongoing research emphasizes the relevance of dynamic encryption solutions that enhance the overall security profile (Marksteiner et al., 2017) (Figure 5.13).

5.4.2. Asymmetric Encryption Methods

Asymmetric encryption methods are crucial to establishing secure communication protocols, particularly for the case of industrial networks. Unlike symmetric encryption, in which a shared secret key is employed by two communicating parties, asymmetric encryption employs a public-key encryption and a private decryption key. The simple design eliminates key distribution vulnerabilities that are common to traditional methods of encryption. In systems with numerous devices communicating with one another, as is true with smart grids employing intricate networks of communication, asymmetric encryption greatly enhances security. With these methods, organizations ensure that confidential data are not leaked out, safeguarding corporations and decreasing threats to improper access (Ghazi et al., 2018). The effectiveness of asymmetric encryption is most prevalent in its application within wireless sensor networks, where different scattered nodes are constantly communicating. The networks accumulate more security threats due to the open exposure of such networks, and hence it is imperative to apply strong encryption mechanisms. Asymmetric technologies enable secure key exchange and data integrity and authentication, addressing threats from malicious devices infiltrating the network without control. They also enable secure key exchange, essential for secure communication protocols in the Internet of Things (IoT). Through the application of public key infrastructures, industrial panels can more effectively reduce computational overhead on devices, particularly in environments where resources are scarce and efficiency is paramount (Dang et al., 2019). While asymmetric encryption adds much to security in industrial networks, it is not without its limitations. The complexity of such methods can lead to increased computational overhead, something that impacts the performance of systems where data is to be moved at high speeds. In addition, having a public key infrastructure creates potential points of failure in itself, such as the need for secure key management and potential vulnerabilities pertaining to key revocation and distribution. Hence, companies are urged to adopt hybrid encryption techniques that combine the strengths of symmetric and asymmetric techniques in order to achieve utmost security and efficiency. By careful consideration of the specific demands of their networks, organizations can implement effective secure communication protocols that take advantage of the merits of asymmetric encryption while eliminating the associated drawbacks (Ghazi et al., 2018) (Figure 5.14).

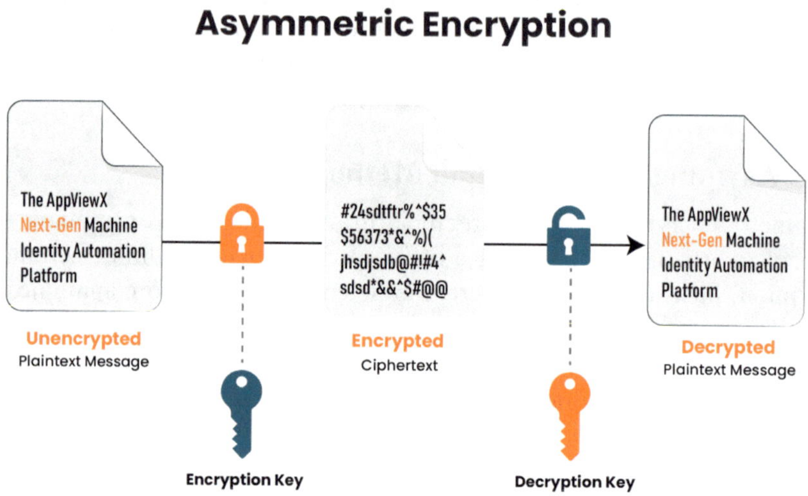

Figure 5.14. Asymmetric encryption methods.

Source: https://www.appviewx.com/wp-content/uploads/2022/02/Asymmetric-Encryption-Works.png.

5.4.3. Hashing Algorithms and Their Applications

Hashing algorithms play an important role in ensuring data integrity and security within secure communication protocols, particularly in industrial networks. By mapping input data into a fixed-size output, or hash value, hashing algorithms make it easier to verify data authenticity without disclosing sensitive information. Their one-way nature makes it computationally infeasible to recover the original input from the hash, providing added security. Applications range from password storage to electronic signing, where the hashed value can ensure the authenticity of data without revealing the content. In industrial networks, where security is also a factor, the employment of hashing algorithms minimizes the threats of unauthorized access and tampering of data, hence establishing confidence in the integrity of data being exchanged on such networks. The increasing intricacy of industry networks necessitates advanced tools to protect communications from the ongoing threat of cyber-attacks. Hashing algorithms fit the bill best, as they have the ability to implement a variety of security protocols like HMAC (Hash-based Message Authentication Code), which combines cryptographic hashes and secret keys to create both data integrity and authenticity. Also, how hashing is applied in blockchain technology, as shown in recent research, depicts how it can be applied to make distributed systems secure. In such a setup, hashing lies at the core of having a tamper-proof ledger, where every block references the previous one through its hash value. This feature enables illegal modifications to be traced easily, thus enhancing the security framework in industrial settings using blockchain technology for record-keeping and data sharing (LeBlanc et al., 2020). The application of hashing functions within smart grid technologies is another obvious use in enhancing secure communication protocols. Smart grids, characterized by two-way communications and utilization of renewable power sources, bring unique security challenges. As required in existing research, cryptographic methods, including hashing, are of utmost significance in ensuring safe access and safeguarding data integrity when being transferred between control facilities and users. With threats ranging from data tapping to illegal access, hashing ensures sensitive information is safeguarded within the interconnectivity system. Furthermore, effective key management methods, in conjunction with hashing mechanisms, are important in preventing unauthorized operations and creating trusted communication connections in smart grids (Alohali et al., 2025). Conclusively, the application of hashing algorithms for secure communication is a significant consideration in improving the reliability and security of industrial networks for technological advancement and operational efficiency.

5.5. IMPLEMENTATION CHALLENGES

When secure communication protocols are applied on industrial networks, several issues emerge that can disrupt their performance. One of the key issues involves the scalability of the protocols as Internet of Things (IoT) devices proliferate at a rapid rate. With industrial setups adopting enormous quantities of sensors and actuators, the need for large quantities of data to be processed without any loss of security becomes paramount. For instance, existing protocols are to be analyzed extensively in terms of latency and network throughput performance in order to serve these devices. As noted by Carpio et al. (2019), the decentralized fog and cloud computing system provides promising solutions, but the choice of proper communication protocol is a critical decision for system architects to guarantee uninterrupted inter-device communication with high security against growing threats. Apart from scalability, the hostile environment characteristically present in industrial setups poses significant implementation challenges in implementing secure communication protocols. Industrial networks are usually installed in hostile environments—exposed to harsh temperatures, electromagnetic interference, or physical stress—that weaken communication integrity. Hostile environments necessitate protocols resilient to handling security and integrity under operational stress. The research in Frey et al. (2018) is about the assessment of content-centric security frameworks that will be robust under such conditions and capable of defending against denial-of-service (DoS) attacks. Accordingly, the protocol flexibility of protocols like CoAP and MQTT should be examined, particularly in their application to safety-critical applications, in order to determine their potential effectiveness and reliability in safeguarding sensitive industrial information. Finally, interoperability between various communication protocols typically complicates the implementation within industrial networks, providing an additional layer to the challenge. With multiple vendors working in the industrial IoT device market that is growing, it becomes more challenging to have smooth communication among different systems. Interfacing between distributed components in space is different in its demands on protocols that not only provide secure data transfer but also allow interfacing between various systems. According to Carpio et al. (2019), interoperability among new and legacy protocols must be addressed in creating a combined IoT-to-fog-to-cloud framework, and hence protocol designers must seek the appropriate remedy that enables secure, harmonized communication across platforms. Eventually, addressing these issues of implementation is critical in realizing the full potential of secure communication protocols towards enhancing industrial network resilience and efficiency.

5.5.1. Cost Considerations for Security Measures

In industrial networks, cost considerations along with security are preeminent to mind because of rising complexity and scale of operations involved. Implementing effective security measures has a cost that is quite varied based on technology utilized as well as the network's specific requirements. Firms must weigh costs of employing sophisticated encryption algorithms, intrusion detection software as well as round-the-clock monitoring technologies against potential financial repercussions of security breaches. Also, due to industrial environments becoming increasingly integrated with Internet of Things (IoT) devices, the cost of security for such connections can increase, primarily due to the sheer volume of devices that possess heterogeneous nature and some varying security requirements. Hence, a deliberate approach towards security costs budgeting is very important, such that business resilience and reliability are maintained by companies while spending money effectively. Incorporation of cybersecurity solutions in today's industrial standards means understanding in detail not only initial costs but also maintenance and upgrade costs in the long run. The introduction of middleware for secure data transfer between different IoT devices, for example, means incurring initial capital limits but leads to lower operation risks and overall financial savings by minimizing the occurrence of security breaches. As signaled, the Cloud of Things (CoT) offers effectively unlimited services that are capable of managing vast volumes of data incurred due to diverse connected devices (Farahbakhsh et al., 2017). Thus, businesses must consider the cost of ownership in such technologies, considering repeat expenses and support to possess productive security postures in the long term while being innovative with their operational frameworks. As the threats to industrial networks increase, there is a need for organizations to be proactive in the way they utilize their resources towards security initiatives. Not only should cost considerations include technology acquisition, but training and employee awareness as well, which can significantly decrease the exposure of human mistakes to security systems. Investment in human capital to build a security-aware culture can have high returns by reducing loss caused by breaches. Organization must also regularly evaluate the cost-effectiveness of security measures and set them according to evolving threats and available new technologies. As the CoT and middleware impacts on IoT debates depict, the above factors are important to ensure operational integrity and long-term investment in industrial communication protocols (Farahbakhsh et al., 2017).

5.5.2. Integration with Existing Systems

The integration of safe communication protocols in industrial networks calls for careful examination of existing systems to secure them against cyber-attacks. By default, industrial control systems were previously isolated, air gapped environments with Programmable Logic Controllers (PLCs) and smart devices isolated from broader networks. This model offered security in the sense of physical access that must be compromised to attack systems. But increasing use of network interfaces and protocols like Profibus and Modbus has dulled these borders, exposing vulnerabilities that must be addressed now. When companies use these systems for efficiency in operations, they inadvertently leave essential infrastructure open to increased cyber threats, necessitating stringent security controls to safeguard communication across such now-interconnected platforms (Evangelopoulou et al., 2017). As industries move toward industrial internet of things (IIoT) and smart grid technologies, the challenge of integrating current systems is exacerbated. The need for real-time communication across heterogeneous grid elements in generation, transmission, and distribution emphasizes the role of interoperability in secure communication. Achieving this integration calls for industrial networks adopting advanced protocols that are able to provide security and privacy but also make communication easy. The existing networking technologies must be implemented to address these challenges, with new systems capable of interworking with legacy systems without compromising security. Such development not only aims to achieve maximum energy efficiency but is also reflective of a broader shift toward intelligent metering and advanced energy management systems that are critical to the current industrial landscape (Chin et al., 2011). Furthermore, incorporation of secure communication protocols within existing systems will also have to consider the ramifications of concerted action towards standardization to address security concerns comprehensively. Standardization is responsible for developing interoperable communications frameworks that facilitate secure interconnections between various components of industrial systems. The collaborative efforts by global organizations to harmonize communications standards will unlock the standards capable of mitigating risks that accompany legacy and innovative systems. Through the alignment of best practices and security requirements, organizations will be able to bolster their defenses, eliminating vulnerabilities susceptible to misuse. Thus, gaining understanding and improving the integration of secure protocols into existing systems becomes imperative not only for conformity but also to preserve industrial network integrity and robustness in a growing interdependent world (Chin et al., 2011) (Figure 5.15).

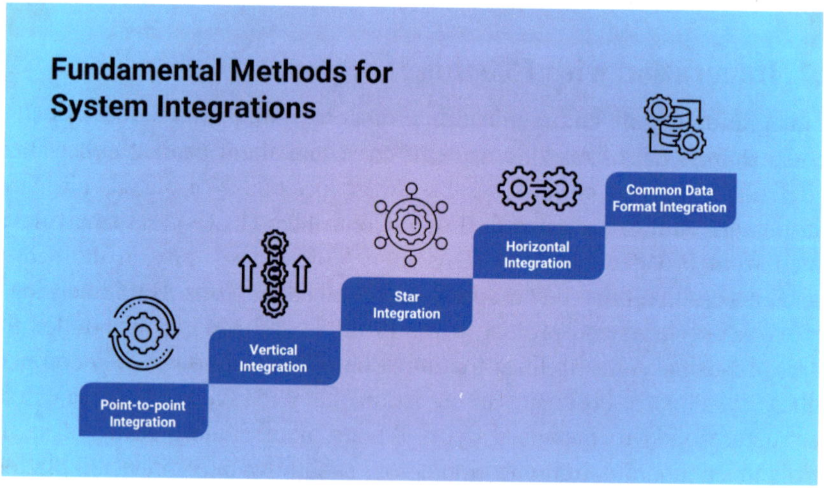

Figure 5.15. Integration with existing systems.

Source: https://dynamics.folio3.com/blog/wp-content/uploads/2023/04/2-1024x576.png.

5.5.3. Training and Awareness for Personnel

Increased network sophistication in sectors demands robust training and awareness mechanisms for employees involved in secure communication protocols. Staff needs to be familiar with both the technological and procedural protocols that govern secure data transfer as the networks evolve. Comprehensive training courses should include not only the fundamentals of safe communications but also the specific threats their surroundings present, such as cyber-attacks on key infrastructure.

This data enables the personnel to identify possible vulnerabilities and take the appropriate measures, thereby strengthening the overall security posture of the company. By promoting an awareness culture in which employees are knowledgeable about the prevalent threats and best practices, organizations will be able to minimize human error-related risk, often the primary source of security breaches. Moreover, tailored training sessions can be able to augment incident response capabilities. When employees are properly trained on secure communication processes, they can react promptly and efficiently against security breaches.

This preparedness is crucial in minimizing potential disruption in critical infrastructures, such as those reliant on water supply and sanitation networks, whose cyber-attacks come with consequential implications that include contamination with chemicals or a breakdown in the network (Gil et al., 2018). Continuous training, which includes simulation exercises and in-transit training, can help keep workers alert and responsive to evolving threats. Such an investment in training not only safeguards the worthwhile properties of a business but also enhances the overall reliability and fault tolerance of industrial networks. Hands-on training methods must also be incorporated with theoretical learning so that personnel can be caused to practically apply what they have been taught. Hands-on training with secure communication units, along with scenario simulation, may provide staff with required skills for effective functioning in complex network contexts. These exposures help close the divide that is found between practice and theory, building a responsive and educated workforce proficient at implementing secure procedures. With advancing technologies through research in new materials, such as carbon fiber composites for the car sector, it is essential that all personnel comprising such technological advancements are highly trained in secure communication processes and how they influence greater operation strategies (Sussmann et al., 1999). By prioritizing training and awareness, organizations enhance their ability to protect sensitive information and provide secure communication through their industrial networks.

5.6. CASE STUDIES OF SECURE PROTOCOLS IN ACTION

In the industrial networking world today, case studies that show the application of secure communication protocols are a goldmine of information about real-world applications and issues. One example of note is the degradation of traditional means of isolation, defined as an air gap, which was a robust method for protecting critical control systems. As systems evolved to interact with broader enterprise networks via protocols such as TCP/IP, exposure to cyber threats has increased. Research of recent attacks on critical infrastructure in Ukraine is an ex post facto warning story, and that makes the need for rigorous forensic inspection of IIoT switches crucial to determine vulnerabilities. The study aims to validate security measures that can protect from such attacks and encourage resilience in Europe and North America, where such infrastructures are common (Evangelopoulou et al., 2017). In like manner, the implementation of secure communication protocols within the smart grid presents a similar persuasive case study, demonstrating the vibrant convergence of energy management and cyber security. Smart grid technology encourages two-way communication to optimize energy distribution and environmental sustainability through improved resource management. Nevertheless, this technological advancement inspires a myriad of security risks, exposing the grid to numerous hazards. Past research highlights the dire need for addressing these issues holistically, with the tendency of focusing on independent attacks at the expense of overall security parameters like accountability. A holistic approach is therefore urgently required to make the security framework of the smart grid robust so that everything is duly secured against potential vulnerabilities (Ghazi et al., 2018). These case studies evidence an immediate need for the development of holistic security policies that can fend off attacks on industrial control systems as well as modern energy infrastructures. The learned wisdom that arises from analytical examination of these systems emphasizes the point that the mere installation of independent countermeasures is not sufficient to ensure long-term security. Rather, measures must consider a global perspective that merges various security fields. Future research should explore emerging strategies that reverse existing vulnerabilities but also adapt to changing threats. Such a shifting narrative with respect to secure communication protocols highlights ongoing awareness and ingenuity in the security of industrial networks, thereby securing global critical infrastructure (Evangelopoulou et al., 2017; Ghazi et al., 2018).

5.6.1. Successful Implementation in Manufacturing

The successful implementation of manufacturing systems relies heavily on the use of secure communication protocols within industrial networks. The use of technologies like the Internet of Things (IoT) can facilitate remote control and monitoring of manufacturing processes, hence increasing productivity and operational effectiveness. However, for these advancements to be achieved, organizations must put a lot of emphasis on cybersecurity interventions guaranteeing data integrity and privacy. Threats such as data breaches can undermine the very foundations of trust required in collaborative manufacturing environments. Industries must navigate through these intricacies by embracing robust security architectures that not only cater to the current needs but also adapt with future advances in manufacturing and communications. This holistic approach guarantees organizations can sustain growth while sidestepping risks of unsafe industrial networks, ultimately leading to enhanced operational success. One of the most important factors for successful manufacturing deployment is the role of collaboration among various stakeholders like technology providers, engineers, and operations managers.

Creating secure communication guidelines is central to ensuring seamless integration between such entities. The advent of IoT technologies turns interconnectivity between systems and devices into the imperative. However, as the spotlight in Abdolee et al. (2019) demands, data privacy, power consumption, scalability, and cybersecurity need to be tackled to the fullest so that the maximum potential of these technologies can be fulfilled. The producers can build reliable networks that not only facilitate operational effectiveness but also secure sensitive information via active debate and adopting standards that put security at the forefront. Hence, collaborative effort in designing and implementing secure protocols can lead to a less volatile manufacturing landscape. As an added point, communication technology diversification has a great contribution towards the success of manufacturing system installation. The evolution of the Internet and its impact on industrial communications is evident in various global contexts, as shown in Licoppe et al. (2025), where the past difficulties experienced in France influenced the development of the Internet.

It is a demonstration that monopolistic systems will smother advancement; therefore, embracing a competitive and open market approach will foster innovation. For manufacturing, embracing various secure modes of communication ensures adaptability and resilience as well as solving certain operational needs. An adaptive manufacturing system which embraces various and secure channels of communication not only enhances productivity but also equips organizations to address market needs in a similar manner. This strategic focus on secure communication reiterates the importance of an active method in modern industrial networks, leading to successful manufacturing deployment eventually.

5.6.2. Lessons Learned from Breaches in Energy Sectors

The energy sector has experienced its share of incidents that revealed critical vulnerabilities, with a focus on putting in place robust secure communication practices in industrial networks. High-profile cyber-attacks, such as the Stuxnet attack on Iranian nuclear facilities, have demonstrated the potential of malicious cyber activity to cause physical damage and impairment of critical services. These incidents highlight the intersectionality between cyber and physical security, and how conventional defense mechanisms must adapt to incorporate the sophistication introduced by cyber threats. According to Marcelo et al. (2008), the need for integrated threat assessment procedures becomes clear; it is essential to consider both man-made and natural accidents while building a secure infrastructure.

Thus, energy corporations must endeavor to implement newer cybersecurity protocols that not only address existing threats but evolve based on new vectors of threats. At the same time, lessons from breaches within this sector point to the importance of energy awareness in cyber-physical systems. As Udoh et al. (2019) observe, use of security mechanisms such as encryption and authentication tends to introduce an overhead in terms of energy on resource-constrained devices, leading to a paradox where stronger security ended up increasing the vulnerability to weaknesses such as denial of sleep attacks. Denial of sleep attacks attack devices' energy management protocols within industrial networks, disrupting operational effectiveness and the lifecycle of hardware as a whole. Thus, energy industries need to sharpen their methods towards cybersecurity by balancing stringent security measures with energy efficiency factors in mind. This two-way vision is essential in the creation of communication protocols that can resist cyber-attacks without impairing the operational capability of industrial systems. The Lessons Learned from all past security breaches in the energy sector also call for the deployment of an adaptive security model with periodic reviews and updates to procedures. Static approaches of the past will no longer suffice; the ever-evolving nature of cyber threats demands to have eyes always open to emerging weaknesses and have response mechanisms that are fast on their feet. National and worldwide efforts focus on hardening critical infrastructures, says Marcelo et al. (2008), but how to make these discoveries functional in a useful manner remains a challenge. Taking preventive action with secure communication protocols, energy organizations can not only secure their systems but also drive innovation through cybersecurity operations. This strategy promises to enhance resilience to subsequent breaches, making the energy sector robust and vibrant in an increasingly globalized world (Figure 5.16).

Figure 5.16. Lessons learned from breaches in energy sectors.

Source: https://www.zmanda.com/wp-content/uploads/2025/04/02-9-industries-that-experience-the-most-data-breaches-r01.jpg.

5.6.3. Comparative Analysis of Different Industries

In examining the comparative analysis of different industries, it becomes evident that the integration of secure communication protocols is paramount, particularly as the Industrial Internet of Things (IIoT) becomes increasingly prevalent. Various sectors, including manufacturing, energy, and transportation, rely heavily on interconnected devices to enhance operational efficiency. However, the unique characteristics of each industry dictate the specific security measures required. For instance, manufacturing would concentrate on real-time data communication within automated operations and would demand secure authentication and encryption solutions particular to counter potential attacks. The built-in vulnerabilities of IIoT devices, for instance, weak key management systems, contribute to this complexity, where adaptive security measures like improved Fiber Channel Password Authentication Protocol (FACP) illustrated in Gu et al. (2024) are inherently required. Effective implementation of such protocols can enhance security very significantly without compromising efficiency. The energy sector presents a special challenge to secure communication protocols due to its critical infrastructure status and increasing Information Technology (IT) - Operational Technology (OT) convergence. This growth in cyber-attacks underscores the requirement to implement advanced security mechanisms that account for the complex interdependencies of IT and OT networks. As cyber threats accumulate, there is a need for protocols that can ensure free communication while safeguarding vital operational data. Technologies such as the real-time ICS cyber test kit described in Mubarak et al. (2021) are excellent examples of how data-driven methodologies can enhance the visibility level of OT traffic and detect potential anomalies. Comparative studies of mature industries show that as industries become more developed, they require industry-specific protocols to guarantee resilience against sophisticated cyber-attacks. Secondly, the transport sector is the best example of the highest importance of security communication protocols, particularly with its reliance on ever more advanced technologies like autonomous systems and smart infrastructure. The application of artificial intelligence (AI) and machine learning (ML) in the transport networks requires the establishment of large security frameworks that are able to react to the evolving threats. Security solutions in this industry must not only ensure data communication integrity but also ensure rapid response to potential breaches, thus the minimum interference with essential services. The demand for proactive security strategy is reaffirmed through the emphasis on machine learning-based anomaly detection systems as recognized in Mubarak et al. (2021). By conducting a comparative analysis of secure communication protocols across these diverse sectors, it becomes clear that a collaborative, innovative approach is vital to fortifying the diverse networks that underpin modern industrial operations.

SUMMARY

In Summary, the implementation of secure communication protocols for industrial networks is paramount for safeguarding system integrity and operational resilience. With industrial networks increasingly based on digital technologies, they are susceptible to a variety of cyber threats that can compromise confidentiality, integrity, and availability. The transition from the traditional infrastructure to more networked, intelligent systems—such as those found in smart grids—highlights the need for general security solutions. The model developed in this essay targets a multi-layered security approach with the ultimate goal of mitigating risks through technology as well as organizational intervention. Through the difficulties posed by evolving cyber threats, organizations are able to ensure a robust defense against potential breaches and the consequent incapacitation of crucial services. The above discussion has also highlighted the critical loopholes in existing security practices, particularly in the areas of accountability and threat detection. As borne out by literature today, the majority of contemporary security solutions do not effectively address the complexity of integrated systems like Building Automation Systems (BAS) that were not designed initially based on proper security architecture. This inadequacy lends support to the updating and fortifying of the security position of industrial networks through embracing new approaches. Integration of findings from various studies suggests that upcoming work needs to focus on developing specific threats, enhancing detection mechanisms, and adapting machine learning methods for effective traffic classification. These enhancements are crucial for assuring security in resource-constrained systems and prevention against emerging threats. Lastly, the future of secure communication protocols in industry networks will call for industry, academia, and government collaboration to develop end-to-end solutions that are effective and dynamic in responding to evolving environments. Because security vulnerability not only has the potential to wreak havoc on the economy but also in public safety and the environment, an integrated effort is a necessity. Projects such as those directed toward research on weaknesses and impacts of cyber-attacks on smart grids, while simultaneously addressing solutions to accountability, are pivotal in fortifying such systems. By ongoing improvement and vigilance in security culture, industries can better prepare for the different dimensions of cyber threats, which will provide stability and reliability to critical infrastructure in the future.

In the study of the terrain of secure communication protocols for industrial networks, there are several key points that highlight the severity and intricacy of the challenges involved. As industrial networks become more advanced in terms of digital communications and control technologies, particularly in areas such as energy management, the vulnerabilities surrounding these developments become even more acute. A fundamental comprehension of the security requirements is a necessity; these are not just the traditional trio of confidentiality, integrity, and availability, but also measures such as accountability and responsiveness to different forms of cyberattacks. The nature of industrial communication networks necessitates an overall strategy and not ad-hoc measures that counter only specific threats. The literature identifies the inadequacy of current approaches, which fail to provide comprehensive protection of the entire network form, making industry participants reason along more holistic security frameworks (Ghazi et al., 2018). Moreover, the shift towards smart grid systems demonstrates the necessity of secure communications infrastructures that place maximum emphasis on security. In conventional power grids, the lack of advanced communication paths may lead to severe threats, particularly with more interface with external networks. Smart grid cybersecurity threats have been researched extensively; surprisingly, the majority of existing work defines threats primarily in terms of their impact on confidentiality, with essential vulnerabilities in accountability and system resilience yet to be fully studied. The fundamental weaknesses of suggested communication technologies necessitate a need for immediate security system upgrading, which enhances reliability as well as protects against potential economic losses and production interference. Implementation of new technologies is not supposed to undermine security, and there ought to be a reassessment of existing protocols as well as initiation of new approaches to protect industrial networks from destructive attacks (Qian et al., 2012). Finally, overall conclusions derived from recent literature point towards a strategic change towards an integrated security model. This involves the development of integrated cyber security strategies that aim not just at near-term industrial network threats but also provide resilience against emerging vulnerabilities in the future. Chief recommendations are the integration of several security solutions into one potent package that involves incorporating continuous monitoring, advanced authentication processes, and active incident response plans. In addition, involving stakeholders from all walks of life in the security debate can help come up with solutions that are both pragmatic and effective. Next-generation research must be focused on creating evidence-based security to ensure that the measures put in place are universally acceptable across industries and are able to ensure the security and integrity of electronic communications against increasingly changing cyber threats (Ghazi et al., 2018).

REFERENCES

1. Abdolee, R., Maroufi, M., & Tazekand, B. M. (2019). On the convergence of blockchain and Internet of Things (IoT) technologies. http://arxiv.org/abs/1904.01936 (accessed on 13th September 2025).
2. Alohali, B. A., Hurst, W., Kifayat, K., & Shi, W. (2025). A survey on cryptography key management schemes for smart grid. https://core.ac.uk/download/42478732.pdf (accessed on 13th September 2025).
3. Apikul, W., Cecchini, S., Clark, M., Cole, M., Crondstedt, M., Gan, D., Loukas, G., et al. (2013). A review of cyber threats and defense approaches in emergency management. https://core.ac.uk/download/42389827.pdf (accessed on 13th September 2025).
4. Bonet Gil, E., Koti, J., & Ugarelli, R. (2018). STOP-IT: Strategic, tactical, operational protection of water infrastructure against cyberphysical threats. https://core.ac.uk/download/224975509.pdf (accessed on 13th September 2025).
5. Brunstrom, A., Damjanovic, D., Fairhurst, G., Grinnemo, K. J., et al. (2016). De-ossifying the Internet transport layer: A survey and future perspectives. https://core.ac.uk/download/80685935.pdf (accessed on 13th September 2025).
6. Carpio, F., Dizdarevic, J., Jukan, A., Masip Bruin, X., et al. (2019). A survey of communication protocols for Internet of Things and related challenges of fog and cloud computing integration. https://core.ac.uk/download/187266039.pdf (accessed on 13th September 2025).
7. Chin, W. H., Efthymiou, C., Fan, Z., Gormus, S., et al. (2011). Smart grid communications: Overview of research challenges, solutions, and standardization activities. http://arxiv.org/abs/1112.3516 (accessed on 13th September 2025).
8. Copeland, R., Crespi, N., Rehmani, M. H., Saleem, Y., et al. (2019). Internet of Things-aided smart grid: Technologies, architectures, applications, prototypes, and future research directions. http://arxiv.org/abs/1704.08977 (accessed on 13th September 2025).
9. Curty, M., Lo, H. K., & Tamaki, K. (2014). Secure quantum key distribution. http://arxiv.org/abs/1505.05303 (accessed on 13th September 2025).
10. Dang, T. K., & Tran, K. T. K. (2019). The meeting of acquaintances: A cost-efficient authentication scheme for light-weight objects with transient trust level and plurality approach. http://arxiv.org/abs/1903.10018 (accessed on 13th September 2025).

11. De Cock, M., Dowsley, R., Fritchman, K., Hughes, J., et al. (2018). Privacy-preserving scoring of tree ensembles: A novel framework for AI in healthcare. https://core.ac.uk/download/237010808.pdf (accessed on 13th September 2025).

12. Evangelopoulou, M., Johnson, C. W., & Pavlova, T. (2017). Applying lessons from cyber-attacks on Ukrainian infrastructures to secure gateways onto the Industrial Internet of Things. https://core.ac.uk/download/111557942.pdf (accessed on 13th September 2025).

13. Farahbakhsh, R., Farahzadia, A., Rezazadeh, J., Shams, Z., et al. (2017). Middleware technologies for cloud of things–A survey. http://arxiv.org/abs/1705.00387 (accessed on 13th September 2025).

14. Ford, B., Kegel, D., & Srisuresh, P. (2005). Peer-to-peer communication across network address translators. http://arxiv.org/abs/cs/0603074 (accessed on 13th September 2025).

15. Frey, M., Gündoğan, C., Kietzmann, P., Lenders, V., et al. (2018). Security for the industrial IoT: The case for information-centric networking. http://arxiv.org/abs/1810.04645 (accessed on 13th September 2025).

16. Ghazi, H. E., Ghazi, H. E., Kaabouch, N., Mrabet, Z., et al. (2018). Cyber-security in smart grid: Survey and challenges. http://arxiv.org/abs/1809.02609 (accessed on 13th September 2025).

17. Gu, H., Shang, J., Wang, P., Mi, J., & Bhattacharjya, A. (2024). A secure protocol authentication method based on the strand space model for blockchain-based industrial Internet of Things. Symmetry. https://www.semanticscholar.org/paper/6c7b2f21aede1ceafacb607ca6e0e7e0400f9617 (accessed on 13th September 2025).

18. Hauser, C. H., Lorenz, D. H., Fudenberg, D., Santacana, E., Kuipers, F., Wu, F. F., Xue, G., et al. (2018). Multi-layer cyber-physical security and resilience for smart grid. http://arxiv.org/abs/1810.00282 (accessed on 13th September 2025).

19. Hong, G. Y., James, H., & Ryu, H. (2006). Quality assessment technique for ubiquitous software and middleware. https://core.ac.uk/download/148639200.pdf (accessed on 13th September 2025).

20. Johnstone, M. N., & Peacock, M. D. (2014). An analysis of security issues in building automation systems. https://core.ac.uk/download/41537512.pdf (accessed on 13th September 2025).

21. LeBlanc, N. (2020). An analysis and enumeration of the blockchain and future implications. https://core.ac.uk/download/322850883.pdf (accessed on 13th September 2025).

22. Licoppe, C., & Benghozi, P. J. (2025). Technological national learning in France: From Minitel to Internet. https://core.ac.uk/download/pdf/6801514.pdf (accessed on 13th September 2025).

23. Marksteiner, S., & Vallant, H. (2017). Towards a secure smart grid storage communications gateway. http://arxiv.org/abs/1710.03017 (accessed on 13th September 2025).

24. Masera, M., & Vamanu, B. (2008). Understanding malicious attacks against infrastructures–Overview on the assessment and management of threats and attacks to industrial control systems. https://core.ac.uk/download/38619766.pdf (accessed on 13th September 2025).

25. Masera, M., Fovino, I. N., & Vamanu, B. (2011). ICT aspects of power systems and their security. https://core.ac.uk/download/38621151.pdf (accessed on 13th September 2025).

26. Mubarak, S., Habaebi, M. H., Islam, M. R., & Khan, S. (2021). ICS cyber-attack detection with ensemble machine learning and DPI using Cyber-kit datasets. 2021 8th International Conference on Computer and Communication Engineering (ICCCE). https://www.semanticscholar.org/paper/ae8cd1dd8ed6f6e29b52fb2ca4417fcccf1687fb (accessed on 13th September 2025).

27. Qian, Y., Sharif, H., Tipper, D., Yan, Y., et al. (2012). A survey on cyber security for smart grid communications. https://core.ac.uk/download/12211131.pdf (accessed on 13th September 2025).

28. Sussmann, M. A. (1999). The critical challenges from international high-tech and computer-related crime at the millennium. https://core.ac.uk/download/62547637.pdf (accessed on 13th September 2025).

29. Udoh, E., & Udoh, E. (2019). An energy aware and secure MAC protocol for tackling denial of sleep attacks in wireless sensor networks. https://core.ac.uk/download/305119264.pdf (accessed on 13th September 2025).

CHAPTER 6

INTRUSION DETECTION AND PREVENTION IN ICS

Contents

INTRODUCTION

The world of Industrial Control Systems (ICS) is increasingly a subject of cybersecurity concerns since these systems are a central element of the functioning of critical infrastructure, including power plants, water treatment facilities, and factories. Previously, the security frameworks surrounding ICS have been weak, primarily due to the support of aged legacy systems that cannot be readily brought up to modern security levels. This vulnerability has a risk of a specific extent, as intruders are going to employ these outdated infrastructures to inject malicious code or trigger paralyzing attacks. Security frameworks that are specifically implemented for ICS must manage special challenges like demands of real-time operational functioning and no ability to introduce changes in existing systems without compromising operational processes. Thus, establishing advanced intrusion detection and prevention mechanisms is of paramount importance while safeguarding such critical systems from exploitation. An important aspect of ICS security pertains to the detection and handling of some threats in the guise of new malware, i.e., ladder logic bombs (LLBs). The attacks can be embedded covertly into the control logic of Programmable Logic Controllers (PLCs), i.e., central units utilized in industrial process control. LLBs have the potential to alter legitimate sensor readings or influence actions based on corrupted data, undermining the integrity and functioning of the systems overall. The emergence of such sophisticated threats calls for a strong call for good detection strategies that can detect unusual behavior in ICS environments. With increasing sophistication of modern-day cyber threats, countermeasures that are able to monitor such emerging types of malware without interfering with operational processes are necessary for system reliability and safety in Realtime systems (Kim et al., 2017). Striking a blow at these security threats, the creation of behavioral analysis tools from electromagnetic side-channel measurements brings to the forefront the potential for new detection mechanisms. By profiling the behavior of PLC-software, security can identify even slight disturbances introduced by rogue changes. What is seen here is the new era of cybersecurity in which traditional defenses must be innovative in keeping pace with the adversaries who ceaselessly update their methods. The solutions proposed aim not merely to enhance the resilience of ICS but also effectively lower operation risks. Since companies increasingly rely on networked systems for automation and control, the integration of sound intrusion detection and prevention methods will be essential to ensuring the integrity and longevities of infrastructures (Beckers et al., 2017).

Definition of Industrial Control Systems (ICS)

ICS play a central function in controlling and automating infrastructure in such industries as manufacturing, energy, water treatment, and transport. These systems are a mix of different devices and software designed to monitor and regulate physical processes in order to provide their optimum performance and efficiency. ICS typically include devices such as Supervisory Control and Data Acquisition (SCADA) systems, Distributed Control Systems (DCS), and Programmable Logic Controller (PLC). As industrial systems increasingly integrate with information technology (IT) and operational technology

(OT), having insight into the architecture and workings of ICS is important. Where vulnerabilities are likely to arise is where the convergence of these technologies happens, leading to potential cyber-attack threats. These threats mandate that organizations put in place robust intrusion prevention and detection measures capable of defending these important systems so that they continue to operate and remain intact. ICS has advanced, integrating cutting-edge technologies such as the Internet of Things (IoT) and data analytics to enhance the capability of these systems. By leveraging real-time data collection and processing, business organizations can monitor system performance and respond to anomalies effectively. Increased connectivity and reliance on digital infrastructure, however, heighten exposure to cyber-attacks. Sophisticated persistent threats (APTs) in such settings are particularly alarming since they deploy sophisticated methods for entering and hijacking industrial networks. The importance of developing and implementing effective monitoring systems is highlighted by the priority of maintaining these critical infrastructures in an operational capacity, as pointed out in recent research on intrusion detection methods. A well-established understanding of ICS and its elements makes security responses and incident response more effective, allowing organizations to defend their systems from potential intrusions (Lemay et al., 2018). Moreover, as cities tap into the power of smart technologies to upgrade life in the urban landscape, the role of ICS in areas such as energy management becomes even more important. Power grids in current times are relying more and more on advanced information and communication technologies (ICT) to facilitate perfect energy generation, transmission, and usage patterns. By embedding such technologies, ICS can fine-tune performance areas such as stability of electricity and improved security in communication. Demand response paradigms reflect the growing trend towards a more interactive approach to energy management, with active prosumer interaction bringing greater efficiency in load balancing and peak shaving. Such innovations, while excellent in themselves, point to the necessity of good intrusion detection systems that can detect and respond to security threats that can benefit from these technological advances. Therefore, proper understanding of ICS, combined with effective cybersecurity measures, is crucial to ensure the continued success and sustainability of modern industrial infrastructures (Bonetto et al., 2017).

Importance of Cybersecurity in Industrial Control Systems (ICS)

Increased reliance on ICS in the management of critical infrastructure heightens the need for robust cybersecurity measures. Because ICS has numerous diverse uses, including power generation to water supply, they are increasingly networked together using Internet Protocol (IP) networks.

Although the interconnectivity, while providing more operational efficiencies and remote management capabilities, also exposes the systems to more cyber vulnerabilities. A prime example was the Ukraine power grid cyber-attack, which demonstrated the manner in which SCADA system vulnerabilities may lead to massive service outages. The consequences of such violations can extend beyond simple outages, posing threats to public safety and national security. Organizations should therefore place highest priority on embracing full cybersecurity frameworks that secure such critical systems from malicious exploitation and ensure reliability and safety of operations (Ginter et al., 2018). Securing ICS is vital, mandating awareness of the technological backdrop as well as future threats. Cybersecurity measures will need to evolve along with the increased sophistication of cyberattacks against critical infrastructure sectors. This challenge is heightened by the shift in paradigm from isolated to networked systems, as exposed by the vulnerabilities introduced by the interfacing of SCADA systems into broader networks. Such vulnerabilities carry significant implications, including economic loss and compromised public safety. Application of an efficient intrusion detection and prevention system (IDPS) has the capability of detecting and preventing threats early enough before they escalate into full-blown incidents. This is in a sense attacker's mentality because it allows organizations to find vulnerabilities that will be exploited and hence strengthen their defense against emerging threats (Almalawi et al., 2017). Furthermore, cybersecurity awareness culture is important in ICS environments. All categories of employees must be educated on the potential risks that cyber-attacks pose and the value of their intervention in system integrity. Regular training workshops and simulation can provide employees with appropriate responses when a cyber-attack occurs, which is essential owing to the human element that is most likely compromised in cyber-attacks. Systemic examination of potential attack vectors and impacts helps to develop a forward-thinking security culture beyond technical security measures. As systems grow more complex, an organizational sense of commitment toward cybersecurity will be essential to risk management. This collaborative effort is paramount to ensuring operational continuity of vital infrastructures and ultimately ensuring the safety and security of communities reliant on the systems (Ginter et al., 2018).

Overview of Intrusion Detection and Prevention Systems (IDPS)

IDPS is a major support in protecting ICS by monitoring traffic in the network for inappropriate activity. An IDPS is designed to identify and respond to probable intrusions by scanning data packets that pass through the network. These two important features are incorporated in this process: intrusion detection, which identifies malicious activity, and intrusion prevention, which actively prevents such activity from passing through before causing harm. In industrial settings, where background interruptions can be so consequential, IDPS integration is a prerequisite. IDPS is necessary because ICS is vulnerable to all sorts of cyber-attacks, including the tampering of control logic, which has the capability to modify operation procedures, as seen in recent malware such as ladder logic bombs (Kim et al., 2017). Effective deployment of IDPS allows organizations to identify threats in real-time, thereby minimizing risks of intrusions. The sophistication of IDPS processes extends beyond detection to encompass prevention processes that prevent critical industrial processes from exposure to vulnerabilities.

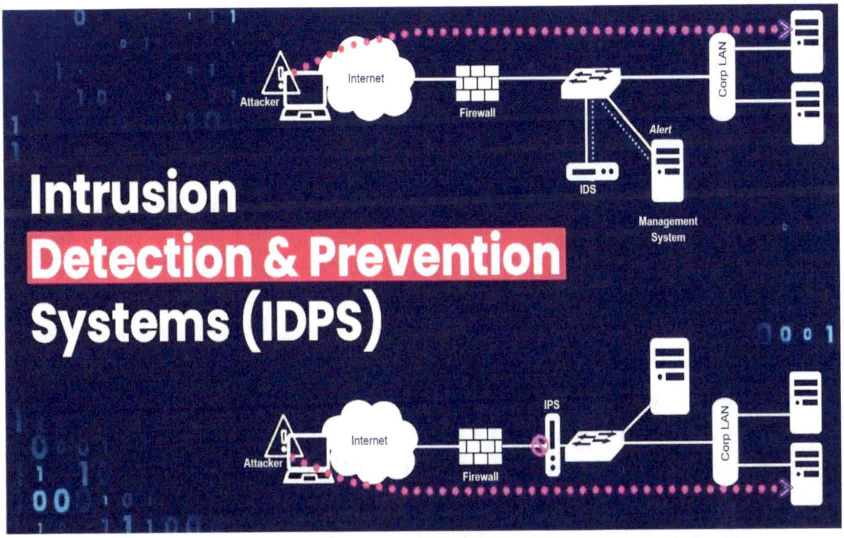

Figure 6.1. The intrusion detection and prevention systems (IDPS).

Source: https://www.goallsecure.com/wp-content/uploads/2024/05/Intrusion-Detection-and-Prevention-System-1024x576.webp.

Because the operating environment within ICS is different, IDPS must be tuned to proprietary protocols and controls used in the system. Legacy network-based intrusion detection systems do not necessarily work well to capture the complexities of ICS, which are based on protocols that are highly susceptible to attacks and tampering. For example, as research reports, industrial protocols have been modified for internet transport, making ICS more vulnerable to external attacks (Nawrocki et al., 2019). Therefore, developing an appropriate IDPS for ICS requires tailored approaches based on these various factors. This may involve establishing detection rules and alert thresholds that match acceptable operational behavior in defining anomalous activities that may indicate an intrusion. Furthermore, the evolving nature of cyber threats necessitates IDPS in ICS to not only respond to known threats but also in accommodating emerging threats. The dynamic nature of cyberattacks means that there are always new intrusion methods being developed, so IDPS has to be ever vigilant and in up-to-date status. With cutting-edge attack vectors such as those against inter-domain communication of ICS becoming a commonality, systems have to utilize cutting-edge analytical means, such as machine learning, to enhance their ability to identify threats (Nawrocki et al., 2019). In addition, enhancing the IDPS entails the implementation of layered security controls coupled with the incorporation of incident response mechanisms capable of quickly containing and responding to potential intrusions. Lastly, a robust IDPS serves as an important security line of defense in the protection of industrial environments against both existing and emerging cyber threats, against operation disruption, and for safety within these critical systems (Figure 6.1).

6.1. TYPES OF INTRUSION DETECTION SYSTEMS

Intrusion detection system (IDS) classification is mostly in two broad categories: host-based intrusion detection systems (HIDS) and network-based intrusion detection systems (NIDS). HIDS executes on individual host computers, monitoring system calls, log files, and application activity for evidences of suspiciousness or unauthorized action. This local orientation gives a close-up view of possible threats unique to a host's configuration and vulnerabilities. In contrast, NIDS considers traffic between multiple devices, and it is therefore feasible to identify odd patterns or signatures that indicate a breach. This system is particularly valuable where there is high data flow, as it has an easy time tracking coordinated attacks that would otherwise be impossible to detect when considered at a host level. For industrial control systems (ICS) specifically, dedicated IDS are needed due to the unique operations procedures and equipment involved. One example of the process is anomaly-based detection, which establishes a standard of what is typical for the system and identifies any variations from this norm. This strategy can identify sophisticated threats, such as those that include ladder logic bombs to change control logic in PLCs, as addressed in Kim et al. (2017). Besides, signature-based detection remains effective, employing a listing of known attack signatures to look for familiar attacks. This is particularly helpful in the ICS setting, wherein some industrial protocols may be subject to extensively documented vulnerabilities, which can be targeted with timely and appropriate countermeasures. Utilization of intrusion detection systems in ICS settings is also influenced by communication protocols used in these settings. ICS normally operate using proprietary, less secure protocols that are vulnerable to eavesdropping and tampering, and hence strong detection is critical, as noted by Nawrocki et al. (2019). The integration of IDS with existing industrial protocols provides real-time monitoring and threat detection that is critical in maintaining the integrity and security of critical infrastructure. With increasingly sophisticated cyber threats, ongoing enhancement of advanced detection methods, including the use of machine learning and monitoring of behavior, will be key to enhancing ICS resistance to intrusions and ensuring an environment for secure operation.

6.1.1. Network-based Intrusion Detection Systems (NIDS)

In Industrial Control Systems (ICS) contexts, NIDS are an essential first line of defense against increasingly sophisticated cyber threats. These systems monitor network traffic and follow unusual patterns or trends, enabling intrusions that may impair operational integrity to be detected. Unlike host-based systems, which are directed at specific devices, NIDS monitor all network communication, providing a broader picture of potential vulnerabilities. As SCADA systems and other ICS designs are more and more interconnected, the need for diligent monitoring of networks increases exponentially, particularly since such systems tend to be based on legacy protocols, which are by their nature insecure. Empirical evidence suggests that intrusion detection mechanisms must evolve in order to successfully counter session hijacking and other network-based vulnerabilities, ensuring the urgent need for robust detection techniques in protecting critical infrastructure (Kleinmann et al., 2017). The effectiveness of NIDS is based on advanced algorithms and analysis techniques that can differentiate between legitimate traffic and malicious traffic. Anomaly-based detection, which seeks deviations from typical traffic patterns, has become the primary approach used in NIDS. The technique is particularly effective in ICS environments, where attackers may employ tactics impervious to conventional signature-based techniques. For instance, sophisticated semantic network-based attacks are capable of manipulating Programmable Logic Controllers (PLCs) and Human Machine Interfaces (HMIs) communication, rendering most conventional detection means useless (Kleinmann et al., 2017). Metadata analysis, such as command sequence and timing, is required to identify such minute forms of attack. Further, integration of machine learning techniques widens the scope for adaptively fine-tuning detection capability to enhance overall system resiliency against emerging threats. Furthermore, with the working environment of ICS more and more making use of Internet of Things (IoT) devices, NIDS's function is even more crucial. These devices introduce a plethora of potential intrusion entry points, which means having a comprehensive framework for detection and monitoring becomes necessary. Existing models have embraced two-layer dimensionality reduction and two-tier classification methods to enhance detection efficacy on high-dimensional data sets, hence enabling detection of multiple attack vectors, including Remote to Local (R2L) and User to Root (U2R) attacks. Through the application of such advanced analytic techniques in NIDS, organizations are now able to build a robust defense system against several threats. In the end, continuous development of NIDS will become even more essential in the continuous protection of ICS from cyber-attacks and ensuring the reliability of critical infrastructure operations (Figure 6.2).

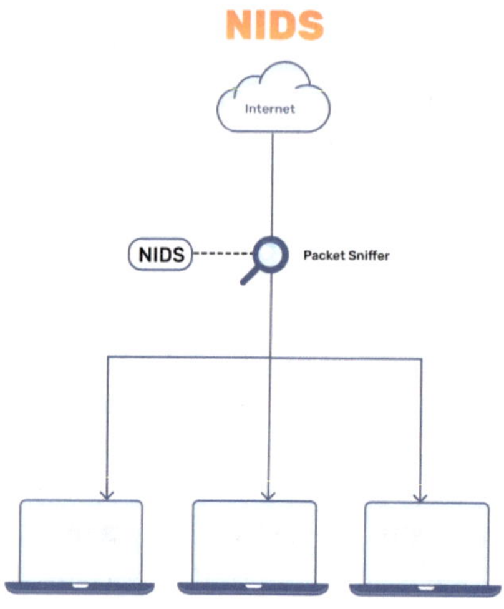

Figure 6.2. Network-based intrusion detection systems (NIDS).

Source: https://bunnyacademy.b-cdn.net/what-is-network-intrusion-detection-nids.svg.

6.1.2. Host-based Intrusion Detection Systems (HIDS)

In industrial control systems (ICS), HIDS is an extremely significant line of defense against increasingly numerous network threats. HIDS operate by monitoring and scanning the internal condition of hosts such as servers, workstations, and network appliances. Compared to their network-based counterparts, HIDS are particularly well-suited at identifying malicious behavior that may be originating from within the system itself and are thus of the utmost importance in environments where sensitive data and operation integrity are of the highest priority. With the use of signature-based detection and anomaly-based identification techniques, HIDS can raise alerts for intrusions, file tampering, and other such malicious attacks jeopardizing the security of ICS environments. Proactiveness is here crucial, especially in the backdrop of increasing vulnerabilities intrinsic to inter-domain ICS traffic referenced in previous studies, underlining the merit of internal monitoring. Moreover, improvements in communication protocols applied to ICS have further complicated things for security, and therefore HIDS becomes even more critical. According to the literature, ICS traffic mostly traverses insecure networks and the Internet, making it more susceptible to threats such as Distributed Reflective Denial of Service (DRDoS) (Nawrocki et al., 2019). Therefore, HIDS must be installed to not only detect the existence of external attacks but also neutralize the danger of less secure protocols. By analyzing host behavior and configurations in real time, HIDS is able to clearly differentiate between normal changes in behavior and potential threats. This is a very critical feature given the results that focus on unguarded inter-domain traffic in ICS. Hence, the role of HIDS is not only detection but also broader preventive controls for safeguarding ICS environments against both internal and external threats. Additionally, designing effective HIDS in ICS requires critical insight into the range of network threats present in the modern digital environment. Observed is that developing a taxonomy for classification of network attacks can be valuable in enhancing the design of specifically designed intrusion detection systems and datasets (Atkinson et al., 2018). Such a methodical approach enables security researchers to focus efforts on refining HIDS techniques that effectively respond to targeted attack vectors typical for ICS. If the listed categorization system is utilized, organizations can have confidence that their HIDS are not only responsive but also capable of adapting to emerging trends in threats. Lastly, integrating robust HIDS into ICS enhances overall security design by offering sophisticated visibility into host-level activity, thereby making the shield robust against increasingly advanced cyber-attacks. While the intersection of cyber threats and industrial processes gets more intense, the importance of HIDS in constructing secure security frameworks cannot be overstated.

6.1.3. Hybrid Intrusion Detection Systems (HIDS)

HIDS is a critical solution in enhancing security for Industrial Control Systems (ICS). By taking advantage of the advantages of signature-based detection and anomaly detection methods, HIDS provide a robust infrastructure to identify known threats as well as support emerging, advanced types of attack vectors. Through this dual-pathed methodology, not only can it quickly react to known intrusion patterns but also identify deviations from routine working patterns. In an ICS ecosystem, data availability and integrity are paramount in real-time; thus, HIDS play a crucial role in safeguarding systems against external and internal threats. Additionally, their adaptability bridges the chasm between traditional IT security controls and the unique requirements of operational technology, giving a general security posture aligned to the unique needs of ICS environments. Incorporating advanced machine learning methods into HIDS greatly improves their detection features. The machine learning algorithms can scan enormous amounts of operational data to learn typical system behavior and then detect anomalies indicative of possible intrusions.

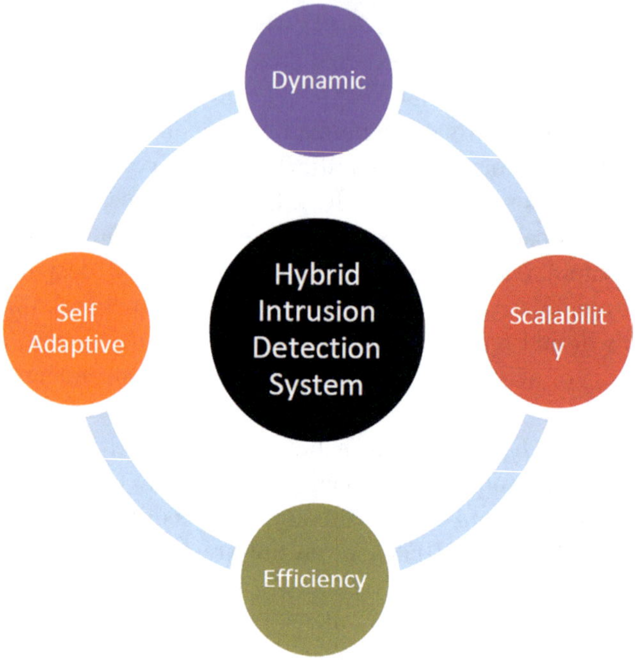

Figure 6.3. Hybrid intrusion detection systems (HIDS).

Source: https://www.researchgate.net/profile/Praveen-Kumar-Rajendran/publication/277949603/figure/fig1/AS:391790973734918@1470421655479/Outline-of-Hybrid-Intrusion-Detection-System.png.

With the development of bringing intelligence to intrusion detection, the feature of HIDS to dynamically adjust their detection thresholds depending on network traffic patterns and contextual information becomes critical. Such intelligent evolution enables HIDS to function efficiently even while attackers develop more sophisticated methods of ICS environment degradation. By virtue of application of these sophisticated algorithms, HIDS not only reduce false positives but also improve the overall reliability of attack detection, and therefore they constitute a foundation of ICS cybersecurity practices in modern times. Besides the above, HIDS in ICS is also enhanced by a multi-layered security approach that considers the particular challenges that happen in these systems.

Advanced infrastructure like the smart grid has intricate interactions involving several components, which could be described using frameworks like game and decision theory. These frameworks help to understand potential adversary actions against the system and thus inform more effective HIDS strategies. By incorporating game-theoretic principles, HIDS can strategize counterattacks to threats and adapt based on anticipated attacker maneuvers. This proactive strategy is pivotal to guaranteeing the dependability and trustworthiness of ICS, in accordance with recent studies that emphasize the necessity of holistic security mechanisms that properly address multi-layer threats (Hauser et al., 2018). In this sense, the future development of HIDS will likely be geared towards enhancing flexibility and robustness against increasingly advanced attack scenarios (Figure 6.3).

6.2. INTRUSION PREVENTION SYSTEMS IN INDUSTRIAL CONTROL SYSTEMS (ICS)

The rapid evolution of ICS reveals a sheer vulnerability profile, bringing into context a compelling need for a robust intrusion prevention system. With persistent threats online, traditional security measures often fail to address the unique needs of ICS networks, which support mission-critical infrastructure operations. An overarching effort through advanced intrusion detection and prevention systems (IDPS) is thus mandated. These systems need to surpass conventional methods by using advanced mechanisms that are capable of observing and analyzing the intricate behavior of the elements of ICS. For instance, MAHIVE: Modular Analysis Hierarchical IDS Visualization Event cybersecurity engine is an advanced solution purpose-built for CPS and ICS environments, indicating a need for developing technologies in sync with the unique environment of industrial networks (Leon et al., 2021). One of the biggest barriers to efficient intrusion prevention system deployment in ICS is the legacy infrastructure that many organizations continue to employ. The inherent challenge of incorporating more recent systems and modern security devices into the older ones can create a significant deficiency in coverage, and the risk of successful cyber-attack increases. Even as the vendors attempt to enhance the security of newer systems, the existing installations persist in remaining vulnerable due to their legacy topology. A proactive strategy is needed to this setting, as evidenced by the use of electromagnetic side-channel analysis—a revolutionary technique that can facilitate detection of behavioral irregularities in software used on ICS platforms like Siemens S7–317 PLCs (Beckers et al., 2017). This new strategy holds much potential in establishing a robust impenetrable barrier against incursions where conventional revamps to security models cannot be achieved. The deployment of advanced intrusion prevention systems within ICS environments is not only about minimizing the cybersecurity risks; it is a paradigm shift towards more secure and dynamic infrastructures. With the implementation of IoT technologies within the industrial scenario, the attack surface expands, and an enhanced security posture is called for. Organizations must embrace a culture of continuous observation and enhancement, with real-time analytics to proactively identify and eliminate threats prior to their ability to cause significant disruption. Furthermore, co-operation between industry participants is essential in the development of harmonized methodologies for ICS cybersecurity, encouraging sharing of best practice and emerging technology. As organizations grapple with these complex challenges, it is evident that an exclusive focus on intrusion prevention is not only advisable but imperative to protect critical infrastructure (Leon et al., 2021).

6.2.1. Definition and Functionality of Intrusion Prevention Systems (IPS)

IPS is a vital component in the realm of cybersecurity, particularly in industrial control systems security (ICS). Situated as network security devices that scan and analyze network traffic to steer clear of potential attacks, IPS technology operates in real-time to identify malicious activity prior to its ability to do harm. With various methods of detection, such as signature-based detection, anomaly detection, and stateful protocol examination, IPS can effectively identify and disarm threats like malware or intrusion attacks. This aspect is vital for ICS installations, which are recurrently hit as they are networked and based on a number of networked devices, thus vulnerable to attacks that could compromise critical operations. Incorporating IPS into ICS provides an active protection framework. Unlike Intrusion Detection Systems (IDS), which merely detect and notify users of potential attack, IPS takes it a notch higher by actually blocking such attacks based on preconfigured security policies. For instance, in an ICS environment where Programmable Logic Controllers (PLCs) control machinery, an IPS can prevent an attacker from injecting malware, e.g., ladder logic bombs, into the control logic of the PLCs. By stopping malware before it can affect operational integrity, IPS protects both the physical infrastructure and integrity of data necessary for maintaining production processes. The operation of IPS thus significantly enhances the security position of industrial environments. Furthermore, IPS's role in ICS is not merely prevention of threats but also continuous monitoring and analysis to adapt to evolving attack paths. Successful IPS products integrate wide-ranging logging and reporting functionality, extremely important for incident post-mortems and compliance with regulations. By constant learning of network traffic patterns, these systems can dynamically refresh their detection rules in response to new forms of cyber-attacks. In those environments employing advanced control logic, as may be found in those controlled to international standards like IEC 61131–3, incapacitating the intruding user becomes increasingly significant. Therefore, as seen in arguments regarding ladder logic bombs and their evasiveness, effective IPS solutions are an essential protective measure against increasingly sophisticated attacks on industrial systems (Alzubi et al., 2015; Kim et al., 2017) (Figure 6.4).

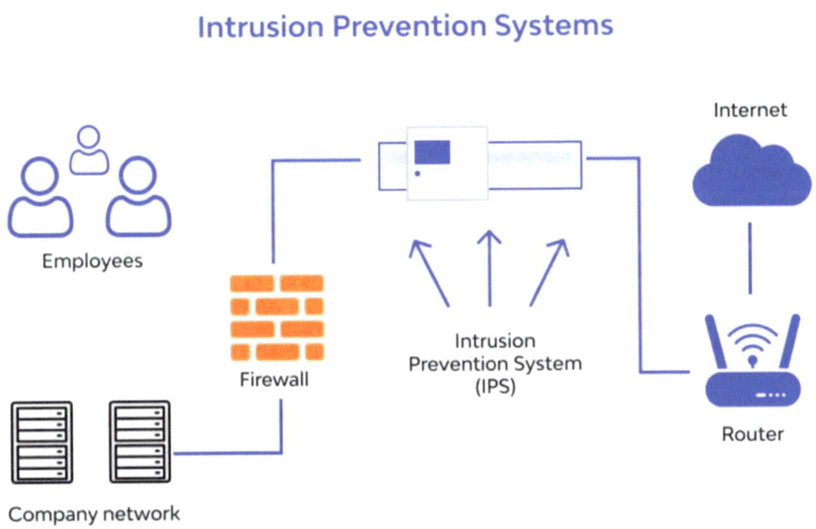

Figure 6.4. Intrusion prevention systems (IPS).

Source: https://cdn.prod.website-files.com/5ff66329429d880392f6cba2/67b4315cd25fd67a4c137911_623d912f786507381fb32bca_IPS%2520at%2520work.jpeg.

6.2.2. Comparison between Intrusion Detection Systems (IDS) and Intrusion Prevention Systems (IPS)

In the world of cybersecurity, specifically in Industrial Control Systems (ICS), the difference between IDS and IPS is pivotal in creating effective security frameworks. IDS is mostly used as a monitoring tool, intended to detect suspicious events by monitoring network traffic and sending notices to administrators when possible intrusions are detected. IDS remains focused on detection and thus offers important information regarding security incidents so that it can be thoroughly investigated after the fact. This reactive method can be particularly important in ICS settings, where an understanding of what type of intrusion has happened is as valuable as preventing it from occurring. Thus, while IDS does bring some insight and analysis, it does not prevent attacks from occurring but enables organizations to improve their defenses by analyzing incidents that have been identified, making it an important component in a comprehensive security strategy. Unlike that, IPS goes one step further by actively blocking the intrusions in real-time. IDS merely alerts on threat detection, whereas IPS instantly takes corrective action to block or restrict malicious activities from happening before they inflict damage. This is especially important for ICS because continuity of operation and security of sensitive systems are both threatened by even low-level security incidents. As the sophistication of cyber-attacks, particularly for interconnected IoT devices in ICS, continues to grow, there is an imperative need for a robust IPS solution. Not only does the solution detect anomalies, but also de-escalates risks using preventative measures, thereby making the defense an effective layers of defense against disruptions as a result of cyber-attacks, as the case of security vulnerabilities in Ahmadinia et al. (2016) argues. But it is also necessary to consider that IDS and IPS are not conflicting but complementary components of an integrated cybersecurity strategy in ICS. The integration of both systems can result in a harmonized environment where the monitoring ability of IDS is supported by the operational effectiveness of IPS. Through analysis of alerts generated by an IDS, an IPS can adapt its action to emerging threats and create a more dynamic and responsive model for managing cybersecurity threats. Further, the hybrid paradigm can support more flexibility in managing the vast range of security problems in ICS environments, particularly in light of existing vulnerabilities in legacy systems, as highlighted in Beckers et al. (2017). In short, by exploiting the strengths of IDS and IPS individually, organizations can build a complete defense mechanism that not only alerts and reacts to an attack but also evolves with the dynamics of the cybersecurity environment.

6.2.3. Role of Intrusion Prevention Systems (IPS) in Mitigating Threats to Industrial Control Systems (ICS)

The integration of IPS into ICS is a critical advancement in cybersecurity, particularly as digital technology increasingly permeates industrial environments. IPS is essentially a first-line defense mechanism that actively inspects network traffic for harmful activity and known vulnerabilities. By analyzing data packets in real-time, IPS can identify potential intrusions before compromising system integrity. This looking-ahead approach is important in ICS, where operation continuity and safety should be maintained. Furthermore, the responsive nature of IPS solutions allows them to respond fast to changing threats, with security measures keeping up with emerging vulnerabilities that the growth in smart systems and networked devices presents, as depicted in Radoglou-Grammatikis (2023). This capacity to immediately act significantly enhances the cyber-resilience of ICS, hence safeguarding critical services. Moreover, the integration of artificial intelligence (AI) in IPS designs enhances them to better detect sophisticated cyber-attacks targeted at ICS systems. With AI technology, the IPS can perform complex anomaly detection and pattern matching, which is essential in detecting and mitigating risk from the Industrial Internet of Things (IIoT). As Radoglou-Grammatikis (2023) explains, the integration of AI into IPS not only improves the effectiveness of threat detection but also enables sounder decision-making regarding incident responses.

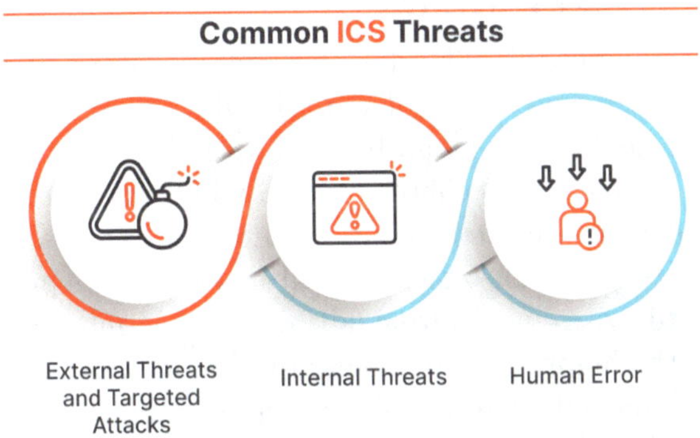

Figure 6.5. Role of intrusion prevention systems (IPS) in mitigating threats to industrial control systems (ICS).

Source: https://www.fortinet.com/content/dam/fortinet/images/cyberglossary/ics-threats.jpg.

Knowledge gained through AI enables IPS to learn from past attacks and adjust its parameters accordingly, making it an adaptive defense system that can refine itself with threats' evolving nature. This added benefit of enhanced detection accuracy and adaptative response ability is invaluable to maintaining the integrity and functionality of ICS, particularly in sectors such as energy, manufacturing, and transport, where security failure will have disastrous effects. Lastly, the role of IPS extends beyond the detection and response to threats; it encompasses the end-to-end protection of critical infrastructure assets in ICS environments. Efficient IPS solutions play a key role in facilitating industry regulations and standards compliance, which are presently more cybersecurity-oriented. As the aviation sector showcases, where there is an emphasis on cybersecurity measures for operational integrity, the same rules apply to ICS in any given sector as reported in. Introducing resilient cybersecurity measures through IPS deployment not only safeguards against cyber-attacks but also raises organizational assurance of operational resilience. Furthermore, real-time monitoring and analysis facilitated by IPS allow organizations to share relevant information about the most current threats, promoting an ethos of cooperation and risk prevention. Such strategic alignment of cybersecurity operations in ICS ultimately leads to a secure and reliable operational environment, vital for protecting both physical resources and stakeholder confidence (Figure 6.5).

6.3. THREAT LANDSCAPE FOR INDUSTRIAL CONTROL SYSTEMS (ICS)

The threat environment for ICS has transformed immensely with the increased digitization and interconnectivity of the critical infrastructure. ICS have, in the past, been isolated systems with little vulnerability to external threats, but advances in communication technology have witnessed them being included in extensive networks. This transition enhances operation effectiveness but simultaneously exposes such systems to all manner of cyber-attacks, among them sophisticated cyber-attacks from adversaries determined to exploit major national assets. The highly specific nature of ICS, e.g., their reliance on real-time data acquisition and control, makes defense strategies more complicated. Protecting ICS, therefore, is the priority with the potentially catastrophic consequences of a successful breach that ranges from disruption to operations to significant economic and safety issues as discussed in Al-Khateeb et al. (2019). With the advanced threats to ICS on the increase, the usual security measures fail. More advanced attack vectors necessitate acceptance of an multi-faceted defensive approach that depends not only on technology but procedural controls as well. Current layered defense solutions are criticized for their inability to prevent sophisticated attackers from successfully breaching systems. A reinforcing of ICS security calls for taking innovative security paradigms like orthogonality to the table. Orthogonality is concerned with having isolated assurance and functional controls at each level of ICS architecture in order to make the system more resilient against a range of cyber threats. This strategic alignment reflects a change of the perception of cybersecurity from absolute defense to proactive defense where potential vulnerabilities are foreseen and resolved ahead of time to avoid exploitation, as discussed in Al-Khateeb et al. (2019). With increasing interconnection among Supervisory Control and Data Acquisition (SCADA) systems, stringent security cannot be overstressed. SCADA systems, which are key to monitoring and controlling infrastructure critical to daily life, are subjected to heightened threats from internet-based cyber-attacks. Evolution of threats necessitates attitude evolution with security audits adopting an attacker's viewpoint to identify gaps and vulnerabilities in the systems. This not only identifies areas of shortfall but also fosters a culture of continuous improvement in security protocol. The consequences of a successful compromise will be disastrous with disruptions to services of a vital kind and losses of enormous financial worth. Thus, proactive security measures are taken coupled with enhancing detection and prevention capacities necessary to eliminate the risk of the new-type attacks targeting ICS, as highlighted in Almalawi et al. (2017).

6.3.1. Common Vulnerabilities in Industrial Control Systems (ICS) Environments

ICS are increasingly integrated into larger network infrastructures, creating entry points for exploitation by third parties. One of the interesting vulnerabilities is the reliance on legacy protocols designed to work on stand-alone networks, now stretched to be applicable over the Internet. Previously, those protocols worked within walled gardens, providing a false sense of security. As ICS networks divert insecure traffic at a worldwide scale, they become more vulnerable to attacks such as eavesdropping and tampering. The shift in the pattern of communications explains the need for potent intrusion detection and prevention measures tailored to counter vulnerabilities intrinsic in ICS environments, effectively altering their posture against mounting cyber threats. Furthermore, ICS's inter-domain communication significantly enhances security threats due to the lack of efficient encryption and authentication processes. According to recent research, the majority of ICS transmit sensitive operation information through unsecured channels, unaware that they are providing chances for cybercrooks to intercept and exploit this information. Data traffic between industrial facilities and remote monitoring stations usually does not get proper filtering and protection, which puts them at risk of Distributed Reflective Denial of Service (DRDoS) attacks that can compromise not only a single ICS component but also the entire Internet ecosystem. Mitigation of such vulnerabilities entails the use of advanced filtering techniques and stronger communication protocols capable of securing ICS data transmissions in a networked environment and facilitating secure interaction with outside systems. Furthermore, game-theoretic modeling application to security for ICS highlights the complexity of securing multi-layered systems from elevated threats. As discussed in the recent literature, the integration of game and decision theory can provide valuable insights into the nature of the adversaries and how the operators of ICS make strategic decisions against them. Based on this analysis, one can better understand possible weaknesses in the operation and management planes of ICS networks. Through the use of a layered security strategy, organizations are better able to anticipate and defend against probable attempts at intrusion and, ultimately, create a more secure operating environment. Addressing the challenges posed by layered security using innovative theoretical models plays a central role in the formulation of robust intrusion detection and prevention capabilities that can adapt to the evolving nature of threats against ICS environments (Hauser et al., 2018; Nawrocki et al., 2019) (Figure 6.6).

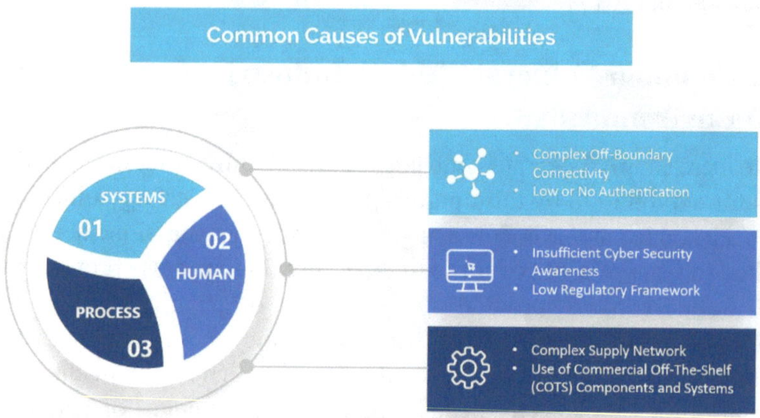

Figure 6.6. Common vulnerabilities in industrial control systems (ICS) environments.

Source: https://www.dts-solution.com/wp-content/uploads/2021/09/Common-Causes-of-Vulnerabilities.jpg.

6.3.2. Types of Cyber Threats Targeting Industrial Control Systems (ICS)

In the ICS sector, cyber-attacks exist in diverse forms, whose overarching similarity is that they are largely characterized by the impacts they have on critical infrastructure. One typical attack is the Denial of Service (DoS) attack, which targets the availability of critical systems by flooding them with excessive requests, rendering them impossible to operate. Such attacks have the ability to disable utilities such as power or water supply, resulting in massive disruption of day-to-day services and threatening public safety. Furthermore, malware specifically targeted at ICS can infiltrate systems, allowing attackers to gain control over physical processes. Such manipulation can lead to catastrophic scenarios, i.e., harm to the equipment or to the environment, underscoring the urgency of the need for efficient intrusion detection mechanisms that can identify and mitigate such types of threats. Another essential type of cyber-attack in ICS is software and hardware component vulnerability exploitation. With systems becoming increasingly networked, they are even more susceptible to intrusions that can compromise their integrity. Attackers are able to utilize exploits in outdated software, employing various tactics to gain unauthorized access to control systems. This danger is heightened by the presence of Internet of Things (IoT) devices in the smart grid infrastructures because these devices mostly lack proper security features, presenting another avenue for attackers. As studies have shown, although communication technology enhances working efficiency, it makes ICS vulnerable to previously unknown dangers (Ghazi et al., 2018). Therefore, extensive measures have to be designed to detect and fix vulnerabilities, along with the use of next-gen intrusion prevention systems. The advent of artificial intelligence (AI) in cybersecurity operations is both a blessing and an opportunity for ICS security. On the one hand, AI-based tools have expanded the capabilities to sense unusual patterns pointing to cyber threats and therefore strengthen defenses against intrusions. They enable strong analysis of long data streams, enabling fast detection of suspected security intrusions and the automation of response algorithms. Simultaneously, the same technologies are being exploited by threat actors who increasingly use AI algorithms to produce advanced threats that outpace conventional cybersecurity controls (Diep et al., 2020). This dual-edged character of AI highlights the necessity of an equilibrium-based approach to ICS cybersecurity where preventive measures are supplemented by a keen sense of emerging offensive methods to make defenses robust enough to counter the ever-evolving pattern of cyber-attacks.

6.3.3. Case Studies of Notable Industrial Control Systems (ICS) Breaches

An examination of notable cases of violating ICS yields salient lessons in vulnerabilities faced by critical infrastructure sectors. One of the most prominent examples of these issues was the 2010 Stuxnet attack, directed particularly at Iran's nuclear facilities. This top-tier worm was designed to disrupt centrifuges, demonstrating both the capability for physical sabotage and the fresh threat landscape in ICS environments. The impact was considerable, not merely generating operational mayhem but also raising questions regarding the intersection of cybersecurity and national security. These case studies underscore the necessity for robust intrusion prevention and detection technology capable of addressing the unique requirements of ICS as threats become increasingly sophisticated and devastating. A second notable example is the 2015 cyber-attack on the Ukrainian power grid, leading to mass power outages affecting hundreds of thousands of citizens. This attack did uncover an orchestrated malware attack on ICS systems that didn't only gain access to the system but also manipulated hardware operations. Even though the hackers avoided security, the attack foregrounded the dual threat of outsider attacks and insider attacks. Lessons obtained from the Ukraine breach emphasize the value of real-time monitoring and incident response mechanisms within ICS as well as the necessity of keeping to established cybersecurity guidelines to prevent risks in advance. The research indicates that not doing so can lead to devastating consequences towards public safety and national security. In addition to this, the energy retail industry itself has also experienced its own challenges, notably the evolution of cyber-attacks fueled by regulatory environments. A European energy retail company case study conducted within the EU NIS 2 Directive framework presents the strict cybersecurity evaluations required in the protection of critical infrastructure.

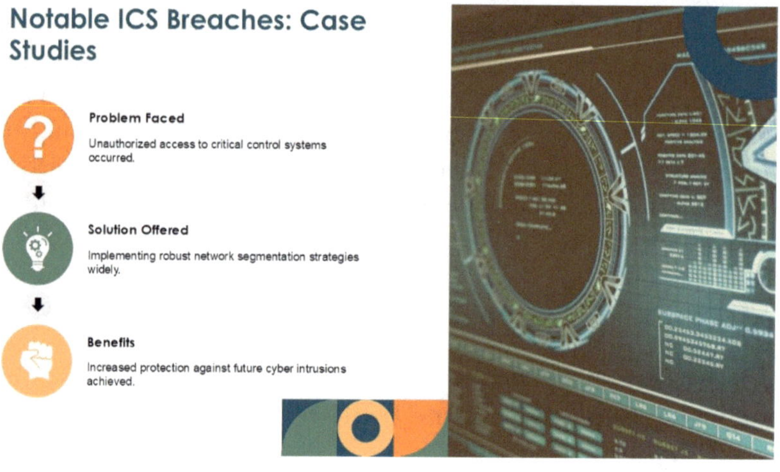

Figure 6.7. Case studies of notable industrial control systems (ICS) breaches.

Source: https://www.slideteam.net/media/catalog/product/cache/1280x720/r/e/real_world_case_studies_of_cybersecurity_breaches_in_ics_ppt_sample_st_ai_slide06.jpg.

The research employs Failure Modes and Effects Analysis (FMEA) and suggests a number of risk categories, which indicate areas of possible weakness and defensive strategies. From the findings, what can be learned will have an impact on the development of sophisticated intrusion detection systems as well as collaborative work to secure energy sectors from cyber-attacks. These case studies are used to illustrate the ongoing development of cyber threats and the necessity for ICS to constantly develop their security controls in an increasingly interconnected world (Helo et al., 2025; Jama et al., 2024) (Figure 6.7).

6.4. IMPLEMENTATION CHALLENGES

Intrusion detection and prevention systems (IDPS) implementation into Industrial Control Systems (ICS) is beset with implementation difficulties that can significantly impede their efficacy. One of the biggest challenges is the unique architecture of ICS, which prefers proprietary protocols and hardware configurations. As noted by analysts, conventional detection methods can be challenging to apply in such specialized environments and are thus less effective for identifying customized threats like ladder logic bombs (LLBs). Such LLBs are typically hidden in control logic and have the ability to direct system behavior surreptitiously, posing a considerable risk to operational integrity. Consequently, designing IDPS capable of appropriately addressing these unique features without compromising on operational continuity is a daunting exercise that security professionals must undertake methodically. Budget limitation is another daunting challenge in deploying IDPS in ICS. Industrial environments prefer performance and reliability over extended security processes due to their primary roles in manufacturing and infrastructure. Most ICS devices, especially those in the Internet of Things (IoT) category, possess limited processing capabilities and memory storage, hence making it hard to incorporate robust security features. Traditional detection approaches, such as signature-based malware detection, are less suitable as they require frequent updates and continuous resource utilization, which could lead to system instability. To overcome these deficiencies, a more resource-aware and smart approach, described in Ahmadinia et al. (2016), can be of significant importance. This approach involves employing adaptive algorithms that are effective at detecting anomalies without placing excessive performance demands on the systems. Additionally, the ever-changing nature of the threat landscape means that IDPS mechanisms in ICS are constantly needed to adapt and get updated. Cyber attackers are now more and more turning to sophisticated mechanisms to exploit control system vulnerabilities, and it is important that security solutions are able to keep pace. The degree of agility required to conduct timely patches and updates can strain running processes, particularly where uptime is critical. For instance, the existence of memory-based vulnerabilities can be a threat to system stability in case they are not quickly resolved by effective IDPS. A general security architecture that incorporates proactive and response mechanisms is key to protecting ICS, as noted in research on detection and prevention being vital in real-time (Kim et al., 2017). By focusing on special responses to special threats while minimizing the effect on operational performance, organizations will achieve a decrease in the risk caused by intrusion attempts while maintaining system integrity.

6.4.1. Integration of Intrusion Detection and Prevention Systems (IDPS) with Existing Industrial Control Systems (ICS) Infrastructure

Incorporating current ICS with IDPS is crucial in enhancing the security status of industrial infrastructures. Legacy systems that have not been fortified with modern security solutions have, on most occasions, predetermined conventional ICS as being vulnerable to cyber-attacks. Incorporating IDPS into such a setup demands a comprehensive approach that combines new technology with existing operational models. This mandates a rundown of existing vulnerabilities while, at the same time, maintaining that the IDPS combination does not disrupt usual operation of the automated functions. Well-specified communication channels between the ICS and IDPS present means for live monitoring and response, thus enhancing defenses against potential attacks and minimizing downtime. Besides, IDPS deployment needs to be synchronized with industry-related safety requirements to guarantee that safety and security controls complement one another and not compete unfavorably. Application of risk-based methodology is essential when deploying IDPS within ICS infrastructures. The approach, already outlined for the Belgian chemical and process industries, can be reproduced by similar industries all over the world (American Petroleum Institute et al., 2014). By adhering to established design principles for security, organizations can leverage IDPS to protect against both cyber and physical threats. Rings-of-protection theory, combining generic security routine with tailored procedure, demonstrates the potential for proactive as well as reactive security measures to be employed effectively within an ICS environment. Correct integration not only enhances the current security landscape but also fosters a safety culture that speaks to employees. Awareness of safety procedures can facilitate smoother implementation and integration of IDPS, ultimately fortifying the protection environment of industrial processes. Moreover, integration of IDPS with present ICS comes with unique challenges and opportunities, particularly given the previous failures of these systems in issues related to security (Beckers et al., 2017). Some organizations face the issue of upgrading legacy hardware while, simultaneously, adding modern security technology. Modern methods, such as the use of electromagnetic side-channel measurements for ICS software behavioral change detection, can offer a new way towards real-time monitoring and anomaly detection without necessitating extreme system overhauls. These methods demonstrate the potential for contemporary security methods to coexist with mature infrastructures without degrading operational efficacy while enhancing protection against new forms of cyber threats. The integration of IDPS must thus be responsive to the diverse technological platform and operational conditions of different industries to provide a generic security solution for the ICS environment (Figure 6.8).

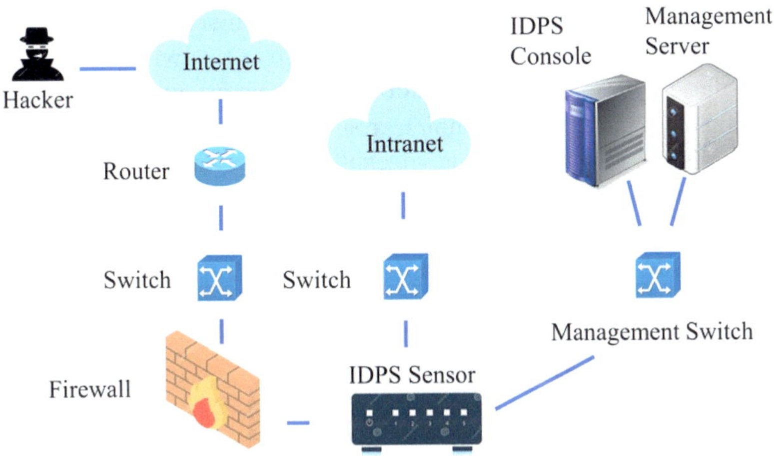

Figure 6.8. Integration of intrusion detection and prevention systems (IDPS) with existing industrial control systems (ICS) infrastructure.

Source: https://www.researchgate.net/profile/Mohammad-Asif-Hossain-2/publication/348269149/figure/fig3/AS:976978349916169@1609941198552/Working-Principle-of-IDPS.ppm.

6.4.2. Resource Constraints and Budget Considerations

Budget considerations are at the forefront of introducing intrusion detection and prevention systems (IDPS) into industrial control systems (ICS). Organizations continually encounter daunting resource constraints that can limit investment in worthwhile cybersecurity measures. Such constraints call for consideration of cost-benefit analysis and budget planning, with an emphasis on high-risk areas that call for high-priority intervention. The problem is not only creating sophisticated technologies but keeping them in the long term because running budgets vary due to unexpected events or shifts in organizational priorities. Thus, decision-makers must balance this fiscal tightrope while keeping their ICS with adequate safeguarding against newly emerging cyber threats that have catastrophic operational impacts. Compounding budget planning refers to the rapid growth of technology within ICS settings. As businesses continue to integrate with digital-controlled and software-driven elements, spending on cybersecurity needs to level up with the growing attack surface. The greater adoption of distributed energy resources (DERs) introduces new threats, calling for more budget to be allocated towards cyber-resiliency activities. Accordingly, organizations must make rigorous studies to ascertain weaknesses in their current systems and invest resources wisely to counteract these vulnerabilities. Effective resource management not only includes investing in prevention and detection tools but also in training human resources that will be responsible for running these systems. This integration of human resources with technology investment is critical to maintaining a strong and responsive cyber defense posture. Against this background of challenges, organizations must adopt a comprehensive budget planning strategy that includes cybersecurity into the overall operation plans. This may be through seeking alternative sources of financing using grants or partnerships aimed at enhancing cybersecurity within ICS. Additionally, organizations can benefit from establishing strong cybersecurity frameworks that adopt industry best practices, potentially as part of their budget considerations. By recognizing that an investment in security measures as a preventative measure is more cost-effective than resorting to reactive action against breaches, organizations can establish a viable cycle of continuous improvement in their cybersecurity frameworks. Addressing resource constraints and budgetary limitations with integrity will eventually enhance their resilience against future attacks, helping to safeguard critical infrastructure in the interconnected environment of today (Chen et al., 2023; Greene et al., 2025).

6.4.3. Training and Awareness for Personnel

Training and sensitization to soundness by personnel are fundamental pillars in safeguarding Industrial Control Systems (ICS) from potential intrusions. Personnel who work in such vital infrastructures must be armed not only with technology savvy but also with a subtle sensitivity to evolving threat patterns. The more sophisticated the technologies are with which ICS get connected, the more intricate potential vulnerabilities turn out to be. Therefore, a structured training program that informs employees about various forms of cyber-attacks, particularly those that exploit system weaknesses, is essential. Training through repeated drills and scenario exercises will equip staff to recognize abnormal patterns of behavior common to breaches, increasing their situational awareness. Through the establishment of the culture of being watchful, organizations are able to make their personnel no longer just passive responders but proactive actors in preventing cyberexploitation and potential terrorist attacks as proposed in recent studies on mitigation strategies and their implications for secure operations within sensitive environments, as indicated in Al-Khateeb et al. (2019). Besides, routine training and sensitization programs should extend to cover more security protocols and contingency planning initiatives. The threat landscape to transportation and other crucial infrastructure assets continues to evolve, and as such, a pro-active approach in sensitizing employees to new tactics must be adopted. For instance, the insidious nature of vehicle-ramming terrorist attacks, as highlighted by the latest research, calls for the sensitization of employees to identify and respond to unconventional threats. Case study training modules will not only define potential threats but also serve as helpful blueprints to adhere to in a crisis.

Figure 6.9. Training and awareness for personnel.

Source: https://fastercapital.co/i/Cybersecurity-measures-The-Role-of-Cybersecurity-Measures-in-Safeguarding-Business-Data--Employee-Training-and-Awareness.webp.

By adding lessons drawn from past attacks to training modules, organizations are able to enhance their employees' capacity to detect patterns of behavior that have been known to precede attacks. This integrated approach not only enhances individual capabilities but also provides overall system resilience as per the available literature on security management practice, citing Edwards et al. (2020). Finally, the establishment of a continuous feedback loop within training programs is essential for relevancy and effectiveness. Recurrent evaluations of the training's efficacy must be done to determine gaps and adjust content on the basis of current intelligence regarding ICS threats. The repeated cycle will make personnel stay updated on new tactics and countermeasures, hence fostering an adaptive security posture. Moreover, development of cooperation with outside organizations that are experts in cybersecurity can equip personnel with a wider appreciation for threats. Capacitating employees through engaging with experts encourages the application of best practices while simultaneously fostering a sensitivity towards mutual responsibilities in preventing threats. Blending training and awareness into organizational culture within ICS better equips employees to handle security concerns, which eventually forms an effective defense system against the complex problems of modern infrastructures, reflecting the holistic strategy advocated in Al-Khateeb et al. (2019) (Figure 6.9).

6.5. BEST PRACTICES FOR INTRUSION DETECTION AND PREVENTION SYSTEMS (IDPS) IN INDUSTRIAL CONTROL SYSTEMS (ICS)

IDPS best practice use for ICS is the heart of protection from increasing cyber threats. Firstly, the organizations need to adopt a risk-based security practice. By assessing vulnerabilities and potential influence of various threats, they are able to prioritize their protection measures effectively. This methodology, as outlined in security risk assessment research particularly for the process and chemical sectors, gathers prime factors concerning management practices, technology, and human communications. The integration of varied layers of security—traditionally known as the rings-of-protection—offers complete protection against incursions. The integration of such an inclusive model enhances the resilience of ICS to different attack vectors and provides a uniform model workers can learn from other safety models (American Petroleum Institute et al., 2014). Integrating Building Automation Systems (BAS) into IDPS for ICS is also a key consideration given that such systems centralize management for sophisticated infrastructures. As they are IP-based communications, BAS are likely to require the robust security measures to ward off any possible cyber-attack. It is noteworthy that traditional security measures usually fail to suffice; therefore, organizations have to focus on enhancing detection and prevention according to the inherent vulnerability of BAS. Recent studies have identified threats to BAS systems and recommended improvement in monitoring approaches, such as machine learning algorithms that provide better traffic classification and intrusion detection capabilities. With the mitigation of such emerging vulnerabilities, companies can implement a multi-layered security system that provides an active defense against cyber-attacks (Johnstone et al., 2014). Finally, extensive training and awareness programs are paramount components in enhancing IDPS in ICS. Staff must not only be instructed in technical aspects of security systems but also in potential human elements which may breach the security protocols. The human factor, overlooked sometimes, may directly influence the efficiency of an IDPS as well. Therefore, periodic training routines must familiarize staff with current security threats and protocol for detecting suspicious activity. In addition, instilling a culture of security encourages the employees to take an active role in protecting ICS environments. By incorporating training with innovation in technology and tailored risk assessments, organizations can build an integrated defense plan that ensures maximum effectiveness of their IDPS and minimizes the threats of intrusion.

6.5.1. Regular Updates and Patch Management

Within the realm of Intrusion Detection and Prevention in Industrial Control Systems (ICS), timely upgrades and effective patch management are crucial in order to maintain a robust security posture. With increasingly more sophisticated ICS infrastructures becoming integrated with national critical infrastructures, the need for continuous software upgrades surpasses typical IT standards. Regular patching seals holes that, if left alone, could end up in the hands of criminal forces attempting to disrupt important services. The constantly changing nature of online threats necessitates a reactive security position that involves patch management as part of its component. Without such focus, ICS remain susceptible to thousands of attacks that, ultimately, would lead to catastrophic effects, which makes this component imperative in organization cybersecurity. In addition, the efficacy of patch management in ICS is supported by the lack of traditional security controls that are normally deployed in common IT environments. As ICS are normally critical components in utilities like energy, water, and transportation, the impact of inadequate security controls can be disasterous. Layered defense strategy is required as existing techniques are questioned more extensively as to their timeliness in responding to advanced persistent threats. Innovations such as the use of orthogonal controls, such as assurance and functional controls across various architectural layers, can support a stronger defense model. By the facilitation of regular updates and robust patch management, organizations can enhance their defenses in a meaningful way so as to significantly increase the bar for attackers to cross their environments (Al-Khateeb et al., 2019). Also, the use of frequent updates and thorough patch management not only enhances security but also promotes a culture of cybersecurity awareness among employees, as proposed in a study on the security awareness of domestic Internet users. The study posits that despite users having base-level awareness regarding essential mitigation methods like patching, there remain gaps in their awareness on more sophisticated security controls (Gibson et al., 2012). This means that businesses must do more than simply replace systems; they must also adopt proactive education and training to ensure that all stakeholders understand the worth of security practices. Building an educated workforce which can recognize threats and value the significance of patch management will lead to an improved secure ICS environment and will be in favor of the overall goal of vulnerability reduction against emergent cyber-attacks (Figure 6.10).

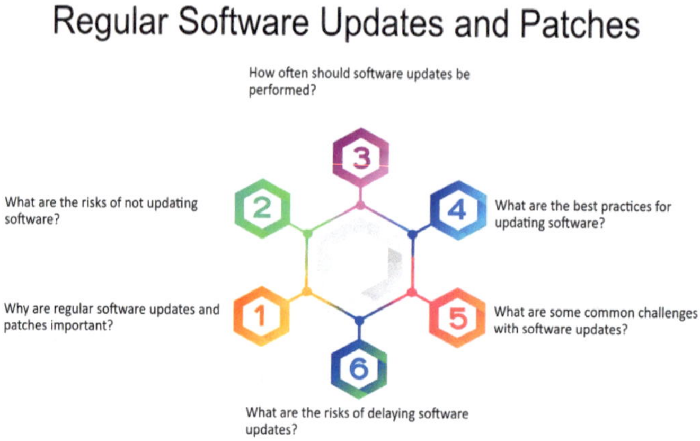

Figure 6.10. Regular updates and patch management.

Source: https://fastercapital.co/i/Cybersecurity--Enhancing-Cybersecurity-Measures-to-Minimize-Data-Loss--Regular-Software-Updates-and-Patches.webp.

6.5.2. Continuous Monitoring and Incident Response

In the Industrial Control Systems (ICS) domain, vigilant monitoring is an absolute necessity to identify potential security threats before they become catastrophic incidents. The ever-vigilant character is crucial in the face of the seriousness of ICS environments charged with managing processes in areas like energy supply and water supply. Real-time monitoring of network activity, system configuration, and user activity is part of ongoing monitoring, by which organizations can detect anomalies that might represent intrusions. The rapid growth of cyber threats rendered static, traditional security tactics ineffective; therefore, dynamic monitoring frameworks need to be set up to maintain strong defenses. By establishing continuous monitoring regimes, ICS operators can facilitate an up-front process to outstanding vulnerabilities, enhancing their overall cyberattack resilience (Al-Khateeb et al., 2019). Incident response, strongly intertwined with continuous monitoring, portrays the formalized process embraced on detecting a security incident. The effectiveness of incident response regimes significantly contributes to the recovery process and general system integrity. In ICS, fast response frameworks should be established to respond to cyber-attacks in real time to limit any resultant damage. An efficient incident response program usually consists of preparation, detection, analysis, containment, eradication, and recovery stages, all of which involve smooth coordination among different stakeholders. The complexity of ICS systems requires incident response plans to be designed specifically to operate requirements, providing timely and comprehensive processes to counteract threats that would bring critical infrastructure operations to a stop (Alzubi et al., 2015). As the nature of cyber threats continues to change, having an effective incident response plan becomes vital in safeguarding the integrity of ICS operations. The synergy of real-time monitoring and effective incident response methodology forms a critical layer in ICS environment defense architecture. Through the cultivation of a culture of awareness and preparedness, it is achievable for organizations to create a synergistic effect that significantly minimizes risks associated with cyber threats. Advanced analytics and machine learning can be employed to enhance monitoring so that quicker detection of suspicious activity is achieved. Moreover, interspersing of industry partners and cybersecurity teams can contribute to enabling information sharing and thus enhancing collective defense against common threats. Existing research suggests that the multi-layer security strategy in ICS plays an important role in combating the complexity introduced by heterogeneous and adaptive attack vectors (Al-Khateeb et al., 2019). Through constant improvement and adaptation, organizations can effectively sustain their cybersecurity controls such that critical infrastructure systems are always secure and reliable.

6.5.3. Collaboration with Cybersecurity Frameworks and Standards

Cybersecurity frameworks and standards must be incorporated into Intrusion Detection and Prevention Systems (IDPS) in order to secure Industrial Control Systems (ICS). With the increased connectivity level of SCADA systems, these standards provide systematic guidelines to detect and avert possible threats. As Almalawi et al. (2017) indicates, interconnectivity of modern SCADA systems exposes them to a variety of cyber-attacks that could have disastrous economic and safety consequences. Organizations are able to establish a good cybersecurity posture in line with best practices by implementing established standards such as the NIST Cybersecurity Framework or ISO/IEC 27001. This pre-emptive approach not only enhances defense mechanisms but also compliance with regulation requirements, thereby contributing to overall resilience to cyber-attacks. Collaboration with industry players, regulators, and research institutions is central to the development of useable cybersecurity interventions to ICS.

The aggregated experience and capacity leveraged through such collaboration has the potential to drive innovations in addressing unique challenges facing critical infrastructures. As noted by Anglano et al. (2018), cooperative effort towards enhancing cybersecurity systems can provide the focal areas of work and actionable strategies that ensure organizational readiness. The coordination provides for sharing of information, making it possible for organizations to stay informed in terms of new risks as well as countermeasures.

Collaboration with other entities also increases the diversity of the information collected in order to do threat modeling, allowing for better understanding of potential vulnerabilities as well as the use of more advanced intrusion detection and prevention techniques. In addition, awareness and training initiatives spurred through collaboration can completely transform the culture of security in organizations. By continuously training personnel in the latest cybersecurity methodology and standards, companies can cultivate a vigilant, alert workforce with respect to potential threats. Almalawi et al.'s (2017) citation emphasizes the importance of adopting the mindset of an attacker to realize weaknesses in defenses, echoing the need for an informed workforce trained in recognizing suspicious activity. Counting training programs based on recognized frameworks ensures that all staff at all levels are equipped to contribute towards the overall security posture. Overall, collaboration in accepting cybersecurity frameworks not only protects the technical defenses of ICS but also protects the human element, crucial in successfully predicting and responding to cyber threats (Figure 6.11).

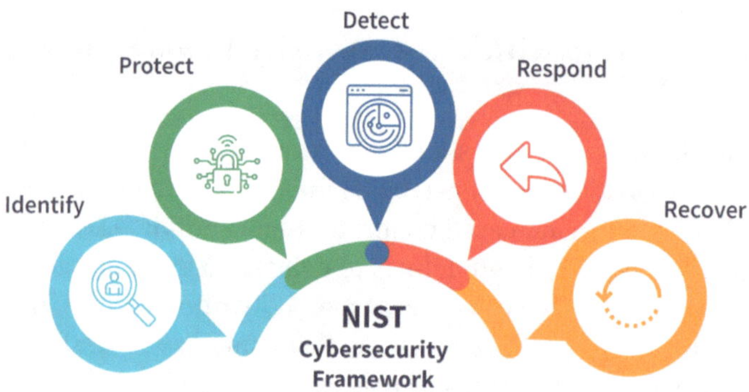

Figure 6.11. Collaboration with cybersecurity frameworks and standards.

Source: https://ats.ae/stagingnew/wp-content/uploads/2022/02/NIST-Cyber-Security-Framework-1024x536.jpg.

6.6. FUTURE TRENDS IN INDUSTRIAL CONTROL SYSTEMS (ICS) SECURITY

As ICS continue to become more sophisticated, future security trends must address the growing complexity and interdependence of the systems. The addition of Internet of Things (IoT) devices to ICS networks makes them more efficient but simultaneously introduces new vulnerabilities. As Romano et al. (2019) explains, the surge of IoT technology not only expands the attack surface but also requires that security engineers develop novel solutions that neutralize innovative threats. One of the forecasted trends is the implementation of adaptive solutions that utilize machine learning algorithms to detect unnormal patterns and behaviors on ICS. By leveraging the potential of advanced analytics, organizations can scale up intrusion detection and prevention to facilitate real-time response to probable security violations. Proactive is not just enhanced resilience, but also addresses growing emphasis on intelligent automation in ICS environments, ultimately safeguarding critical infrastructure. Moreover, human factors in ICS security are being considered an integral part of future strategy increasingly. Research indicates that human behavior is likely to be a critical source of utility network vulnerability, making extensive training and awareness campaigns crucial (Busby et al., 2016). The future directions must emphasis the development of a security culture in an organization to complement technological advancements.

The transformation entails engaging employees through ongoing training and simulation that includes social engineering methods usually employed by cyber attackers. These efforts have a significant influence on an organization's general security by offering a warranty that everyone is able to recognize threats and react as needed. In the process, organizations not only bolster their technology interfaces against abuse but also foster a complete security strategy that takes into consideration the human behavior facet as a component of risk management. Lastly, the future of ICS security will require more emphasis on regulatory compliance and use of global standards. As more cyber-attacks take place, governments and industry groups will start implementing strict regulations requiring organizations to adopt standardized security practices. This is evident with the growing applicability of frameworks that integrate risk management practices into operation security practices. Such structures will encourage organizations to monitor their cyber hygiene on a continuous basis and adopt best practices that are a reflection of the latest threat landscapes. Moreover, regulators can mandate reporting of security incidents, which can become a more transparent culture of accountability within the industry. As businesses become increasingly aligned with such regulations, not only will they enhance their security position, but they will also contribute to making the overall integrity and reliability of essential infrastructures, thus create even greater public confidence in ICS systems.

6.6.1. Advancements in Machine Learning (ML) and Artificial Intelligence (AI) for Intrusion Detection and Prevention Systems (IDPS)

The evolution of ML and AI has significantly transformed the IDPS landscape for industrial control systems (ICS). Traditional security measures lag behind in automatically adapting to changing threats and sophisticated attack channels and thus need the integration of smart mechanisms. In these, ML algorithms can analyze colossal amounts of network data and identify patterns that indicate cyber threats. These advancements enable IDPS to monitor for anomalies and potential threats in real-time, thus enhancing the overall ICS security position. With the complexity of attacks increasing, the deployment of machine learning models based on approaches such as decision trees and support vector machines becomes paramount in facilitating proactive defensive strategies that actively respond to neutralize breaches and vulnerabilities, emphasizing the need for a paradigm shift from conventional solutions to next-generation AI-driven solutions (Abu-Salih et al., 2024). Furthermore, the incorporation of advanced deep learning frameworks within AI-fueled security products supports instant detection of known as well as unknown attack patterns. By adopting models like recurrent neural networks (RNNs) and convolutional neural networks (CNNs), organizations can deploy IDS that can learn from historical events and adapt based on evolving malicious trends. This capacity is more than just threat identification; it enables quick countermeasures to large-scale attacks, such as Distributed Denial of Service (DDoS) attacks. Effective metadata collection and preprocessing are necessary in order to allow efficient deployment of such technologies by minimizing bias and interpretability problems associated with them, which are necessary to achieve trustworthy AI performance in cybersecurity use cases. Addressing these problems increases trust in AI models, ultimately to higher organizational adoption of sophisticated defense systems (Ali et al., 2025). While AI and machine learning are promising in terms of enhancing IDPS for ICS, several challenges discourage their massive adoption and implementation. Data reliability issues, model explainability, as well as built-in biases in AI-based decision-making, pose major stumbling blocks. Organisations could be deterred from adopting AI technologies to the very essence due to a lack of faith in the trustworthiness of outcomes presented by AI, alongside repeated demonstrations of adversarial attacks on vulnerabilities in AI. Hence, while machine learning developments open wide doors towards deepening protection against novel threats, there needs to be a steady dialogue on the ethical issues and actual-world limitations of AI in the cybersecurity space. By alleviating these issues, the cybersecurity community can more effectively harness the potential of AI, enhancing defenses against increasingly sophisticated cyberattacks and the integrity of the systems they aim to protect (Ali et al., 2025) (Figure 6.12).

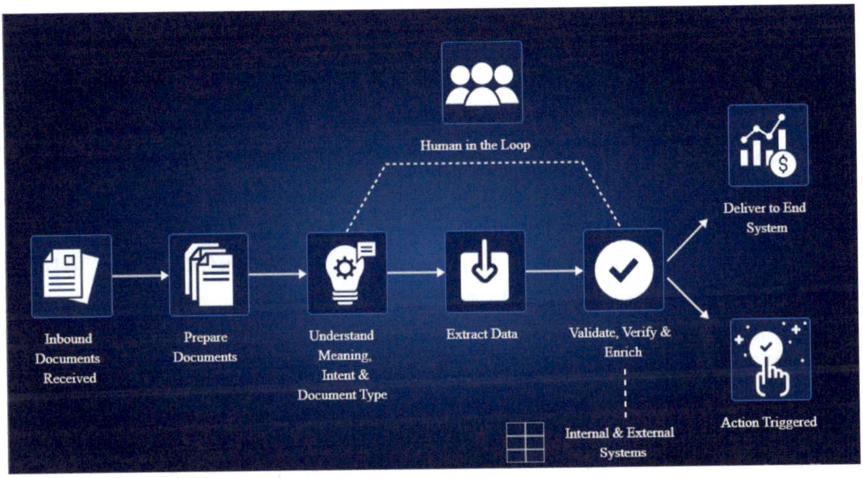

Figure 6.12. Advancements in machine learning (ML) and artificial intelligence (AI) for intrusion detection and prevention systems (IDPS).

Source: https://d3lkc3n5th01x7.cloudfront.net/wp-content/uploads/2023/05/ 2505 5959/Intelligent-Document-Processing-IDP-Banner.svg.

6.6.2. The Role of Cloud Computing in Industrial Control Systems (ICS) Security

Cloud computing in ICS security is a significant innovation in critical infrastructure protection. Previously, ICS environments were usually isolated and possessed limited connectivity due to legacy approaches. Utilizing cloud technology, however, offers more flexibility, scalability, and resource optimization. This makes real-time monitoring and data correlation easier, which are essential for successful intrusion detection and prevention. With the support of cloud services, enterprises are able to utilize advanced analytics and machine learning techniques to identify patterns that can indicate future threats. This shift not only enhances security but also promotes proactive threat management, so it becomes important to identify the underlying vulnerabilities of cloud infrastructure as the attack surface evolves. Therefore, cloud computing use in ICS security is not necessarily about data storage; it becomes an integral part of obtaining a lasting security posture in a dynamic threat landscape. Furthermore, cloud data security and privacy are of utmost importance, especially in strategic sectors such as energy and finance, where failure would have widespread consequences. As highlighted in recent researches, cloud computing environments can be made safer with comprehensive solutions encompassing prevention and detection of attacks. Intrusion Detection Systems (IDSs) are crucial to this end, giving visibility into possible infringements while allowing organizations to respond swiftly. The deployment of IDS within cloud-based ICS not only avoids unauthorized access risks but also enhances the integrity and availability of critical operational data as a whole. It is essential to deploy cooperative IDS systems based on shared intelligence between various environments to ensure effective protection for sensitive information, safeguarding it from relentless evolution in cyber threats (He et al., 2014). In addition, cloud computing introduces new challenges, particularly in terms of regulatory compliance and data sovereignty. As cloud services are being increasingly adopted in ICS, organizations must deal with varying policies and regulations that govern data security across different jurisdictions. This variety highlights the need for robust intrusion detection and prevention mechanisms that can learn to resist varied regulatory requirements while maintaining data confidentiality. The possibility of cyber disruption, i.e., temporary or permanent loss of service, emphasizes the importance of continuity of operation within cloud systems. As explained in earlier research, it is crucial to design methods that preserve confidentiality, integrity, and availability of data within such an environment. The use of strategic cloud security frameworks thus responds not just to compliance concerns but also fortifies ICS against the uncertainty of cyberattacks (Kifayat et al., 2025).

6.6.3. Emerging Regulations and Compliance Requirements

Increased dependence of Industrial Control Systems (ICS) on digital infrastructures has created the demand for developing regulations and requirements for compliance that aim to improve cybersecurity. With advanced attackers in increasing numbers, international regulating bodies are developing standards that necessitate stronger protection of ICS. These new standards are meant to address the vulnerabilities inherent in such systems because their significance positions them at the forefront of national infrastructure. Laws such as the Cybersecurity Improvement Act and recommendations of the National Institute of Standards and Technology acknowledge the need for proactive measures in security, compelling businesses to incorporate comprehensive intrusion detection and prevention systems in their operations procedures. Not only do these laws enable resilience against cyber assaults but also aim to establish standardized best practices across industries. Increased compliance requirements are increasingly founded on an understanding that existing layered-defense methods may be insufficient against advanced cyber threats. Security design orthogonality, the deployment of assurance and functional controls in all ICS infrastructure layers, is in high demand as a new method to transcend traditional security system limitations. By adopting this multi-dimensional strategy, organizations are able to fortify their defenses against advanced attack vectors, as posited in recent studies highlighting the need for a robust security architecture in ICS (Al-Khateeb et al., 2019).

Figure 6.13. Emerging regulations and compliance requirements.

Source: https://linfordco.com/wp-content/uploads/2013/06/a-guide-to-regulatory-compliance.jpg.

This paradigm shift is an indication of a broader recognition that merely checking boxes for compliance is insufficient; nowadays, companies must conduct constant compliance testing involving novel security technologies while transforming to adapt to evolving threat developments. Also, the establishment of global standards for electronic security is crucial in constructing a harmonized compliance approach in ICS environments. Through considerations related to the elements needed to possess a secure electronic environment, seven pillars of greatest concern have been established that underpin these standards to address issues ranging from legal frameworks to risk-management styles (Glaessner et al., 2025). These pillars create pillars upon which organizations may develop their cybersecurity policies in an effort to be compliant with requirements that promote operational integrity. By focusing on an all-encompassing approach to electronic security, like the requirement for certification, private insurance, and better incident reporting procedures, organizations will better address the nuances of regulatory compliance. Lastly, the convergence of emerging standards and stringent compliance requirements makes it possible for a culture of accountability and resiliency to become a necessity to safeguarding critical infrastructure in an age of expanding cyber threats (Figure 6.13).

SUMMARY

In Summary, the implementation of robust IDPS in ICS is an imperative requirement against expanding cyber threats. The long history of poor security technologies of such systems, particularly legacy systems, underlines the necessity of innovation in security measures more than ever before. As acknowledged in the current climate of cybersecurity challenges, manufacturers are compelled to up their defensive measures. Forward-thinking technologies, such as those utilizing electromagnetic side-channel measurements to profile behavioral change in software, demonstrate how much more secure ICS can be made. Such methods not only address real-time intrusion detection but also address the long-term issues introduced by legacy systems Beckers et al. (2017), thus providing a baseline for greater levels of operational integrity. Moreover, the combination of cybersecurity and risk management in ICS calls for a multilateral approach to adequately address vulnerabilities. Organizations must place utmost importance to thorough analysis to identify specific risks and respond proactively. This is particularly applicable to sectors like healthcare, where widespread fraud and abuse have incurred substantial financial losses. Historical studies demonstrate that best practices of electronic fraud detection in other industries can prove crucial in enhancing security measures in ICS (Hillegersberg et al., 2011). With these best practices being implemented, organizations are better positioned to counter threats of unwanted intrusions, thereby maintaining the integrity and reliability of critical infrastructure. Finally, the dynamic nature of cyber-attacks necessitates a commitment to periodic improvement and upgrading of intrusion detection practices. As technology progresses, methods of potential intrusion will consequently become more complex, and proactive steps must be taken in order to stay ahead of threats in the making. Cooperation between manufacturers, security experts, and regulatory bodies could result in innovation in IDPS to allow the development of more resilient systems. In so doing, ICS-reliant industries can not only strengthen their security posture but also comply with requirements for regulatory standards and protect critical national interests. Accordingly, as the dynamic nature of ICS security continues to evolve, commitment to robust, responsive, and durable intrusion detection systems must remain a top priority.

The extensive spread of cyber threats to ICS calls for the implementation of efficient intrusion detection and prevention systems. As discussed, the traditional security measures applied to ICS have proven inadequate, often due to the legacy systems that dominate the landscape. These outdated systems were not designed with stringent security protocols and thus lack the resilience required to combat sophisticated attacks. Manufacturers are beginning to address these vulnerabilities, yet the sheer scale of existing installations complicates rapid updates and implementations of state-of-the-art security solutions. The challenge is not only to enhance security but also to manage the integration of modern technologies with established infrastructures. Researching new intrusion detection systems is a major advancement for safeguarding ICS environments. It can be done by utilizing electromagnetic side-channel measurements in identifying variations in software behavior executed in ICS. This approach is capable of providing precise profiling and differentiation of program modifications, including in complex setups such as Siemens S7-317 PLCs. This type of approach is novel since it integrates traditional security techniques with cutting-edge cryptographic analysis and thus enriches the range of means accessible for industrial cybersecurity. The ability to detect minor changes in program activity by monitoring these electromagnetic emissions suggests an avenue of opportunity for research and implementation that could revolutionize the way organizations evaluate the integrity of their control systems (Beckers et al., 2017). Essentially, the topics of intrusion detection and prevention in ICS underscore the need for ongoing innovation in security practice. The application of advanced methodologies, such as side-channel analysis, offers promising avenues for protecting key infrastructure against burgeoning threats. Because ICS are a core part of industries such as energy, transport, and manufacturing, securing them cannot be overstated. As the dynamics of cyberattacks change, so too should the methodology applied to protect such systems. The future R&D activities have to work to not only improve existing technologies but also find new ones that have the capacity to co-exist with existing systems and also offer overall security (Alzubi et al., 2015).

REFERENCES

1. Abu-Salih, B., Al Hwaitat, A. K., Al-Sharaeh, S., Alassaf, M., et al. (2024). Overview of mobile attack detection and prevention techniques using machine learning. https://core.ac.uk/download/616997972.pdf (accessed on 13th September 2025).
2. Ahmadinia, A., Javed, A., Larijani, H., Saeed, F., et al. (2016). Intelligent intrusion detection in low power IoTs. https://core.ac.uk/download/293881756.pdf (accessed on 13th September 2025).
3. Ali, M., Akimjonov, A., Bakhronkulova, L., Khabirova, N., et al. (2025). Artificial Intelligence in Cybersecurity: Threats, Defenses, And Future Directions. https://core.ac.uk/download/658857303.pdf (accessed on 13th September 2025).
4. Al-Khateeb, H., Burnham, K. J., Epiphaniou, G., Hammoudeh, M., et al. (2019). Preliminaries of orthogonal layered defense using functional and assurance controls in industrial control systems. https://core.ac.uk/download/196185875.pdf (accessed on 13th September 2025).
5. Almalawi, A., Baecher, A., Boldea, B., Bouchti, N., Brand, M., Byres, E., et al. (2017). Assessing and augmenting SCADA cyber security: A survey of techniques. https://core.ac.uk/download/293885085.pdf (accessed on 13th September 2025).
6. Alzubi, J., Hamdoun, H., & Nazir, S. (2015). Cyber-attack challenges and resilience for smart grids. https://core.ac.uk/download/30276016.pdf (accessed on 13th September 2025).
7. Anglano, C., Aniello, L., Antinori, A., Armando, A., et al. (2018). The future of cybersecurity in Italy: Strategic focus area. https://core.ac.uk/download/188830442.pdf (accessed on 13th September 2025).
8. Arata, C., Dowell, E., Ellis, F., Fennelly, M., Fontaine, M., et al. (2014). Security risk assessment and protection in the chemical and process industry. https://core.ac.uk/download/30731359.pdf (accessed on 13th September 2025).
9. Atkinson, R., Bayne, E., Bellekens, X., Hindy, H., et al. (2018). A taxonomy of malicious traffic for intrusion detection systems. https://core.ac.uk/download/228178450.pdf (accessed on 13th September 2025).
10. Beckers, A., Cui, A., Archambeau, C., Peeters, E., Heyszl, J., Longo, J., Msgna, M., et al. (2017). Side-channel based intrusion detection for industrial control systems. http://arxiv.org/abs/1712.05745 (accessed on 13th September 2025).
11. Bonetto, R., & Rossi, M. (2017). Smart grid for the smart city. https://core.ac.uk/download/84922332.pdf (accessed on 13th September 2025).

12. Busby, J. S., Gouglidis, A., Green, B., Hutchison, D., et al. (2016). Threat awareness for critical infrastructures resilience. https://core.ac.uk/download/76959109.pdf (accessed on 13th September 2025).
13. Chen, J., Deng, R., Ge, P., Liu, Y., et al. (2023). Enhancing cyber-resiliency of DER-based smart grid: A survey. http://arxiv.org/abs/2305.05338 (accessed on 13th September 2025).
14. Choo, R., Dehghantanha, A., Haddad Pajouh, H., Javadian, M., et al. (2019). A two-layer dimension reduction and two-tier classification model for anomaly-based intrusion detection in IoT backbone networks. https://core.ac.uk/download/74220285.pdf (accessed on 13th September 2025).
15. Conte De Leon, D., Oyewumi, I., & Steiner, S. (2021). MAHIVE: Modular Analysis Hierarchical Intrusion Detection System Visualization Event Cybersecurity Engine for cyber-physical systems and Internet of Things devices. https://core.ac.uk/download/479741396.pdf (accessed on 13th September 2025).
16. Diep, Q. B., Truong, T. C., & Zelinka, I. (2020). Artificial intelligence in the cyber domain: Offense and defense. https://core.ac.uk/download/323112801.pdf (accessed on 13th September 2025).
17. Edwards, F. L., & Goodrich, D. C. (2020). Transportation, Terrorism and Crime: Deterrence, Disruption and Resilience. https://core.ac.uk/download/289097215.pdf (accessed on 13th September 2025).
18. Eleimat, M., & Őszi, A. (2025). Cybersecurity in aviation: Exploring the significance, applications, and challenges of cybersecurity in the aviation sector. https://core.ac.uk/download/642633291.pdf (accessed on 13th September 2025).
19. Ghazi, H. E., Ghazi, H. E., Kaabouch, N., Mrabet, Z., et al. (2018). Cyber-security in smart grid: Survey and challenges. http://arxiv.org/abs/1809.02609 (accessed on 13th September 2025).
20. Gibson, K. (2012). The Great New Zealand botnet: Broadband to the door an asset or security issue? (Unpublished manuscript).
21. Ginter, A., Ten, C. W., Vasilakos, A. V., Yamashita, T., et al. (2018). Impact assessment of hypothesized cyberattacks on interconnected bulk power systems. http://arxiv.org/abs/1801.01048 (accessed on 13th September 2025).
22. Glaessner, T., Kellermann, T., & McNevin, V. (2025). Electronic security–Risk mitigation in financial transactions: Public policy issues. https://core.ac.uk/download/pdf/6644802.pdf (accessed on 13th September 2025).
23. Greene, B., Olsen, W., Romeo, J., Sorkin, D., et al. (2025). Remediation

Technology Collaboration Development–A Compendium. https://core.ac.uk/download/pdf/162000436.pdf (accessed on 13th September 2025).

24. Hauser, C. H., Lorenz, D. H., Fudenberg, D., Santacana, E., Kuipers, F., Wu, F. F., Xue, G., et al. (2018). Multi-layer cyber-physical security and resilience for smart grid. http://arxiv.org/abs/1810.00282 (accessed on 13th September 2025).
25. He, X., Liu, R., Nagar, U., Nanda, S., et al. (2014). Enhancing big data security with collaborative intrusion detection. https://opus.lib.uts.edu.au/bitstream/10453/34094/5/d3hux-proof1_Thomas%20%281%29.pdf (accessed on 13th September 2025).
26. Helo, P., & Suorsa, M. (2025). Cybersecurity risks and defense for a European energy retail business: A case study using FMEA and Bowtie incident analysis. https://core.ac.uk/download/664416763.pdf (accessed on 13th September 2025).
27. Jama, Y. (2024). Adopting machine learning-based IDS systems in SMEs. https://core.ac.uk/download/638720787.pdf (accessed on 13th September 2025).
28. Johnstone, M. N., & Peacock, M. D. (2014). An analysis of security issues in building automation systems. https://core.ac.uk/download/41537512.pdf (accessed on 13th September 2025).
29. Kifayat, K., Mac Dermott, A. M., & Shi, Q. (2025). An elastic scaling method for cloud security. https://core.ac.uk/download/96564887.pdf (accessed on 13th September 2025).
30. Kim, D. Y., John, K. H., Goldenberg, N., Zonouz, S., Zonouz, S., & Liu, Y. (2017). On ladder logic bombs in industrial control systems. http://arxiv.org/abs/1702.05241 (accessed on 13th September 2025).
31. Kleinmann, A., Kleinmann, A., Kleinmann, A., Mukherjee, B., Alcaraz, C., Pasqualetti, F., Liang, G., et al. (2017). Stealthy deception attacks against SCADA systems. http://arxiv.org/abs/1706.09303 (accessed on 13th September 2025).
32. Lemay, A., Lin, C. T., Watts, D. J., Hutchins, E. M., Pagani, G. A., Kneis, J., Rubio, J. E., et al. (2018). Tracking advanced persistent threats in critical infrastructures through opinion dynamics. https://core.ac.uk/download/214841416.pdf (accessed on 13th September 2025).
33. Nawrocki, M., Schmidt, T. C., & Wählisch, M. (2019). Uncovering vulnerable industrial control systems from the Internet core. http://arxiv.org/abs/1901.04411 (accessed on 13th September 2025).

34. Radoglou-Grammatikis, P. (2023). SecureCyber: An SDN-enabled SIEM for enhanced cybersecurity in the industrial Internet of Things. https://core.ac.uk/download/586549556.pdf (accessed on 13th September 2025).
35. Romano, N. (2019). Securing our future homes: Smart home security issues and solutions. https://core.ac.uk/download/213463704.pdf (accessed on 13th September 2025).
36. van Hillegersberg, J., Müller, R. M., Thornton, D., Travaille, A., et al. (2011). Electronic fraud detection in the U.S. Medicaid healthcare program: Lessons learned from other industries. https://core.ac.uk/download/pdf/11481166.pdf (accessed on 13th September 2025).

CHAPTER 7

RISK ASSESSMENT AND SECURITY POLICIES

Contents

INTRODUCTION

Widespread use of technology in infrastructures has profoundly altered the landscape of risk analysis and security policy. With businesses increasingly reliant on Commercial Off The Shelf (COTS) software, including Linux and other VOIP software, they unknowingly expose themselves to potential vulnerabilities that can be used by cyber threats. This situation emphasizes the need to design powerful security frameworks that are able to address the unique needs of safety-critical systems. The employment of these technologies is alarming and poses significant concerns regarding the effectiveness of existing security policies, particularly when the policies fail to address the intricacies and interdependencies of modern-day infrastructures. Thus, there is a critical re-examination of risk management procedures necessary in order to reinforce the bridges between safety protocols and cybersecurity initiatives, since traditional methods may be unable to continue working in our fast-paced age of rapid technological shifts and higher levels of sophistication of cyberspace threats. Apart from the complexities of technological integration, the effectiveness of risk assessment measures is compromised by the frailties of regulatory frameworks that are failing to harmonize security and safety. The failure of public-private partnerships to coordinate sufficient cybersecurity measures adds excessively to the burden on safety-critical applications, having a tendency to make them susceptible to exploitation. Existing safety standards such as the ISO27k series struggle to match the pace of cybersecurity guidelines, resulting in gaps and inconsistencies that may lead to ineffective defense measures. Moreover, hybrid standards such as IEC 62443 are under-valuated and lack sufficient support for the enforcement of broad risk management measures. This tension necessitates a shift towards new approaches that not only address the technical requirements of safety but also incorporate a deep understanding of cybersecurity challenges, thereby bringing about resilience to potential threats in networks. The ever-changing online business environment brings in another dimension of risk management and security policy enforcement complexity, with organizations grappling with increasingly complex threats that will undermine essential functionalities and degrade consumer confidence. To counter these threats, it is absolutely essential for businesses to proactively manage their online security. Continuing evaluation and adjustment of security controls in conformity with set information technology and information security standards become necessary in managing risks. Benchmarking frameworks can assist entities in performing this by ensuring that security measures are not only developed in conformity with prevailing standards but also dynamic enough to handle the dynamic nature of cyber threats. Such frameworks should promote continuous improvement, hence enabling organizations to regularly examine and enhance their security policies over time, becoming progressively more resilient to new threats and reinforcing stakeholders' confidence (Johnson et al., 2016; Pye et al., 2007).

Definition of Risk Assessment

Risk assessment is the foundation in security policy development that provides a systematic approach to determining, analyzing, and controlling potential risks that might be a threat to an organization. Risk assessment is essentially a detailed analysis of weaknesses in an organization's infrastructure, processes, and systems. Through systematically identifying these weaknesses, organizations can classify them by the likely impact and probability of occurrence, hence making decision-making easier. The relevance of this method grows in the framework of rapidly evolving IT technologies, which demand constant adaptation of security steps to repel new threats. The inclusion of risk assessment in security policy drafting effectively allows organizations to utilize resources more effectively and formulate a robust mechanism for guarding their assets and activities. Risk assessment process is typically structured into several stages: identification, analysis, evaluation, and treatment. At identification, organizations enumerate potential threats that might compromise their objectives, ranging from cyber-attacks to inefficiencies in the workflow. Subsequently, at analysis, the threats are all investigated to ascertain their likelihood and potential effect, enabling organizations to quantify their overall vulnerability. The assessment stage enables organizations to prioritize risks, which serves as the foundation for effective risk avoidance strategies. Treatment programs are therefore developed to neutralize risks identified and make security policies not only reactive but proactive. Risk assessment therefore becomes a central organizational governance aspect that triggers resilience and adaptable measures in uncertainty, particularly in IT governance contexts where Gheorghe (2025) describes the need for comprehensive risk assessment. Moreover, the emphasis on best practice risk assessment reiterates the value of its contribution to increased security in schools and throughout the broader public sector. Innovative safeguarding practice employed by top-performing schools has been noted, for instance, by Her Majesty's Inspectors, demonstrating how sound risk assessments contribute to building the quality of safety environments. Such evaluation, besides bringing increased security, also provides a framework for continuous improvement and accountability in security practice in most organizational environments. With policies undergoing change, lessons derived from prudent risk analysis come to play an important role in informing effective security strategy that is sustainable as well as responsive to new threats and vulnerabilities. Accordingly, the synergy between risk assessment and security policies cannot be overemphasized, as it provides the basis for developing robust systems that are able to withstand emerging threats while ensuring the well-being of stakeholders across various contexts (N/A, 2011).

Importance of Security Policies

Implementing effective security policies is essential for organizations aiming to reduce risks and protect their assets securely. Security policies are a framework that outlines the processes and guidelines necessary to provide protection for sensitive information and operation integrity. For instance, by outlining well-defined acceptable use policies, employee roles, and incident response processes, organizations can create a culture of security awareness. This is not fire-fighting in nature but encourages continuous monitoring and frequent renewal, enabling organizations to respond rapidly to emerging threats. Therefore, creating comprehensive security policies is needed to minimize vulnerabilities and comply with industry standards and regulations, ultimately enhancing the organizations overall resistance to likely security incidents.

Figure 7.1. The implementation of robust security policies.

Source: https://fastercapital.co/i/Labeling-Service-Security-Ensuring-Data-Integrity--Labeling-Service-Security-for-Startups--Implementing-Robust-Security-Measures.webp.

Furthermore, the character of the threat landscape in the current era requires a superior understanding of security policies, particularly in the field of technological advancements and the increasing sophistication of cyber threats. Organizations must realize that security is not only an IT issue but a vital risk management element that involves all facets of the organization. Policies must be informed by regular risk assessments that identify vulnerabilities and evaluate the potential impact of threats to effectively counter future risks. As noted, security policy alignment with risk assessment models ensures that the organization can prioritize resource allocation and develop targeted strategies to respond to captured threats. This alignment is essential in developing a security posture that can adapt through time, maintaining the importance of having current and relevant security policies. Besides this, success of security policies also heavily relies on the active participation and engagement of all organizational employees. When security policies are well formulated and made public, they give a sense of shared responsibility to staff members, ultimately making the organization's defense system more effective. Employees' training workshops and regular awareness campaigns are required to ensure that each employee clearly knows his responsibility in keeping these policies intact. As individuals become attuned to the implications of their actions upon security, they become critical in avoiding potential security threats. With this in mind, building a security-awareness culture, supported by robust policies, allows organizations to successfully mitigate the complexities of the risks, achieving firm protection against a broad range of emerging threats (Figure 7.1).

Overview of the Relationship Between Risk Assessment and Security Policies

The risk assessment and security policies relationship is central to creating a solid foundation for dealing with potential threats. Risk assessment forms the basis upon which efficient security policies are created, as it involves identifying, analyzing, and classifying risks that an organization is likely to face. By methodical approaches, organizations are in a position to evaluate their current vulnerabilities and potential risks, and thus develop targeted policies to address these determined issues. By way of illustration, determining the likelihood and scope of potential risks allows stakeholders to better allocate resources, thereby leading to an improved security position. Such an active and well-consulted approach not only enhances the resilience of the organizations but also instills security-conscious culture penetrating all organizational layers. Furthermore, the effectiveness of security policies depends on their compliance with the outcomes taken from extensive risk analyses. If security policies are constructed without proper attention to risks recognized, they can be rendered ineffective, leading to protection gaps to be filled by adversaries. It is, therefore, critical for organizations to incorporate constant risk assessment processes and policy formulation alongside. This will ensure security measures evolve with time due to changing threat environments, emerging technologies, and evolving organizational priorities. The interdependent interlinkage between risk analysis and the development of security policies facilitates the dynamic security strategy that not only responds to attacks but also anticipates them. Such interlinkage eventually makes the defenses of the organizations stronger against probable security attacks and increases their ability to respond to incidents more efficiently. Furthermore, the interdependent interlinkage between risk analysis and security policies emphasizes the importance of stakeholder engagement in organizations. Management, IT professionals, and end-users are all stakeholders that play a role in risk assessment as well as implementing security policies. These individuals are important in identifying potential vulnerabilities that may not be noticed by outside auditors or analysts. Effective communication and coordination among these different groups offer a full understanding of risks and encourages shared involvement in security projects. Following the views of certain research, as it has been highlighted, strengthening collaboration among state authorities and educational establishments can be achieved through establishing partnerships, demonstrating the potentiality of working together in creating improved security systems (Atkinson et al., 2004; at Virginia Commonwealth University et al., 2014). Developing a common responsibility for security can therefore lead to stronger organizational procedures as well as forward-looking countermeasures against the various challenges presented by various kinds of threats.

7.1. THE PROCESS OF RISK ASSESSMENT

Grasp the elementary guidelines that govern the process under risk assessment in order to create sound security policies. Risk assessment enables organizations to identify possible weaknesses and threats that can influence their operations, assets, or persons involved. With proper examination of these elements, companies can classify risks based on probability and potential impact, thereby distributing resources in the best manner. This preliminary risk identification phase not only involves identification and review of risks but also requires updating periodically in response to changes within internal and external environments. Organisations can, in iterative fashion, ensure that their risk management strategies remain up to date and effective, aligning with new challenges as and when they arise. Therefore, a robust risk assessment process forms the bedrock of good decision-making and the development of effective security policies that can effectively address identified risks. Following identification and analysis of risks, the assessment and prioritization of such risks is the next extremely vital step in the risk assessment process. This step is fundamental in deciding which among these risks need to be treated on an urgent basis and which need to be monitored in the long term. By quantifying risk as a matter of probability and potential impact, organizations can strategically allocate their resources. This process of assessment also often involves the participation of stakeholders, which can be used to bring additional insight and foster a shared culture of responsibility for security and risk management. A good example is the European Commission's Privacy Impact Assessment (PIA) framework, which specifically highlights involvement by diverse stakeholders, particularly in cases with emerging technologies like RFID. This kind of engagement ensures that varied perspectives enter into the risk assessment process to produce more efficient security policies that consider stakeholder concerns and social values (Spiekermann-Hoff et al., 2020). Ultimately, the success of any risk assessment process is determined by its integration into overall organizational security policies. By ensuring that risk assessment results directly influence security measures, organizations are in a position to ensure a culture of proactive risk management. This synthesis not only enhances compliance with the law, including human trafficking and other criminal laws but also ensures the focus on protecting vulnerable populations. As can be observed from recent trends of evolution in U.S. policy interventions against human trafficking, the identification of emerging risks and the deployment of focused response measures are critical in the evolution of a robust security system. Therefore, the continuous cycle of feedback between the process of assessing risk and security policy formulation is crucial to the balance between risk protection and operation efficiency (Siskin et al., 2012). By continued activity in this field, organizations are able to adapt to emerging threats while maintaining regard for ethical considerations and stakeholder engagement (Figure 7.2).

Figure 7.2. The process of risk assessment.

Source: https://63sats.com/wp-content/uploads/2024/03/Risk-Assessment-Process-Master-Image-Thumbnail.webp.

7.1.1. Identifying Potential Risks

Identifying prospective dangers in today's organizational context is an imperative task, a mere regulatory compliance but rather situating organizations in a better position in the midst of competition. Organizations are flooded with volumes of information capable of informing strategic choices. To effectively utilize this information, managers need to identify a range of risks that may jeopardize organizational goals. Through systematic risk assessment, companies are not only able to identify probable weaknesses but also formulate resilient measures that reduce such risks negative impacts on overall operational integrity. Such a proactive approach to risk management summarizes a dual agenda: strengthening business processes while maintaining flexibility towards an ever-changing marketplace (RUSU et al., 2025). Thus, determination of potential risks becomes a key feature of larger risk management procedures, driving organizations into higher levels of efficiency and competitiveness in the market. Also, in sectors like energy, intricacies of determining potential risks are further governed by the intricate web of supply chains and their significance in supporting operational efficiency.

The dynamics of such supply chains, commonly characterized by interdependencies across fuels, technologies, and infrastructures, necessitate an appreciation of the attendant risks. Supply chain disruptions, for instance, can result from geopolitical tensions or environmental concerns, leading to significant widening of energy security gaps. Mitigating the risks entails appreciation of how strongly connected ecosystem drivers tend to blur boundaries between short-term threats and long-term vulnerabilities (Arent et al., 2014). Hence, a systemic review of supply chain risks must take into account not only the current operating context but also anticipated changes towards low-carbon systems since these developments introduce additional uncertainties that must be closely watched. Effective risk identification is then followed by existing practices, which are mostly reactive and not proactive in nature. Without a structured way of doing things, organizations are likely to overlook early warning signs that portend impending threat before it gradually becomes crises. Risk management frameworks should entail both qualitative and quantitative methods in order to formulate an integrated knowledge of vulnerabilities that cover operational, technological, and regulatory areas. Integrating these kinds of frameworks will enable organizations to anticipate risks from market volatility, technological redundancy, and dynamic legal climates, thereby fostering a culture of resilience and flexibility. As companies grapple with the intricacies of the modern connected realms, risk identification practices and strong security policies blend perfectly, reinforcing sustainability in the face of paradigm changes and competition (RUSU et al., 2025).

7.1.2. Analyzing Risk Impact and Likelihood

Where risk analysis is concerned, knowledge of the impact and likelihood of threats is vital when making security policies. The risk impact analysis takes into account the potential consequences arising from a threat, enabling organizations to coordinate response efforts. For instance, with human trafficking, a problem with significant implications for the victim, law enforcement, and general public safety, risk impact analysis can guide the allocation of resources into prevention and intervention activities. As stipulated in the international and domestic human trafficking report, U.S. policy measures to counter such threats must be founded on a sound conceptualization of the magnitude of the threat and the effectiveness of countermeasures. Such cooperative effort not only creates awareness but also fosters stakeholders' solidarity in combating endemic ailments, ultimately resulting in enhanced communities' immunity to such threats (Siskin et al., 2012). Risk likelihood analysis involves quantitative measurement of the chances of a specific threat occurring and assessing their potential frequency. This is particularly useful in developing an overall picture of weaknesses in different contexts, such as organizations or communities. With the use of models that incorporate qualitative and quantitative data, organizations can assess risk likelihood better and sequence activities accordingly. For instance, information security awareness studies reveal that it is beneficial for organizations to be aware of the actions of employees toward security practices. Drawing from a broad array of theoretical frameworks, varying from protection motivation theory, organizational culture and individual behaviour can make the prospects of the enforcement of security policy a success. Thus, a clear understanding of risk impact and likelihood allows organizations to develop more effective and resilient security policies that ward off looming threats appropriately (Hutchinson et al., 2020). Additionally, the inclusion of risk impact and likelihood assessment results in an eloquently worded risk management plan. This two-way assessment forms the grounds for the development of well-informed security policies that enhance current threats and looming future problems. By embarking on an iterative process of assessment that constantly examines these factors, companies are able to respond adaptively to their emerging threats. As highlighted in numerous policy studies, such as those for human trafficking, ever-changing security models are of primary importance in a fluid risk environment. Finally, an integrated approach to risk impact as well as likelihood analysis ensures that security policies are not merely reactive but proactive, building long-term resilience and protection against sophisticated threats. This process supports the necessity of sound design risk assessment practices as one of the elements of good security policy design (Siskin et al., 2012) (Figure 7.3).

Likelihood and Impact Model

			Likelihood				
			Rare	Unlikely	Moderate	Likely	Almost Certain
			1	2	3	4	5
Impact	Severe	5	**5**	**10**	**15**	**20**	**25**
	Major	4	**4**	**8**	**12**	**16**	**20**
	Moderate	3	**3**	**6**	**9**	**12**	**15**
	Minor	2	**2**	**4**	**6**	**8**	**10**
	Negligible	1	**1**	**2**	**3**	**4**	**5**

Figure 7.3. Analyzing risk impact and likelihood.

Source: https://thealsgroup.com/wp-content/uploads/2024/03/Likelihood-and-impact-model.png.

7.1.3. Prioritizing Risks for Mitigation

Risk prioritization for mitigation is an all-important practice in the realm of risk management and security policy because it helps improve overall resistance to potential threats. Picking out some risks for intervention intentionally allows organizations to spend their resources more effectively and better. By identifying vulnerabilities in their operating environments, decision-makers can target high-impact risks that pose direct threats to their objectives. This ordering not only does the most with limited resources but ensures that issues of greatest importance are being addressed first and therefore reduces opportunity for serious disruption or loss. Learning Alliance for Adaptation in Smallholder Agriculture is a prime example of the worth of intelligent decision-making in risk prioritization. Through recording and cross-program comparison of experience lessons from existing programs, the Alliance illustrates the way the application of evidence-based approaches can lead to more successful adaptation in response to threats from climate change. This, in turn, identifies the inherent value of ranking risks in the context of farm policy, especially since the stakeholders are cooperating to aid the smallholder farmers under increasingly uncertain climate. Building on systematic knowledge production, the Alliance upholds the argument that interventions ideally targeted, based on empirical evidence, can holistically enhance resilience and sustainability in risk-exposed areas and thus aid closely risk management best practices (Ekong et al., 2019). Even further, with the new adaptation landscape, it is increasingly important to realize that prioritizing risks is not a static process. Research breakthroughs typically reveal emerging vulnerabilities associated with climate change, triggering continuous assessment of risks. For instance, highlighting agricultural adaptation strategies can lead to foraying into new approaches closing knowledge gaps in resilience. Addressing the social vulnerabilities associated with climate change allows organizations to implement anticipatory measures that reduce exposure to such emerging threats. This cyclical pattern of risk assessment, reduction, and adaptation, highlighted in the United States climate change adaptation reports, forms the central template for furthering security policy intended to lead toward a sustainable future (Snover et al., 2013).

7.2. TYPES OF RISKS IN ORGANIZATIONS

In organizational risk analysis, the distinction between risk types is of utmost importance in the formulation of effective security policy. Strategic risks are decisions which can impact an organization's long-term objectives and competitiveness position. Poorly managed strategic choices, either in terms of market entry or asset allocation, can lead to huge monetary losses. Secondly, since organizations depend increasingly on technology, they are open to operational risks that relate to internal procedures and system failures. Such types of risks, such as supply chain failures to machine malfunctions, identify the necessity for robust operational systems. Therefore, an understanding of such forms of risks provides a context for constructing targeted risk management frameworks that underpin organizational resilience. Financial risks, another type of organizational risk category, arise as a result of uncertainties affecting profitability and asset valuation negatively. Volatility in the market, credit risks, and liquidity problems highlight the need for organizations to adopt end-to-end financial governance processes. The emergence of regulation standards, such as Basel II, mandates companies to not only think about their financial structure but also evaluate their adherence to prescribed governance processes (Önal, 2025). This highlights the need to incorporate risk evaluation into business processes at its core, such that financial decision-making remains in line with overall risk management policies. Through identification of financial weaknesses, entities are able to create better responses to avoiding future crises and bolstering their stability. The growing intricacy of organizational contexts has amplified the relevance of cybersecurity threats, a top concern for most entities currently. Cyber threats like malware attacks and data breaches pose substantial threats to both operational integrity and stakeholder trust. These represent potentially cascading risks to an organization's reputation and customer relationships with grave long-term consequences. It is in this context that research like the Cetifs project identifies the imperative for organizations to close there currently existing gaps in critical infrastructure modeling and analysis (Bloomfield et al., 2009). Through proactive cybersecurity, organizations are better able to safeguard their assets, imposing compliance with well-defined security policies aligned with strategic priorities and business needs (Figure 7.4).

Figure 7.4. Types of risks in organizations.

Source: https://fastercapital.co/i/Risk-management--Mitigating-Risks-through-the-Topping-Up-Clause--Exploring-Different-Types-of-Risks.webp.

7.2.1. Physical Risks

Physical risks, within the context of risk assessment and security policies, refer to a variety of dangers that may pose risks to organizational operations and well-being of staff. These risks could be caused by environmental factors, health conditions, or intentional violence. Businesses must recognize that physical threats such as nature-based calamities or malevolent attacks can have the potential to disrupt business continuity in a considerable way. Businesses need to have comprehensive risk assessment policies to identify these vulnerabilities. For example, businesses that conduct their activities in disaster-swept regions may have to put more priority on emergency response plans based on their functional necessities. As per the investment risk study of Ukraine's renewable energy sector, the overall need for state intervention and effective policies to counteract physical risks is crucial in creating stable infrastructure (Tkachuk et al., 2025). Additionally, physical risk effects transcend immediate safety concerns; they have an influence on stakeholders' perception and trust towards an organization. For instance, a serious incident of workplace violence could result in high anxiety among employees and can bring discredit to the institution. Physical risk management therefore has to include regular risk assessment with staff training programs to enhance safety and awareness culture. As emphasized by the results of a recent research in Ugandas Fintech industry, understanding how security actions relate to incidents is crucial as organizations struggle to secure their mitigation efforts. The successful management of physical threats by means of well-informed security policies not only protects staff but also complements the overall operating resilience of businesses under different kinds of threats (Mackenzie et al., 2024). In altering security policies against physical threats, companies must employ advanced monitoring systems and establish solid communication networks for incident reporting. The evolving state of threats necessitates continuous review of security arrangements, responding to environmental as well as organizational changes. Through active measures, such as comprehensive security policies and employee training, companies can almost certainly reduce their susceptibility to physical threats. The necessity for regular analysis is also underscored in literature by the necessity to build a risk management culture that can respond to changing challenges. For the Ukraine energy industry, as noted, the interactive relation between policy systems and physical risk mitigation must be continually scrutinized to strengthen the stability and security of operations in a rapidly changing world (Tkachuk et al., 2025).

7.2.2. Cybersecurity Risks

The contemporary digital world is increasingly plagued with cybersecurity risks that tend to undermine organizational solidity and data security. The risks come from numerous sources, from external cyberrisks to insider compromises and unintentional leakage of sensitive information. For instance, as organizations get increasingly networked by advanced technology, the attack surface expands, making them vulnerable targets for cyberspace criminals. The consequence is sometimes severe, from economic loss to reputational blemish. As emphasized through initiatives like the Trusted CI project, aimed at providing centralized cybersecurity services, the effectiveness of good risk assessment becomes a priority to ensure data integrity across industries, like academia and research, where data breaches are impactful (Adams et al., 2019). Organizations must recognize vulnerabilities in advance and implement general security policies to address these prevalent threats. Moreover, the intersection of cybersecurity and artificial intelligence (AI) introduces new risks and possibilities that must be considered in any valid risk assessment methodology. The Centre for European Policy Studies, through its Task Force on AI and Cybersecurity, has created that while AI may enhance cybersecurity measures—e.g., making detection and response to threats smarter—so too can it be exploited by criminal elements. This double purpose of AI necessitates current security architectures to be redesigned to address not just the way AI systems can be made secure but also how they might inadvertently increase threats. For example, AI-driven systems are able to introduce new vulnerabilities that can be exploited by attackers, thereby complicating risk management (Fantin et al., 2020). Security policy must therefore include provisions for AI technologies, acknowledging their potential to enhance and erode cybersecurity. Lastly, the continuously evolving pattern of cybersecurity threats necessitates constant monitoring and revision of security policies to ensure they remain effective and relevant. With the rapid pace of technological trends and attack methodologies employed by cyber attackers, organizations must develop a culture of vigilance that enables continuous employee education and training. Reviewing and updating security policies regularly aligns not just with regulatory requirements but also fosters an organization's immunity to evolving threats. Further, creating integrated collaboration among multiple stakeholders—government, private sector, and academia, as highlighted by initiatives like the HLEG on AI Ethics and Trustworthy AI—can assist in creating a better picture of actual risks and adaptive security measures. This mutual engagement lies at the heart of creating an integrated cybersecurity model that can respond and anticipate evolving threats within the virtual world (Fantin et al., 2020) (Figure 7.5).

Figure 7.5. Cybersecurity risks.

Source: https://marvel-b1-cdn.bc0a.com/f00000000310757/www.fortinet.com/content/dam/fortinet/images/cyberglossary/cybersecurity-risks-of-working-from-home-(1).png.

7.2.3. Operational Risks

Operational risks are one of the most interesting areas of attention for organizations wishing to enhance their resilience and sustain competitive advantage. Such risks are predominantly linked to process, system, and people failures, among a broad spectrum of possible events that could affect operations. In the advanced economic environment of the present day, where the adoption of business process automation and reliance on technology become increasingly prevalent, attention to the management of operational risks cannot be emphasized enough. Managers need to be acutely sensitive to the several potential vulnerabilities arising from both internal and external sources, necessitating the uptake of a proactive approach in risk assessment. Organizations stand a better chance of reducing the impact as well as maintaining operational continuity when they are informed and aware of possible threats, thereby guaranteeing their reputation and financial soundness (RUSU et al., 2025). Sound risk management practices are essential for the organization to be able to navigate the complex character of operational risks. The adoption of a systematic approach enables the managers to review, analyze, and rank the risks to ensure efficient allocation of resources to counter potential losses. One of the most important aspects of the process is ongoing monitoring and revisiting of the current risks because the changing face of the business environment has the potential to introduce new threats. In addition, utilization of information technology governance models has the ability to improve the ability of organizations to control operational risks of information technology and systems greatly. Integration of risk management practices in organizational culture increases proactiveness in the employees and fosters an environment in which awareness of risk is a priority (Önal, 2025). Apart from this, the role of operational risk management is not just to remain in compliance with regulatory guidelines; it further entails continuous improvement in business processes. Utilizing an iterative process, organizations are able to learn from past experiences and apply the knowledge in future risk assessments. This embedding of best practices not only reinforces the overall risk profile but also supports operational efficiency. As the reliance on digital technology increases, organizations specifically must focus on the control mechanisms of their information systems. Not only does a strong operational risk management system reduce financial losses, but it also helps build stakeholder confidence, which enhances the organization's market position and long-term sustainability (RUSU et al., 2025).

7.3. DEVELOPING EFFECTIVE SECURITY POLICIES

Developing sound security policies is a serious task that requires a definite understanding of the risks to which an organization is exposed. Risk assessment is the cornerstone of developing the policies because it identifies the possible threats and vulnerabilities that can be exploited by such risks. Examining them in an orderly manner enables organizations to focus their security drive and invest the required resources. This is achieved by establishing an active response to possible security breaches instead of a passive one. Effective policies need to address not only the technical aspects of security but human factors as well, including training of personnel and incident handling procedures. Lastly, the integration of risk assessment in the formulation of security policy makes the organizations themselves more able to withstand threats, encouraging a culture of security among stakeholders.

Figure 7.6. Developing effective security policies.

Source: https://media.licdn.com/dms/image/v2/D5612AQGiRA2s4Ymwrw/article-cover_image-shrink_720_1280/article-cover_image-shrink_720_1280/0/1678920649520?e=2147483647&v=beta&t=9q6fv0M2JEtSkrUv4nAA9ZSRcblUWQWoiQ-_uSzR6mY.

Moreover, the effectiveness of security policies also depends on their dynamism in tracking evolving risk paradigms. As organizations are presented with multitude emerging threats, ranging from cyber threats to ecosystem threats, security policies must be constantly evaluated and updated. A static policy instantly becomes obsolete and leaves organizations vulnerable to emerging challenges. Through the incorporation of dynamic methods of risk evaluation through constant monitoring and scanning vulnerabilities, organizations can stay ahead of the curve. By setting up a cycle of periodic updates and reviews, organizations are able to keep their security controls up to date and effective. This cyclical process not only demonstrates a devotion to security, but it also places organizations in a position to react quickly to unexpected events, thereby minimizing resultant losses and building stakeholder trust. Finally, effective implementation of security policies requires sound communication and cooperation at all organizational levels. Well-documented policy is ineffective if it is not disseminated to appropriate employees who would implement it. Engaging stakeholders in the design of policy helps elicit buy-in as well as enhance compliance. Training programs for staff that educate employees on security procedures and the necessity thereof can also minimize risks based on human error. Additionally, the inclusion of a system for giving feedback allows employees to raise concerns and recommend new policy. By establishing a security awareness culture across the organization, organizations can be in their best position to utilize their resources not only to protect against attacks but also to maximize their nimbleness in having to respond to a more sophisticated security landscape as emphasized in the need for integrated strategies in addressing the risks of climate change and other global issues as discussed in Kahiluoto et al. (2009) and O'Brien et al. (2008) (Figure 7.6).

7.3.1. Key Components of Security Policies

Having clear goals and scope in security policies is the core of organizations that are ready to manage risk effectively. Such policies outline the function of security controls so that it becomes evident for everyone what their responsibilities and roles are. By establishing well-defined goals, e.g., minimizing data breaches or protecting sensitive information, organizations can make their defenses support their business objectives. Not only does this make the defenses stronger for resisting potential attacks, but it also attain compliance with relevant regulations. Security policies must address an explicitly stated scope in order to address the unique problems a company faces, from physical security to cyber-attacks. As the evolving environment of threats presents new challenges, policies must evolve constantly in a bid to incorporate emerging threats. Organisations must then review and update these components as part of their risk management policy from time to time, developing the foundation for effective security culture. Risk assessment is the cornerstone while developing security policies as it enables organisations to identify, evaluate and rank potential threats. With proper risk assessment, organizations can identify vulnerabilities that would make them vulnerable to victimization by threats such as human trafficking and data breaches. Such evaluations include a determination of likelihood of different kinds of attacks and an understanding of the likely impact on stakeholders and business operations. Proper policy draws on knowledge from risk assessments to inform security measures, offering a realistic response to resource allocation. It enables the development of targeted action to counteract the most severe threats while still ensuring compliance with mandated standards and regulations. For instance, a security policy founded on detailed risk analyses could mean that training programs or technological upgrades are implemented as measures to prevent, thus raising the overall security posture of the company. Implementation and enforcement are also crucial components of effective security policies, ensuring that guidelines established are translatable into practice. Organizations must have certain procedures implemented for employee training, monitoring, and enforcing any infractions against the policy. Training programs help aid in teaching staff about security best practices and the importance of adhering to policies. Further, ongoing monitoring and evaluation are required in order to quantify the effectiveness of security measures, thus providing a feedback mechanism for continuous improvement. As organizations grow with developing threats, the enforcement of security policies should include stated penalties for non-adherence, promoting a culture of responsibility among the personnel. This not only strengthens the organizations ability to combat risks but also puts personal responsibility into the hands of each and every employee. By doing so, organizations can develop a forward-looking security environment that reduces vulnerabilities and fosters resilience to prospective threats.

7.3.2. Aligning Policies with Organizational Goals

Proper alignment of security policies with organizational aims is fundamental to risk assessment and security policies in contemporary enterprises. Organizational aims are usually entangled with performance results, and it is therefore crucial that security policies be aligned with these aims. Whenever organizations introduce new technology, the management is challenged at the same time with optimizing value and reducing information technology (IT) risks. Current governance frameworks such as those described in Gheorghe (2025) provide procedures for organizations to perform audits that are risk assessment-based. By adopting risk-based methodologies, IT auditors can enhance the effectiveness and efficiency of security policies such that they directly reflect the organization's overall goals. Aligning security policies from a strategic viewpoint presents a foundation for a more adaptive organizational framework capable of supporting change in risk environments. Incorporating stakeholder feedback in the development of security policies also acts as a deciding factor in goal alignment in the organization. By an embracing of new processes, such as those followed by the Department of Homeland Security (DHS) in its Quadrennial Homeland Security Review, organizations can facilitate a better model of security policy formulation. Stakeholder participation not only yields diverse perspectives but also instills transparency and collaboration within the organization, thus enhancing the implementation of security policies. It is by including the stakeholders in the policy-making process that the organizations are able to identify the critical risks and fill existing gaps more precisely, as the

Cigler et al. (2010) describe. This alignment in this way makes the organizational mission robust as well as ensuring the security measures are comprehensive and sensitive to both internal and external risks. To successfully align the security policies with organizational goals, constant evaluation and fine-tuning are required. Security environments are not static, and regular inspection of security policies in line with shifting organizational objectives is therefore mandatory. A proactive risk assessment procedure ensures that organizations are able to identify potential vulnerabilities and amend their policies in response. According to Gheorghe (2025), implementation of a systematic audit procedure provides organizations with the tools to methodically examine their IT governance, as included in effective risk management. Besides, having built-in measures to decide on the impact of policy changes ensures that consistency with organizational goals in the future will be maintained. By prioritizing flexibility and responsiveness, organizations can create security policies that not only protect their assets but also drive their strategic agenda (Figure 7.7).

Figure 7.7. Aligning policies with organizational goals.

Source: https://fastercapital.co/i/Organizational-Alignment--Fostering-Unity-through-MBO--Aligning-Employee-Goals-with-Organizational-Goals.webp.

7.3.3. Involving Stakeholders in Policy Development

Involving stakeholders in policy-making is a requirement to enhance the power and effectiveness of risk assessment and security policies. Having diverse participants—business executives, government organizations, and members of society—guarantees that different perspectives are included in the decision-making process. The participatory process can yield more representative policies of the publics' needs and issues and therefore generate greater compliance and trust among constituents. One notable example of such engagement can also be identified in the Department of Homeland Security's Quadrennial Homeland Security Review (QHSR), in which the review engaged feedback from thousands of stakeholders. Through transparency and collaboration, the DHS not only gained constructive feedback but also forged a strong stakeholder engagement capability in its various agencies, in which policy strategies were aligned with public priorities and overall policy effectiveness was enhanced (Cigler et al., 2010). The process of engaging the stakeholders also creates a feeling of ownership and accountability among the stakeholders, with greater opportunities for the policies to be enforced successfully. If the stakeholders are involved in the discussion and development of security policies, they are more likely to understand the need for such policies. It is particularly significant in risk assessment, where compliance with complex policies depends on the support of the public. Also, stakeholder interaction can identify and minimize the potential for future conflicts to enable a smoother process from policy making to action. The experience gained from stakeholder interaction can further guide the development of quantifiable measures that can be utilized to measure outcomes, such as levels of trust within communities and the effectiveness of communication campaigns (Cigler et al., 2010). But it is necessary to evaluate if existing governance structures are adequately capable of dealing with the complexities of stakeholder participation in crises or extraordinary events, as foregrounded in the food system governance. Standard stakeholder structures may be inadequate when unforeseen threats arise that threaten to ruin the same policies intended to protect communities. Policymakers must remain vigilant and nimble in response so they can make the frameworks for stakeholder engagement dynamic enough to adapt to the evolving environment of risks and security threats. Constant dialogue among stakeholders must be facilitated in order to enable constant development of policies and practices to address new challenges. By creating an adaptable governance structure that engages stakeholders proactively, security policies can endure the pressure of crisis environments and safeguard people's interests with better effectiveness.

7.4. IMPLEMENTING SECURITY POLICIES

Security policy enforcement is an integral component of effective risk management in organizations. Through the imposition of explicit rules and guidelines, organizations can adopt a rigorous framework that minimizes vulnerability to threats and prevents potential risks. This systematic approach allows organizations to anticipate hazards and institute countermeasures that are suitable for their working conditions. According to RUSU et al. (2025), organizations are required to utilize information within their possession and analyze threats that can impact their mission. Not only does this establish a safety culture, but also provides authority to employees to recognize and report security threats in advance. Ultimately, the efficiency of security policy not only protects physical and digital assets, but significantly enhances the organizations overall immunity to disturbances. Furthermore, effective deployment of security policies necessitates ongoing assessment and adaptation to shifting security dynamics. With threats emerging or developing, organizations must be prepared to update their policies to maintain their effectiveness. Regular audits and reviews of the policies guarantee that they flag areas of weakness and room for improvement, keeping the security measures up-to-date and robust. Input from employees and lessons drawn from past breaches can also drive policy changes. As pointed out, effective actions ought to be grounded in a comprehensive understanding of the security threats an organization is faced with, as outlined in Cannon et al. (2015). Such a feedback loop encourages organizational culture that treats security as an ongoing process and not a compliance exercise. Overall, security policy implementation is a central practice in safeguarding an organization's assets and reputation. Through active involvement in risk analysis and developing thorough, dynamic policies, organizations are able to build resistance against threats. Through the encouragement of the culture of security first by employees, organizations foster shared responsibility for keeping things secure and for plugging gaps in defenses. This approach not only supports operational integrity but also organizational credibility in the business world. As organizations pursue long-term growth and stability, the importance of ensuring strict implementation and continuous improvement of security policies cannot be overemphasized, reaffirming the link between effective risk management and long-term success.

7.4.1. Training and Awareness Programs

Under the context of risk analysis and security policy, Training and Awareness Programs are core factors that have immediate effects on the efficiency of a security infrastructure within an organization. The intent of such programs is to heighten employees' attentiveness to security threats and measures, thus inculcating an organizational culture of vigilance and preparedness. By employing workshops, simulations, and briefing sessions, organizations can equip their employees with the skills that enable them to detect and respond to probable threats. This proactive approach ultimately makes the organization more resistant to security breaches, as better-aware employees are the first line of defense. The integrated nature of such programs is brought out through programs like the cybersecurity protocols issued under Executive Order 87 N/A

(2016), which stresses the need for continuous education to effectively address ever-evolving threats. Apart from this, raising awareness about potential security vulnerability points also contributes meaningfully to creating an informed staff in responding to incidents efficiently. Training exercises on the basis of scenario learning assist the workers in thinking more rationally regarding their actions and implications of their choices in security aspects. Phrasing, for example, can be duplicated to simulate the types of threats present in modern cyberspace. Such an inclusion of risk assessment protocols not only prepares the personnel for the real event but also inspires them towards a sense of realizing the significance of their respective roles in the overall security scenario. Adherence to an official awareness program is witnessed in collaborative practices documented in policies like EO87, which inspires collaborative interagency efforts aimed at reinforcing cybersecurity awareness (N/A, 2016).

Figure 7.8. Training and awareness programs.

Source: https://prolink.insure/wp-content/uploads/2022/05/Security-Awareness-Training-02.png.

Apart from that, the effectiveness of Training and Awareness Programs relies on their ability to respond to the specific needs of the workforce and operating environment of the organization. With evolving technology and new risks emerging, specialized training modules handling relevant risks must be addressed. Organizations must constantly assess the usefulness of their awareness programs so that they can make them current with today's risk profiles and incorporate people's feedback. This dynamic process ensures the continued evolution of security policies and measures, reinforcing commitment to security excellence and fostering a culture of collective responsibility. As the nature of security threats continues to shift with advances in emerging technology, the resultant lessons learned from these programs will inform future risk analyses and security planning, demonstrating that training and awareness are paramount to the continued battle against cyber-attack (Figure 7.8).

7.4.2. Monitoring Compliance and Effectiveness

Successful monitoring of compliance and effectiveness is essential for any organization seeking to adhere to risk assessment and security policy. Monitoring compliance involves thorough checks on procedure, process, and behaviour to ensure that they are in line with policies put in place and regulatory requirements. This attention not only helps in the identification of areas of non-conformity but also in developing a culture of responsibility within the company. Organizations utilize various methodologies, for example, audits and performance reviews, in order to obtain credible data on their compliance status. As these evaluations inform administrative decisions, they provide a basis for risk reduction strategies, thus enhancing effectiveness in overall security measures. The addition of continuous monitoring enhances an organization's ability to catch up with new threats and emphasizes active engagement in maintaining compliance within the increasingly broadening regulatory landscape (Fair Labor Association, 2012). Monitoring compliance takes additional complexity in the context of IT governance with the rapid pace of technological evolution. The challenge is to redefine IT elements to construct value with reduced risks using sound management practices. Here, current security policies not only need to address technology-related risks but must align with broader organizational objectives. Improvement in IT governance must have a process that integrates risk assessment within the auditing process. This approach enables organizations to properly measure risks that come with their IT governance frameworks and identify areas of high risk that may need high-level scrutiny (Gheorghe, 2025). Subsequent continuous and proactive monitoring for compliance is needed in dynamic security policies that are responsive to the dynamic technology environment. Monitoring mechanisms' effectiveness impacts organizational security policies, particularly in the context of compliance. Companies that are actively engaged in effective monitoring are more likely to respond to compliance breaches and address any form of weakness in their security systems. Furthermore, the feedback loops established via monitoring activities allow iterative enhancement of risk assessment procedures, which can lead to an improved security system. As companies collect and analyze data on compliance and impact, they ought to emphasize transparency and communication across all levels. Periodic reporting can allow stakeholders to understand risk dynamics and ascertain that compliance is everybody's responsibility. This mutual character of monitoring brings into existence a more enlightened organizational culture, which fosters a proactive mindset in regards to not just compliance but also the effectiveness of security policies in general (Fair Labor Association, 2012; Gheorghe, 2025).

7.4.3. Adapting Policies to Changing Risks

In an increasingly unpredictable world, policy adjustment to changing threats has become a necessary component of risk analysis systems in security policies. The rapid evolution of environmental, social, and technology-driven threats demands policy proactivity in both policy-making and policy implementation. Policymakers must understand that static frameworks are insufficient to address dynamic and composite threats. An effective policy adaptation cycle begins with comprehensive risk analyses that identify probable vulnerabilities and threats across various sectors. Quality data and ongoing monitoring form the bedrock so that these analyses remain current with respect to the ground reality so that policy guidance can be realigned in a timely manner. This adaptive policy framework not only encourages resilience but also encourages stakeholder engagement, fostering a setting that is conducive to enabling varied voices to contribute to decision-making processes. In addition, there are some sectors, such as agriculture and urban development, that display the demand for particular approaches within the process of recasting policy in terms of meeting particular issues arising from hazards. For instance, according to recent studies, urban food security is highly vulnerable to climate change, particularly in Southern Africa, where poverty levels are high. Policy makers have to reflect on how shifting climate patterns impact on the urban system and food chain in order to develop sustainable interventions.

By incorporating scientific climate projections into policy planning, the stakeholders are better able to coordinate resources and strategic initiatives to reverse food insecurity. A targeted approach in this manner guarantees that policies are not only responsive but proactive as well, addressing the root causes of vulnerability while promoting sustainability at grassroots levels and thereby gaining the required food resources amid variable climatic circumstances. Lastly, climate change adaptation and risk management should be mainstreamed into security policy in order to address the growing global concerns. It is crucial to understand how climate change risks impact societal stability and the availability of resources. Adaptive strategy development under continuous climate assessment can deliver more resilient communities capable of better adapting to environmental change. Policymakers need to invest in education, technology, and infrastructure in order to enhance adaptive capacity while capturing the interdependence of risks in borders. Multidimensional policies will not only be a response to the present threats but also a stepping stone for sustainability and security in the future, thereby recommitting adherence to the adaptation of policies to evolving risks in an ever more interdependent world (Figure 7.9).

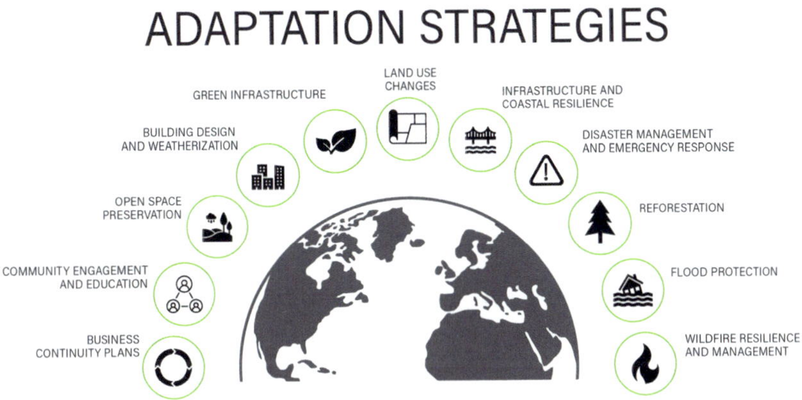

Figure 7.9. Adapting policies to changing risks.

Source: https://eaest.com/wp-content/uploads/2022/01/Climate-Change-Adaptation-Strategies-1-1024x461.png.

7.5. RISK MANAGEMENT FRAMEWORKS

As part of creating effective security policies, extensive knowledge of risk management frameworks is required. These frameworks provide structured methods of approaching risk identification, analysis, and reduction in organizations. By addressing potential hazards through a structured process, stakeholders are better positioned to prepare for threats that may impact operational integrity and regulatory compliance. A key part of these frameworks is their ability to adapt to an ever-evolving risk environment, enabling organizations to adjust accordingly. In addition, a good risk management framework also includes factors like legal compliance, stakeholder engagement, and operational efficiency, making it a complete approach that reinforces both security policies and operational procedures. This multi-layered strategy ensures organizations are proactive, not reactive, in determining risk, implementing a strong framework that is aligned with organizational objectives and stakeholder aims. The integration of advanced technologies, such as artificial intelligence and data analytics, has become a key feature of modern risk management frameworks. All these technologies have the capacity to streamline the process of risk identification and enable real-time analysis of large volumes of data.

Automation, as established by recent studies, enhances risk assessment and mitigation capacities across healthcare, finance, and transport industries. For instance, the cross-EU and U.S. comparative discussion of AI risk management strategies determines the emphasis on compliance and transparency in the EU and innovation through decentralized regulation in the U.S. These differences are representative of the varying perceptions toward balancing effective risk management and the facilitation of technological development. Understanding such frameworks allows companies to modify risk strategy to meet specific industry demands with adherence to relevant legislation and guidelines ultimately enhancing overall security policies. Additionally, in addressing complex geopolitical challenges, such as with border management frameworks, good risk management must consider the interplay of economic, legal, and technical factors. The study of customs and border regimes demonstrates the way risk analysis needs to be synchronized with operational objectives in a way that maintains customs procedures flowing smoothly while ensuring security. A three-dimensional strategy that harmonizes various regulations and technological advances, particularly in the environment of high risk like international trade, is needed. Automated document verification system technology brings out the requirement for dynamic policies that can accommodate changing security threats along with allowing cross-border commerce. By harmonizing such conflicting constituents, risk management frameworks can significantly enhance the resilience and responsiveness of an entity during sudden short-term as well as long-term threats, ultimately establishing a secure operating environment that is effective and efficient.

7.5.1. Overview of Common Frameworks (e.g., National Institute of Standards and Technology (NIST) and International Organization for Standardization (ISO))

Compliance with standards is crucial for organizations that seek to balance operating efficiency against security in the context of risk assessment and security policy. The NIST and the ISO are the best-known standard frameworks among these. NIST provides a comprehensive set of guidelines that are known for their intimidating and applicability in various industries, particularly in the United States. The NIST Cybersecurity Framework, for instance, promotes a risk-based method, assisting organizations in identifying, defending against, detecting, responding to, and recovering from cyber threats. This framework not only describes best practice but also promotes an integrated process for managing risk so that organizations can define their security policy through individual risk profiles successfully. On the international stage, global standards, particularly those addressing information security management systems (like

ISO/IEC 27001), play a key part in building security policies to international best practice. These standards encompass an integrated view of risk management with emphasis on the evaluation and reduction of risk by context and purpose of the organisation. By applying ISO standards, organisations acquire credibility and trust, since compliance with these standards indicates dedication to security and risk management processes to acceptable standards. Such frameworks also establish a base for communication among stakeholders, thereby encouraging a security-conscious organizational culture.

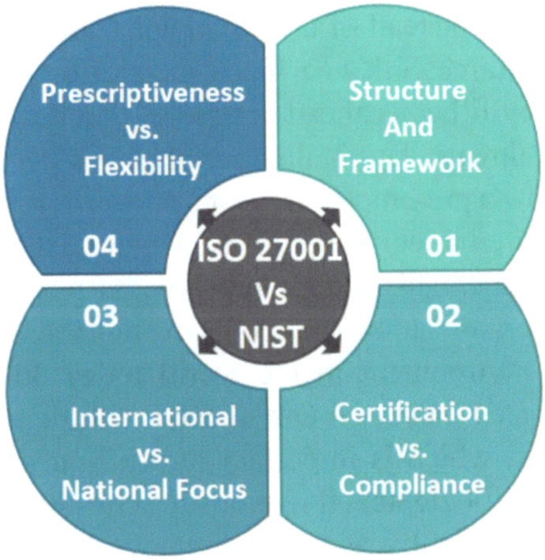

Figure 7.10. Overview of common frameworks (e.g., National Institute of Standards and Technology (NIST) and International Organization for Standardization (ISO)).

Source: https://cdn.shopify.com/s/files/1/0564/9625/9172/files/3.1_eca6cf37-3b7a-40ab-89b5-8375d9495966_600x600.png?v=1729164909.

The effectiveness of these models is augmented further when organizations integrate them into ongoing risk analysis and security policy formulation. Through an iterative process, security solutions remain adaptive to evolving threats and regulatory landscapes. Stakeholder participation and cultural compliance with accountability are central elements in such a process, as elaborated in the context of professional responsibility in policy formulation (Feltus et al., 2025). For instance, the enactment of privacy impact assessments (PIAs), for example, as outlined within frameworks like the RFID PIA Framework sponsored by the European Commission, illustrates the benefit of iterative evaluation and stakeholder involvement in addressing privacy matters and technological revolutions (Spiekermann-Hoff et al., 2020). Not only does such convergence strengthen security policy but also enables an environment within which risk management becomes everyone's responsibility across all organizational hierarchies (Figure 7.10).

7.5.2. Benefits of Using a Framework

Constructing a framework is important in the field of risk assessment and security policy because it provides a structured method of safeguarding organizational assets. A structured framework enables organizations to find, analyze, and prioritize risks effectively, and thus make informed decisions regarding the distribution of resources. Instead of taking a haphazard method, organizations that operate within a framework can focus their efforts on the most critical risks, ensuring that they fill gaps ahead of time so that they do not develop into more comprehensive threats. By facilitating effective knowledge of probable risks, frameworks allow a pre-emptive response towards risk management, with staff at every level invited to engage with security implementations and adhere to established procedures. This systematic process not only strengthens the security posture of an organization but also its overall resistance to the changing and dynamic threat environment in the world today. In addition, having a framework also promotes increased compliance with regulatory requirements and industry standards, which is important in the context of risk assessment and security policy. With heightened regulatory scrutiny by regulatory bodies, the application of a universally accepted framework assists organizations in meeting mandated compliance expectations on a level playing field. For example, these frameworks typically have provisions that ensure conformity with existing laws and regulations, which minimize the likelihood of offending regulators and resulting penalties. Once again, clearly documented processes and procedures within these frameworks make it possible for organizations to handily demonstrate compliance in audits or reviews. By adopting frameworks consistent with legal guidelines, organizations will avoid fragmented and defensive approaches that may lead to loopholes in their security policy and ultimately provide a safer and accountable platform for stakeholders and customers. Lastly, advantages from the use of a framework are not only in terms of operational effectiveness and compliance but also hugely contribute to improved communication and coordination within organizations. An efficient common framework translates to common language and risk comprehension across departments, breaking down organizational silos that prevent proper risk management. Common stakeholders like management, IT professionals, and compliance personnel can collaborate more efficiently if they are operating within the same framework. Cross-functional identification of risks is made easier, and overall security policy practice is promoted. As organizations continue to become more integrated, it is essential that there is an atmosphere of cooperation, where all the members play collectively in unity of purpose towards overall security goals. The inclusion of frameworks within the organizational culture ensures collaboration and assists in developing a united response to emerging threats, ultimately securing an enhanced security posture that is strategically aligned relative to the overall mission of the organization.

7.5.3. Challenges in Framework Implementation

The application of frameworks such as the NIST Risk Management Framework (RMF) in the cybersecurity domain is plagued with serious challenges that make organizational compliance and security effectiveness very difficult. Through the complexity of the modern threat landscape, organizations will unintentionally leave their systems in unauthorized status, hence becoming non-compliant with the prescribed regulations. Most of these professionals struggle to apply the RMF in real-world contexts, leading to severe vulnerabilities that can be easily targeted by cyber threats. Such a prevalence of such issues indicates the extent to which there is a need for effective methodologies that can facilitate smoother integration of security policies and risk assessment methods, enabling organizational assets to be kept sufficiently secured. Thus, surmounting these implementation challenges becomes not only important for individual company protection but also essential for enhancing the collective security infrastructure within the extended cybersecurity environment. Among the most promising paths toward optimizing RMF implementation has been through the application of Model-Based Systems Engineering (MBSE) principles. This research brought into the limelight of present literature acknowledges that MBSE can be a valuable technique, streamlining cybersecurity control integration and improving risk assessment processes. Employing modeling techniques, businesses may avoid traps concerning defective structures, which ultimately equate to smarter and more streamlined processes. This practice enables greater openness in determining Authorization-to-Operate (ATO) statuses, which is essential when adhering to standards. Thus, embedding MBSE within framework implementation not only addresses some of the inherent difficulties that confront organizations but also enables them to respond more effectively to evolving cyber threats, further entrenching their defense systems against potential vulnerabilities. In addition, the difficulties associated with framework implementation are usually compounded by the dynamic nature of the threats they face in cybersecurity, which calls for a transition towards innovative approaches. For instance, the integration of blockchain technology with dynamic access control models has proven to be a durable solution for enhancing security and reliability in distributed systems. Leveraging concepts such as zero trust and real-time updating of trust values, organizations can redesign their approach towards access control, eliminating threats caused by untrusted nodes. As is argued in recent studies, blockchain-enabled architectures display a remarkable improvement in successfully managing access permissions, hence system integrity. Thus, research on different methodologies such as MBSE and blockchain technology provides viable options in the disintegration of problems in implementation, with organizations capable of maintaining strong cybersecurity positions in a time of rapid technological advancement and increasing complexity in cyber-attacks (Figure 7.11).

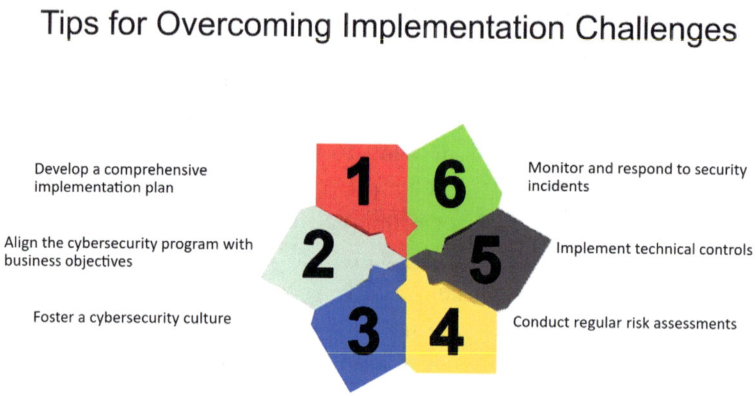

Figure 7.11. Challenges in framework implementation.

Source: https://fastercapital.co/i/NAIC-s-Cybersecurity-Framework--Safeguarding-Insurance-Data--Tips-for-Overcoming-Implementation-Challenges.webp.

7.6. CASE STUDIES IN RISK ASSESSMENT AND SECURITY POLICIES

The study of case studies in security policy and risk assessment takes a central role in improving our knowledge of security threats and their control. A superficial look at the terrorist attacks of 2024 on President Trump, as described in the literature under review, identifies immense lessons about the efficacy of extant security measures. The research brings into perspective extensive flaws, such as the absence of a coordinated command structure within response agencies that weakened the potency of incident management. Furthermore, the inefficiencies of insufficient timely communication and readiness in the face of imminent threats underscored the need for reform. Through an analysis of the meaning behind these weaknesses, policymakers will be able to design more potent security policy actions aimed at curtailing threats. Such case studies not only provide indications of past failures but also draw out potential pathways for advancement in enhancing national security measures in the context of high-profile figures and events (Kim et al., 2024). Technological advancements in the field of preliminary security risk analysis (PSRA) create revolutionary potentialities for the improvement of response measures. The fast velocity at which probable risks are formed necessitates current assessments that are challenging for human actors to keep up with under high-stress environments. The progressive case study of fine-tuned models (FTM) in PSRA validates an outstanding enhancement of the speed and accuracy of risk analysis, vital under mission-critical operations. Organisations have documented dramatic error reduction coupled with the acceleration of security risk identification through the use of advanced language models. The ability to quickly summarize complex risk analysis allows experts to avert repetitive interventions that will sap resources and deflect attention away from lengthy risk analyses. These technological advancements highlight the relevance of integrating advanced methodology in formulating comprehensive security policies that respond suitably to emerging threats (Esposito et al., 2024). Case studies allow for multi-disciplinary examination of risk assessment mechanisms across various contexts, creating potential for improved security policies. Findings in incidents such as the foiled plots against high-profile individuals can not only influence government responses but also response modes applicable to private sectors, making them more secure against potential threats. The development of a multi-security system, involving the public, private, military, and police organizations, is a quintessential example of a proactive reaction to counter-terrorism dynamics. Furthermore, the establishment of a participatory governance structure for countering new forms of terrorism, such as drone terrorism, is an example of how case studies can stimulate innovation in security policies. Through integrating the lessons from past failures and the advancements of technology in measuring risk, policymakers can create an innovation-driven security policy that is not only responding to threats, but also pro-actively addressing them (Esposito et al., 2024; Kim et al., 2024).

7.6.1. Successful Implementation Examples

Successful implementation of security policy and risk assessment is crucial to safeguarding confidential data in various sectors, especially e-commerce. An ideal example is organizations that actively perform vulnerability scans to identify potential threats before they happen. Such proactive measures are likely to include deep audits of existing security systems, hardware and software configuration. By such focus on preventing design defects that could lead to breach, organizations are not only addressing today's risks but also establishing a security-aware culture among the workforce. This approach is critical because it states that security cannot be appended to a product but must be part of the entire software development process. As Glisson et al. (2005) observes, most of the security flaws are a result of basic design errors and not security weakness, so the need to make security inherent from the outset. Another successful deployment example is in organizations that have flexible methods for security policies. This flexibility allows for tailoring to suit the specific needs of the organization and, concurrently, keeps pace with industry standards.

Figure 7.12. Successful implementation of risk assessment and security policies.

Source: https://media.geeksforgeeks.org/wp-content/uploads/20240530144323/Steps-of-a-successful-security-assessment-660.webp.

By using a modular framework, organizations are well placed to respond rapidly to changing threats and integrate improvements in security technologies. Such frameworks make it possible to continuously revise security protocols so that they remain relevant in a rapidly evolving digital landscape. Furthermore, the integration of security components in the web engineering process during the initial stages can significantly reduce the likelihood of costly breaches. Highlighted earlier, security violations and design errors are related, confirming that security measures taken in the initial stages can avoid loss of money as well as overall improved business resilience, as cited in Bailey et al. (2002). Lastly, the success of security policy is also shown by regular employee training and awareness drives. These activities are instrumental in creating an organizational culture that concerns data security and risk management. Regular training exercises equip staff to spot probable security threats, such as phishing or social engineering attacks, and respond accordingly. Organizations that promote an awareness culture not only lower their vulnerability to external threats but enhance their incident response capability. Lastly, successful execution of risk assessment policies relies on achieving a balance between technical controls, adaptive methods, and human factors. This combined strategy ensures that security is naturally built into all layers of the organization, balancing technological safeguards with readiness and vigilance of workforces to safeguard precious information assets (Figure 7.12).

7.6.2. Lessons Learned from Failures

Risk assessment and security policy failures are typically rich learning experiences, both revealing inherent systemic vulnerabilities and the necessity for adaptive responses. When organizations do not take into consideration the threats, the outcome can be catastrophic, leading to compromised infrastructure and lost public trust. An example is the inability to correctly assess risks in critical infrastructure, which can yield distant effects, and this explains the importance of having strong risk management systems. Such systems should have not only technological assessments but also socio-organizational factors, as explained in existing literature. By learning from past failures, organizations can extrapolate important lessons that amplify resilience and responsive adaptation to unforeseen adversity, cultivating a culture for continuous improvement in risk evaluation procedures, exemplified through the guiding trade-off with risk principles formulated in previous studies (Procházka et al., 2018). Furthermore, emphasizing resilience as a significant aspect of risk management demonstrates how failure learning may feed forward to shape proactive action. As firms are faced with an increasingly sophisticated threat environment, particularly in relation to the environment and climate change, the necessity for adaptive security policy increases. The Rockefeller approach to building resilience through neighborhood planning illustrates the need to establish a learning culture in the event of failure. By designing frameworks that not only react to threats but embed adaptive learning in their own design, organizations are able to satisfactorily anticipate and defend against anticipated security breaches or infrastructure failure. This spiraling approach, built on the lessons from previous failures, is a pointer towards wanting to learn in the midst of failure and develop sustainable security strategies that protect human life and public welfare (N/A, 2009). Lastly, by embracing the knowledge acquired from errors in the framework of risk management and security policy, organizations can sharpen their operational intelligence and enhance their response frameworks. The volatility of threats leaves organizations with no choice but to veer away from fixed policies and towards adaptive, learning-oriented strategies. Positioning collective risk management through the implementation of feedback loops makes organizations' defenses more potent against looming failures. Furthermore, deconstructing historical case studies of security failure reiterates the assumption that failure, when properly examined, is part of being a member of the broader corpus of knowledge on the dynamics of risk. Through this reflective practice, organizations are able to build a culture that not only accepts accountability but also promotes innovation and forward-looking activities, so that the lessons of the past are transformed into the building blocks for a secure future (N/A, 2009; Procházka et al., 2018).

7.6.3. Industry-Specific Considerations

Industry-specific considerations in security risk assessment and security policies are increasingly being understood as key to effective management of organizations. Different industries face unique risks depending on their operational environments, regulatory environments, and stakeholder expectations. For example, healthcare industries must have patients' information and privacy as their first priorities due to strict laws such as HIPAA, and banks must grapple with the complexities of fraud control and capital market threats. Understanding these particular threats critically enables organizations to create tailored security policies that are compliant and proactive in risk mitigation, not reactive. Therefore, adapting risk assessment methods to the specific nuances of an industry allows organizations to deploy resources more efficiently and counteract anticipated weaknesses before they inflict extensive harm or monetary loss. As industries evolve, the influence of dynamic external forces such as economic shifts, technological shifts, and environmental concerns necessitates a dynamic response to risk management. Rapid climate change is a good example, showing how environmental changes are able to generate imminent as well as long-term challenges for various sectors (Young et al., 2025). Industries such as fisheries and agriculture will have to include disturbances caused by climate change and revise their risk assessment accordingly. In these cases, an understanding of the interdependence between environmental change and operational risk allows firms to develop effective adaptation frameworks encompassing physical risk and social and economic components. By consequence, an industry-specific strategy that facilitates flexibility in risk management can yield adaptive systems with the ability to resist shocks of both expected and unexpected changes. Aside from these issues, one must also appreciate the merit of getting the stakeholders involved in shaping risk assessment and security policy. Experts in the industry must collaborate with government, academic, and community stakeholders to get different perspectives and insights (Beveridge et al., 2000). The collaborative nature ensures that security policy not only captures the best practices but is also grounded in the realities that most stakeholders face. Participatory method promotes shared accountability for risk mitigation, ultimately generating more effective and sustainable outcomes. By embedding stakeholder feedback into ongoing evaluation, organizations are able to iterate strategies and remain sensitive to changing industry landscapes. In total, industry-specific considerations should place priority on nimble processes and shared frameworks to fully address the multi-faceted risks confronting diverse industries (Figure 7.13).

Figure 7.13. Industry-specific considerations.

Source: https://fastercapital.co/i/Return-on-assets--Comparing-Return-on-Equity-and-Return-on-Assets--Industry-Specific-Considerations-for-ROA-and-ROE.webp.

SUMMARY

In Short, the role of risk assessment in security policies is crucial for businesses that wish to make it in a dynamic world. Businesses, particularly those in high-risk sectors such as oil and gas, demonstrate the necessity for an efficient Enterprise Risk Management (ERM) system. Corporations like ExxonMobil and Chevron demonstrate this premise by instituting thorough risk analysis that betrays a sharp perception of the complex dangers that confront them, ranging from financial uncertainty to issues of regulation. The availability of specialized risk management functions, e.g., Chief Risk Officers, reflects an intention to prioritize risk in corporate governance arrangements, thereby ultimately allowing such firms to better manage the intricacies of contemporary market conditions (Ethridge et al., 2013). In addition, the ever-changing nature of internet business environments suggests a need to regularly update security policies to address changing threats and vulnerabilities. Organizations must adopt a proactive strategy for risk management to secure both their operational integrity and consumer trust. This necessitates a continuous review process based on set benchmarks derived from sanctioned information technology and information security standards. Adopting a continuous improvement strategy, organizations are able to enhance their security systems and watchfulness towards newly emerging threats and thereby sustain their online presence and consumer trust (Pye et al., 2007). Lastly, the relationship between risk assessment and security policies is the core of an organization's strategic objectives. Through sufficient analysis and anticipatory management of threats, organizations not only protect their assets but also employ their threat knowledge to create competitive strengths. The findings from sound risk management practices observe that effective security policies are not merely defensive measures but essential components that enable businesses to grow and flourish. As businesses continue to encounter new challenges, the unproblematic integration of risk management into strategic thinking will be key to their long-term success and ability to withstand a changing world marketplace.

In risk assessment, detection and evaluation of potential risks to an organization are rudimentary concepts that determine security policies. One key is the application of frameworks specialized in specific technologies or methods, like in the case of the Privacy Impact Assessment (PIA) framework sponsored by the European Commission. This framework, initially developed to address privacy concerns associated with radio frequency identification technology, places emphasis on proactive risk discovery brought about by emerging technologies associated with the Internet of Things. By defining the PIA process, companies can better anticipate issues and enhance security measures, building a culture of sensitization for privacy upkeep and risk management in their operating models (Spiekemann-Hoff et al., 2020). To that effect, insider threats constitute a significant security risk element that companies are facing more and more. Unlike external threats, insider attacks can also become very hard to detect and ward off since the authorized users have the right access and permissions. The emergence of a broad base of research aimed at quantifying and addressing such threats is testifying to the importance of understanding insider dynamics. By categorizing attack strategies through structured decomposition trees and correlating them with defense strategies, organizations can create stronger, goal-based requirements to counter potential insider threats effectively. This proactive approach facilitates awareness and empowers personnel to contribute to the organization's overall security posture, thereby bridging gaps in traditional security measures (Franqueira et al., 2006). Finally, good security policies and risk analysis require continuous strategy development based on both technological advancement and the evolving threat landscape. The use of systems like the PIA for technologies such as RFID, and good insider threat evaluation methods, helps organizations examine their vulnerabilities holistically. Furthermore, employee training on the implications of insider threats and organizational culture that considers security as a collective responsibility can lead to proactive action. Lastly, the development of an adaptive strategy involving both external and internal threats will enhance an organization's resilience against possible security breaches and an overall defense strategy that continually adapts to the evolving nature of risks (Franqueira et al., 2006; Spiekermann-Hoff et al., 2020).

REFERENCES

1. Acosta Alba, I., Andrieu, N., Chia, E., Gamba-Trimiño, E., et al. (2019). Co-designing climate-smart farming systems with local stakeholders: A methodological framework for achieving large-scale change. https://core.ac.uk/download/275689779.pdf (accessed on 13th September 2025).
2. Adams, A., Avila, K., Basney, J., Brunson, D., et al. (2019). Trusted CI experiences in cybersecurity and service to open science. http://arxiv.org/abs/1904.05383 (accessed on 13th September 2025).
3. Al-Maamari, A. (2025). Between innovation and oversight: A cross-regional study of AI risk management frameworks in the EU, U.S., UK, and China. https://www.semanticscholar.org/paper/76b02882f2265accd210bdce7f17e2deab6b12ec (accessed on 13th September 2025).
4. Arent, D., Bustamante, M., Mitchell, C., Candelise, C., Kern, F., Genus, A., et al. (2014). Supply chains and energy security in a low carbon transition. https://core.ac.uk/download/20267316.pdf (accessed on 13th September 2025).
5. Atkinson, R. C., Dunlop, J., & Irvine, J. (2004). A personal networking solution. https://core.ac.uk/download/9039076.pdf (accessed on 13th September 2025).
6. Bäckman, S., Kahiluoto, H., Rötter, R., Sumelius, J., et al. (2009). Sustainable rural development with emphasis on agriculture and food security within the climate change setting: SARD-climate final report. https://core.ac.uk/download/10929656.pdf (accessed on 13th September 2025).
7. Bailey, C., Field, D., Froggatt, K., Krishnasamy, M., et al. (2002). Web development evolution: The assimilation of web engineering security. https://core.ac.uk/download/pdf/1393957.pdf (accessed on 13th September 2025).
8. Beveridge, M. C. M., Bulcock, P., & Hambrey, J. (2000). The improved management of small-scale cage culture in Asia: Final technical report. https://core.ac.uk/download/11018591.pdf (accessed on 13th September 2025).
9. Bloomfield, R. E., Chozos, N., & Salako, K. (2009). Current capabilities, requirements and a proposed strategy for interdependency analysis in the UK. https://core.ac.uk/download/9559636.pdf (accessed on 13th September 2025).
10. Bultman, J., Kanywanyi, J. L., Maarifa, H., Mtei, F., et al. (2012). Tanzania health insurance regulatory framework review. https://core.

ac.uk/download/pdf/11307330.pdf (accessed on 13th September 2025).

11. Burger, C. P. J. (2009). Food calamities and governance: An inventory of approaches. https://core.ac.uk/download/pdf/29246635.pdf (accessed on 13th September 2025).

12. Cannon, A., & Ueberall, S. (2015). Assessing New York City's youth gun violence crisis: Crews–Volume II–CompStat for violence prevention programs: Collecting program specific data to manage performance and inform policy. https://core.ac.uk/download/75781298.pdf (accessed on 13th September 2025).

13. Chackiewicz, M., Orłowska, M., & Orłowski, D. (2025). Customs and border systems on the external border of the EU on the example of the border crossing in Medyka–Legal and economic aspects. Zeszyty Naukowe SGSP. https://www.semanticscholar.org/paper/92b0ffdde8392edc3dd8429f7b9973d4579b500c (accessed on 13th September 2025).

14. Cigler, B. A., Stephens, D. W., Stanley, E. M. Sr., & Reeder, F. S. (2010). The National Dialogue on the Quadrennial Homeland Security Review. https://core.ac.uk/download/71345948.pdf (accessed on 13th September 2025).

15. Ekong, J. (2019). Review of the Learning Alliance for Adaptation in Smallholder Agriculture. https://core.ac.uk/download/220111438.pdf (accessed on 13th September 2025).

16. Esposito, M., & Palagiano, F. (2024). Leveraging large language models for preliminary security risk analysis: A mission-critical case study. Proceedings of the 28th International Conference on Evaluation and Assessment in Software Engineering. https://www.semanticscholar.org/paper/fcbb801ab7fad852c06b3638eb7a8962ab9c6b0e (accessed on 13th September 2025).

17. Ethridge, J. R., & Rogers, V. C. (2013). Enterprise risk management in the oil and gas industry: An analysis of selected Fortune 500 oil and gas companies' reaction in 2009 and 2010. https://core.ac.uk/download/289123530.pdf (accessed on 13th September 2025).

18. Fair Labor Association. (2012). Assessment Report 2012 Russell Brands, Pakistan AA0000000108. https://core.ac.uk/download/33605839.pdf (accessed on 13th September 2025).

19. Fantin, S., Ferreira, A., & Pupillo, L. (2020). CEPS Task Force on Artificial Intelligence and Cybersecurity Technology, Governance and Policy Challenges: Task Force evaluation of the HLEG Trustworthy AI Assessment List (pilot version). CEPS Task Force Report, 22 January 2020. https://core.ac.uk/download/287647893.pdf (accessed on 13th September 2025).

20. Feltus, C., Georges, A., & Petit, M. (2025). Definition and validation of a business IT alignment method for enterprise governance improvement in the context of processes-based organizations. https://core.ac.uk/download/pdf/6465762.pdf (accessed on 13th September 2025).

21. Franqueira, V. N. L., & van Eck, P. (2006). Defense against insider threat: A framework for gathering goal-based requirements. https://core.ac.uk/download/9632612.pdf (accessed on 13th September 2025).

22. Frayne, B., & Ziervogel, G. (2011). No. 08: Climate change and food security in Southern African cities. https://core.ac.uk/download/159478643.pdf (accessed on 13th September 2025).

23. Gampel, A., & Eveleigh, T. (2025). Model-based systems engineering cybersecurity risk assessment for industrial control systems leveraging NIST risk management framework methodology. Journal of Cyber Security and Risk Auditing. https://www.semanticscholar.org/paper/be4acdab9f94609b0d760858eea432bb139f6873 (accessed on 13th September 2025).

24. Gheorghe, M. (2025). Audit methodology for IT governance. https://core.ac.uk/download/pdf/6673187.pdf (accessed on 13th September 2025).

25. Glisson, W. B., & Welland, R. (2005). Web development evolution: The assimilation of web engineering security. https://core.ac.uk/download/1393957.pdf (accessed on 13th September 2025).

26. Gooden, S., Graham, F., Martin, J., et al. (2014). Bridging the data divide: Understanding state agency and university research partnerships within SLDS. https://core.ac.uk/download/215487233.pdf (accessed on 13th September 2025).

27. Hutchinson, G., & Ophoff, J. (2020). A descriptive review and classification of organizational information security awareness research. https://core.ac.uk/download/288653621.pdf (accessed on 13th September 2025).

28. Johnson, C. W. (2016). Why we cannot (yet) ensure the cybersecurity of safety-critical systems. https://core.ac.uk/download/46664182.pdf (accessed on 13th September 2025).

29. Kahiluoto, H., & Rötter, R. (2009). Implications of and possible responses to climate change. https://core.ac.uk/download/10929655.pdf (accessed on 13th September 2025).

30. Kim, T. Y., & Song, J. H. (2024). Study on the development of national security policies through the analysis of the terrorist incidents involving President Trump in the United States. The Korean Association for Terrorism Studies. https://www.semanticscholar.org/paper/6f83e5032d4c07908c0059e3abab84f14b5c7de4 (accessed on 13th September 2025).

31. Mackenzie, D., & Njunwamukama, S. (2024). Strengthening fintech security in Uganda: An analysis of insider threats and effective risk management strategies. International Journal of Technology and Systems. https://www.semanticscholar.org/paper/e164cc9fb47028e71832f58ef3db69416de66ea3 (accessed on 13th September 2025).

32. Mohajan, A., & Jahan, S. (2025). Embedding security awareness into a blockchain-based dynamic access control framework for the zero-trust model in distributed systems. Electronics. https://www.semanticscholar.org/paper/4e1fdcaa74a636085d3b0546821c1efdb43abda2 (accessed on 13th September 2025).

33. Mohamed, M. Y. (2021). Simulation on off grid hybrid PV-battery system. https://core.ac.uk/download/42955345.pdf (accessed on 13th September 2025).

34. N/A. (2009). Building climate change resilience. https://core.ac.uk/download/71349675.pdf (accessed on 13th September 2025).

35. N/A. (2011). Safeguarding in schools: Best practice. https://core.ac.uk/download/4159700.pdf (accessed on 13th September 2025).

36. N/A. (2016). State of Iowa cybersecurity strategy, July 2016. https://core.ac.uk/download/76640486.pdf (accessed on 13th September 2025).

37. O'Brien, G., O'Keefe, P., Rose, J., Wilson, T., et al. (2008). Climate change and variability, energy and disaster management: Produced risks without produced solutions: Rethinking the approach. https://core.ac.uk/download/4146915.pdf (accessed on 13th September 2025).

38. Önal, M. Z. (2025). An aggregated information technology checklist for operational risk management. https://core.ac.uk/download/pdf/6299795.pdf (accessed on 13th September 2025).

39. Procházka, J., Procházková, D., & Rusko, M. (2018). Principles for management of risks of critical infrastructure. https://core.ac.uk/download/287889332.pdf (accessed on 13th September 2025).

40. Pye, G., & Warren, M. (2007). A model and framework for online security benchmarking. http://hdl.handle.net/10536/DRO/DU:30007062 (accessed on 13th September 2025).

41. Rusu, A. C., & Stroie, E. R. (2025). Security risk management–Approaches and methodology. https://core.ac.uk/download/pdf/6612749.pdf (accessed on 13th September 2025).

42. Siskin, A., & Wyler, L. S. (2012). Trafficking in persons: U.S. policy and issues for Congress. https://core.ac.uk/download/19958958.pdf (accessed on 13th September 2025).

43. Snover, A., Hansen, L. J., Jackson, L., Gregg, R. M., Ellworth, S., & Arroyo, V. (2013). The state of adaptation in the United States: An overview. https://core.ac.uk/download/71361521.pdf (accessed on 13th September 2025).
44. Spiekermann-Hoff, S. (2020). The RFID PIA–Developed by industry, agreed by regulators. https://core.ac.uk/download/286370697.pdf (accessed on 13th September 2025).
45. Tkachuk, L., & Pidhorodetskyi, M. (2025). State regulation as a tool for minimizing investment risks in Ukraine's renewable energy sector: Challenges and prospects. Economic Scope. https://www.semanticscholar.org/paper/7f250bde16aa187373337898b0aa09b7201e26f6 (accessed on 13th September 2025).
46. Young, C., Handmer, J., & Jones, R. N. (2025). Valuing adaptation under rapid change. https://core.ac.uk/download/pdf/30676891.pdf (accessed on 13th September 2025).

CHAPTER 8

ENDPOINT AND DEVICE SECURITY

Contents

INTRODUCTION

In the dynamically evolving world of technology, device and endpoint security is today a topmost priority for organizations and individuals as well. With more and more devices being internet-enabled from personal smartphones to complex medical devices, so too does potential exposure to security risk from these endpoints rise. The need to protect these devices is emphasized by the increasing sophistication of cyber threats, which can result in data breaches, business disruption, and significant financial loss. Also, with companies becoming more digital and cloud-dependent, the necessity for effective endpoint security is exacerbated. Securing devices is not a technological endeavor alone but an essential need to maintain trust and safeguard sensitive information in an increasingly connected world. Endpoint and device security dynamics are more than technical solutions; they are integrated solutions addressing human, procedural, and technological aspects. As emphasized by current studies, there are various barriers hindering effective installation of security systems in different industries. Factors hindering the application of contemporary security technologies include cost constraints, measurability reliability, and the lack of acceptable protocol guidelines applicable everywhere. Such are the complications represented by examples of device integration with devices like continuous glucose monitoring (CGM) systems, wherein in spite of their potential to enhance healthcare outcomes, user resistance and operational reliability are barriers to widespread use (Bergenstal et al., 2017). Creating effective guidelines and systemic solutions for their operation is therefore essential to enhance their security in actual use. Furthermore, the intersection of endpoint security and cloud computing offers unique opportunities and challenges for organizations. With increasing organizational uses of cloud infrastructures, they must grapple with data integrity, privacy, and security vulnerabilities peculiar to such environments. The emergence of open-source platforms like OpenStack provides a foundation for building flexible and secure cloud solutions, as seen in the formation of the BradStack project (Al-Shaidy et al., 2017). This flexibility allows organizations to integrate legacy systems and third-party applications with enhanced cybersecurity controls. Ultimately, the pursuit of end-to-end endpoint and device security solutions necessitates an interdisciplinary initiative involving technological innovation, best practices, and regulatory compliance to mitigate the myriad threats found in today's digital landscape.

Definition of Endpoint and Device Security

The concept of device and endpoint security entails the defensive solutions that are used to protect various devices connected within an environment, including computers, smartphones, tablets, and IoT devices. The larger the number of endpoints in any given organizational system, the greater is the necessity for robust security solutions. These devices are most likely to be the focal points of cyber-attacks as they are prone to being vulnerable. Thus, endpoint security is best defined as the implementation of security solutions involving hardware along with software components designed to prevent unauthorized entry and protect data integrity. This element of cybersecurity proves crucial in consideration of the threat sophistication in today's world that takes advantage of the vulnerabilities intrinsic in device connectivity, rendering traditional security ineffective in an evolving landscape. Also, the advent of mobility and remote work necessitates re-thinking of traditional security perimeters. Traditional security models are normally characterized by a hard perimeter, hoping to see a demarcation between internal trusted assets and outside threat. With de-parameterization, organizations must re-think how they establish and defend their digital security environment. This emerging trend represents a shift toward a more dynamic security stance, where workers engage with sensitive data from multiple endpoints outside the confines of a secure organizational periphery. Device and endpoint security therefore have to adapt to more dynamic security measures such that all devices, regardless of location, are adequately monitored and protected from imminent threats posed by heightened connectivity (Cleeff et al., 2008). Besides, the challenges of embedded devices make endpoint security discourse even more complicated. As most organizations rely on connected devices for operational efficiency, the potential for security exploits is greatly improved. For instance, vulnerabilities can be from protocols such as USB, which, as highly flexible as it is, can be employed to negative ends (Butler et al., 2017). To offset such risks, security procedures have to incorporate in-depth firmware examination and evaluations to recognize malicious behavior prior to compromising network integrity. By combining proactive and response options, organizations can reduce risks of vulnerabilities with endpoint devices. Such multi-layered approach is essential in establishing a sound security framework that is capable of coping with the continuously changing threat landscape, such that all endpoints are well secured against changing and advanced cyber threats (Figure 8.1).

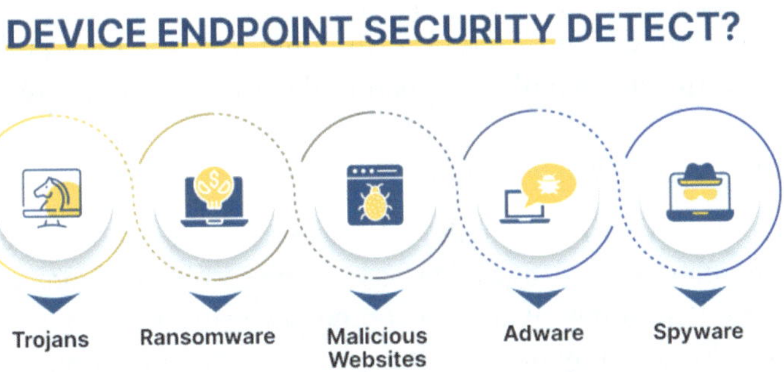

Figure 8.1. The concept of endpoint and device security.

Source: https://www.extnoc.com/learn/wp-content/uploads/2022/05/What-is-Device-Endpoint-Security.png.

Importance in the Modern Digital Landscape

With the constant rise in technology and devices being connected together in the modern digital era, endpoint and device security has become extremely vital. As more people and organizations become reliant on digital platforms, avenues for weaknesses in the form of cyber-attacks increase manifold. The infiltration of smart devices into individuals' daily lives has resulted in enormous amounts of personal and sensitive data being collected, often without the users even knowing the risks. This offers rich ground for nefarious users who seek to capitalize on vulnerabilities, and thus robust endpoint security is not just a necessity but a core necessity for data safeguarding. Also, the increase in the demand for privacy is a demonstration of the need for strict policies and shield mechanisms for data privacy, particularly for endpoints of residential contexts. The research explained in Bronfman et al. (2016) shows the focus on balancing between benefits of smart home technologies and the risks that come with them. The elderly and disabled communities, who increasingly rely on these devices, need a concerted effort towards ensuring that their data is safe. This twin narrative works to emphasize the necessity of device security encompassing a greater understanding of moral data handling and the essentiality of surveillance today. Finally, the necessity to establish a culture with security as its focus extends beyond users to encompass developers and policymakers as well. As argued once more in Hallnäs et al. (2006), research into issues at the root of interaction design is crucial to the design of effective security and methods for designing devices. At a stage where technology and design have interlinked connotations, coming up with an understanding of the potential risks at the nexus of technology and user interaction is important. By prioritizing endpoint security, stakeholders can ensure a safer digital world, where innovation need not come at the expense of privacy and protection of the user. In doing this, not only is the end user protected, but trust in technology as part of daily life is also established.

Overview of Key Challenges and Threats

With the rapidly changing technology landscape, device and endpoint security are facing numerous challenges that have to be resolved by organizations for protecting vital information and operation integrity. One of the significant challenges is presented by the growing use of cloud computing, its agility, and flexibility, which also brings in potential threats. Integration of cloud platforms such as OpenStack demands potent security measures to protect against eventual breaches. Institutions must tread carefully when deploying secure cloud platforms like the BradStack Multi-node cloud infrastructure, particularly through penetration testing to verify security controls are sound. As an increasing number of organizations rely on cloud-based services, institutions are likely to balance the convenience provided by such technologies against the need for sound security controls to effectively manage the dynamic threat environment (Al-Shaidy et al., 2017). With the continuous expansion of the Internet of Things (IoT), it adds more complexities to the ones that are currently there, which are linked to endpoint security. The emergence of an open Internet of Things, which has grown from a closed Intranet model, brings issues regarding how secure resource-restricted devices that are connected to the public Internet are. The complexity the openness introduces has the potential to put organizations at greater risk, particularly in relation to data integrity and unauthorized access. The convergence of restricted networks demands strict control mechanisms to manage exchanges securely. Without adequate solutions, vulnerabilities in these networks expose endpoints to malicious usage, compromising organizational data as well as user privacy (Demeester et al., 2014). Moreover, the lack of universal global standards and protocols for device and endpoint security exacerbates the issues mentioned above. With varied devices manufactured by varying manufacturers, seamless interoperability without any compromise on security becomes increasingly difficult. Enterprises typically end up with patchwork systems that fail to address the entire spectrum of security threats. This fragmentation not only complicates incident response planning but also compromises the mechanisms for adopting complete security across devices. As the environment continues to evolve, it will be essential to maintain cooperation between stakeholders in building effective frameworks for endpoint security. All of these intricate problems will eventually become of utmost significance in mitigating risk and enhancing the overall robustness of organizational cybersecurity efforts.

8.1. TYPES OF ENDPOINTS

Endpoints are integral points of access within networks that serve a variety of distinct roles depending on type. Common examples of endpoints include mobiles, laptops, desktops, and Internet of Things (IoT) devices. Laptops and desktops have typically served as the primary user interfaces, enabling communication with databases and applications. They are the typical objectives in attack due to their integration into organizational networks, on which sensitive data is being processed. Contrarily, mobile devices have taken over, with the associated unparalleled security concerns due to their mobility and varying operating systems. All endpoint categories possess special vulnerabilities that need to be addressed with special security to ward off unauthorized access, data loss, and malware utilization attacks. Accordingly, understanding the different categories of endpoints is essential to create suitable security strategies that ward off the peculiarities of each one. For IoT devices, heterogeneity creates opportunities as well as challenges for endpoint security. IoT devices are characterized by their multi-faceted means of connectivity as well as protocols, which can complicate security. As a basic example, a smart thermostat can communicate through Wi-Fi, while an industrial sensor communicates through a proprietary protocol. As outlined in recent research, the proliferation of IoT devices calls for a comprehensive integration strategy to service, particularly since these devices operate in environments of limited resources that bound their computing power. Security requirements must account for the varied communication methods used by such devices to enable their security. The use of a cloud platform to manage these disparate systems has proved useful, beneficial to service developers and generally the IoT system's security stance, as emphasized in seminal research on the subject (Demeester et al., 2015). Furthermore, ongoing endpoint security innovation is such that new vulnerabilities must be considered, namely those of USB devices. The extensive set of abilities within the USB protocol can be gamed by attackers, such as with attacks like BadUSB that take advantage of the very openness of the USB connections. This is particularly pertinent to embedded devices that utilize microcontrollers, which are frequently overlooked during security audits. Emergent models such as FirmUSB have subsequently arrived to analyze USB device firmware, identifying potential malicious behavior without source code exposure, equivalent to dynamic analysis tools targeting embedded architectures. As endpoint security matures, handling such nuanced vulnerabilities becomes the highest priority within organizations attempting to safeguard their digital valuables and neutralize threats associated with various endpoint types (Butler et al., 2017) (Figure 8.2).

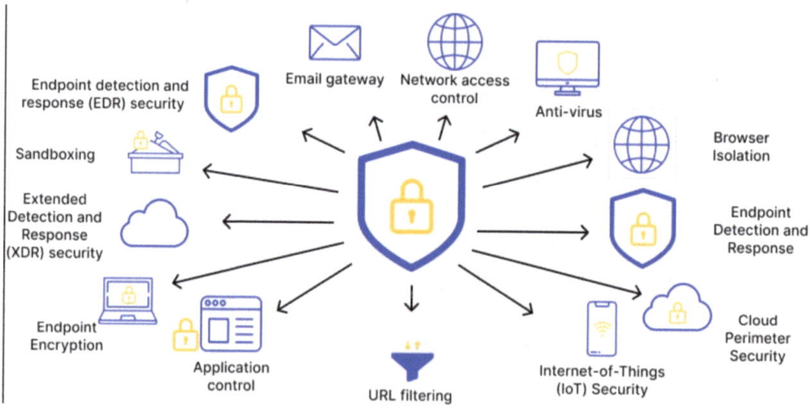

Figure 8.2. Types of endpoints.

Source:. https://www.facebook.com/groups/2408899619294901/posts/2576088369242691/

8.1.1. Desktops and Laptops

The high penetration of desktops and laptops in the home and business environments necessitates a solid approach to endpoint and device security. Laptops and desktops serve as key entry points into organizational networks and sensitive data. With increasing mobility of laptops, laptop threats become even more sophisticated; an improperly secured, stolen, or lost device can result in significant data breaches. Old security solutions that were so perimeter-control dependent are now unable to keep up with the vulnerabilities that exist in today's computing environment. This has prompted organizations to adopt a more agile security solution that entails hardware and software for the purpose of securing both laptops and desktops wherever they may be. As businesses move to cloud-based applications, they must bolster endpoint security to protect sensitive information accessible through these devices and thereby render defenses more paramount than ever. Endpoint security is an expansive array of measures for minimizing risk at desktops and laptops.

These steps include the installation of antivirus software, firewalls, and encryption mechanisms designed to safeguard one of the most exposed areas of organizational infrastructure: user endpoints. Adding advanced forms of authentication, such as two-factor authentication or biometrics, also adds additional security. But installation alone is not sufficient; organizations must be actively involved in real-time monitoring and management of their endpoint security posture. With the evolving nature of cybersecurity threats is the necessity for the strategies employed to combat them to evolve equally. The de-perimeterization issues, as put forward by the Jericho Forum, challenge organisations to reconsider security practices and integrate preventative and reactive policies in their endpoint device security (Cleeff et al., 2008). Apart from this, the desktops and laptops' technology compatibility with a variety of remote communication aids contributes to the intricacies in security practice. New approaches will aim to manage automation systems via mobile and tablet integration, highlighting a move toward more agile IT environments. This flexibility, although desirable, could inadvertently enhance the attack surface for potential attackers. The increasing use of software applications for remote control of devices is a prime example of the fine line organizations need to walk between adding functionality and adding security. As set forth in recent research on human-machine interfaces (HMIs), investments in secure but user-friendly systems are essential to the preservation of operational integrity while keeping potential vulnerabilities to a minimum (Demaiter et al., 2016). As organizations persist in exploring the digital frontier, an integrated approach to endpoint and device security that involves both technology and staff training will be pivotal in the protection of sensitive information on desktops and laptops.

8.1.2. Mobile Devices

Mobile devices in the current cybersecurity landscape are both precious opportunity and potent threat to business organizations. The explosion of the smartphone and tablet has revolutionized how employees communicate and do business, offering more convenience than ever before that enhances productivity. However, this convenience comes at a cost; mobile devices are more and more appealing targets for cybercriminals, which makes endpoint security the leading cause of concern for businesses. The study of such devices and their connectivity offers a greater attack surface, and so organizations are forced to rethink traditional security models. With mobile devices increasingly the primary access method to confidential information, legacy models founded on hard perimeters cannot ensure their security since they have been proved insufficient in replicating the dynamic nature of threats in today's world of computing (Cochran et al., 2018). Compounding on this complexity are the innovations of the Internet of Things (IoT), where mobile technology is more and more incorporated into its architectures. The incorporation of mobile technology into IoT infrastructures offers more functionalities while simultaneously introducing security concerns.

For instance, IoT-capable mobile applications integrated into IoT architectures can facilitate real-time data mining and operations management, as presented in research comparing various architectures of IoT and how their performance is measured (Cochran et al., 2018). Thus, organizations need to use robust security solutions specifically architected for mobile device management that can handle both the in-built vulnerabilities and sophisticated attack mechanisms employed by cyber threats. As both mobile devices and IoT systems continue evolving, so do the means employed to safeguard sensitive data and offer security compliance. To effectively protect mobile devices, organizations must adopt a multi-layered security system with organizational controls and technical controls. These may entail the implementation of strong access controls, stringent encryption methods, and instituting frequent security training for staff members operating these devices. Additionally, organizations must take into consideration the de-perimeterisation practices, which demand adaptive security bubbles that can shape themselves according to the changing technology landscape. As described in security literature, these steps are required to cope with the difficulties stemming from increased connectivity between devices and the delicate balance to cope with potential security breaches (Cleeff et al., 2008). Placing emphasis on these integrated approaches, organizations can create a security culture that understands the intrinsic difficulties of mobile devices without losing functional effectiveness in a more mobile, connected environment (Figure 8.3).

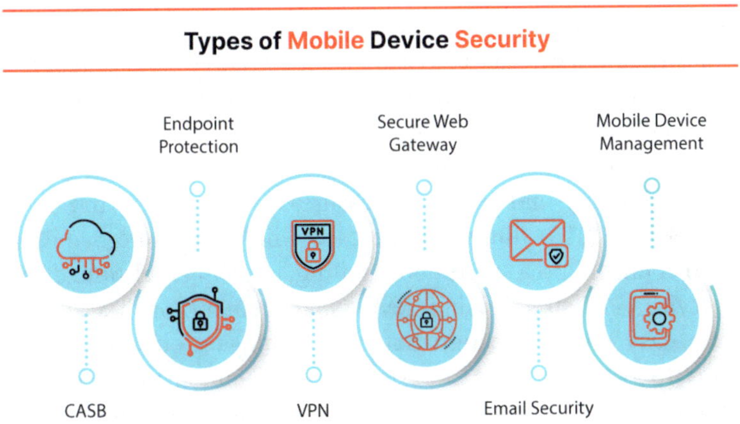

Figure 8.3. The mobile devices represent both significant opportunity and notable risk for organizations.

Source: https://www.fortinet.com/content/dam/fortinet/images/cyberglossary/mobile-device-security.png.

8.1.3. Internet of Things (IoT) devices

The popularity of IoT devices has transformed the world of technology, with both enormous opportunity and enormous security concerns. With such connected devices covering homes, enterprises, and cities in larger numbers, the sheer volume of connections is a cause for concern when it comes to endpoint security. Each device, whether a smart thermostat or a factory sensor, is an entry point that can be exploited. Implementing robust security solutions is crucial not only to protect sensitive data but also to keep platforms trustworthy and intact. The sophistication of the issue is also exacerbated by the sheer number of devices available, with each potentially employing different standards and communication protocols, which can further conceal areas of vulnerabilities. To manage the intricacies of IoT devices with efficiency, solutions based on cloud infrastructure have proven to be a potential approach in handling endpoint security. A cloud platform that can connect heterogeneous IoT devices makes it easier for different systems to communicate and integrate, thus making the security process more straightforward. This method enables service developers to innovate without getting entangled in the complexities of device interoperability. For instance, Demeester et al. (2015) illustrates how such a platform minimizes the operational load on resource-starved devices as well as encourages service integration. This is particularly significant as service developers struggle with an increasingly fragmented IoT landscape based on varied standards and connectivity means that require convergent security strategies. Finally, a unique application of this cloud infrastructure is in smart home settings where connecting multiple devices to cloud services significantly increases functionality and security measures. For instance, the architecture in Cochran et al. (2018) illustrates the case where Arm microcontrollers are employed to connect into a smart home setting, enabling real-time data transfer to cloud services and streamlining device management. Such assimilation not just enables easy communication but also employs cloud machine learning mechanisms to optimize user experiences, e.g., anticipating comfort levels from environmental data. Through the guarantee of such devices sharing data securely with the cloud, end users are able to enjoy greater convenience without compromising their security, thereby demonstrating the critical balance between connectivity and protection within the paradigm of IoT (Figure 8.4).

Figure 8.4. The proliferation of internet of things (IoT) devices.

Source: https://www.techtarget.com/rms/onlineImages/iota-devices_mobile.png.

8.2. COMMON THREATS TO ENDPOINT SECURITY

As the majority of businesses now rely on multiple digital channels, the extent of cyber-attacks, particularly ransomware, has increased significantly. Ransomware attacks are specifically configured to lock critical data and demand ransom in exchange for decryption keys and therefore remain an ongoing source of anxiety for endpoint security. The sudden expansion of ransomware has resulted in notable repercussions like financial loss and system downtime on businesses from various industries such as healthcare, finance, and manufacturing. Understanding the history of these attacks, which started from malware and Trojan codes, is necessary. Learning about the encryption algorithms used in these cyberattacks, such as AES and RSA, will aid organizations in securing their systems. It is proactive threat detection and a good security infrastructure that are required to thwart ransomware, as knowing how it operates enables measures to be put in place to limit system vulnerabilities (Aziz et al., 2016). Another prevalent endpoint security threat is the utilization of Universal Serial Bus (USB) protocols, which have been notoriously found wanting because of their widespread deployment and very nature. USB peripherals can be used to exchange data between systems but can also serve as vectors for malicious activity. The BadUSB attacks underscore the threat potential of infected USB peripherals, whereby the openness of the protocol is exploited by attackers to insert malware into innocuous-looking peripherals. These threats highlight the problem of correctly ascertaining whether USB devices in a network are safe, as the potential for malicious activity is still clouded in uncertainty. Utilizing a company such as FirmUSB can enhance USB firmware analysis, enabling organizations to detect likely threats more effectively. Increased caution and innovative analysis tools are thus needed to ensure that devices are not exploited at endpoints (Butler et al., 2017). Other than USB exploits and ransomware, phishing attacks are one of the biggest threats to endpoint security, striking at the human element in cybersecurity. Phishing attacks mislead users into divulging sensitive information, such as personal data and login credentials, by masquerading as legitimate sources of authority. As the sophistication of phishing techniques increases, standard measures may prove insufficient. Businesses must therefore give priority to raising awareness among the users to inculcate a culture of awareness and vigilance among the employees. Anti-phishing software and robust email security controls are also critical in securing endpoints from such attacks. Regular simulated phishing attacks training can help users recognize and respond to these threats more effectively. With enhanced user education and implementing appropriate technological controls, organizations can diminish the risks associated with phishing, thus making endpoint security overall more robust (Aziz et al., 2016).

8.2.1. Malware and Ransomware Attacks

The spread of malware and ransomware attacks is one of the major problems in endpoint and device security. Within the last couple of years, particularly during the COVID-19 pandemic, these kinds of cyberattacks have substantially risen with the unprecedented switch towards remote work and online interactions. This adjustment opened up a convenient setting for cybercriminals, with companies scrambling to accommodate new realities of conducting business, often neglecting good security practices. Different endpoints, particularly those in domestic settings, were not properly secured, making them vulnerable targets for hackers. These were further reinforced by the influx of new users to the virtual space, who may unwittingly expose their systems to vulnerability through recourse to risky habits, like clicking on malicious links or installing unknown applications. In this manner, identifying and mitigating these threats has become essential for safeguarding an organization's cyber space. Ransomware attacks are particularly sinister in the sense that they can bring operational systems to their knees by encrypting critical data, rendering it inaccessible until ransom is paid. The healthcare sector has been hit particularly hard, with many reports citing that ransomware attacks targeted hospitals and health organizations during the pandemic. After a consideration of global and domestic experiences, such attacks increased as health systems were bogged down by heavy patient volumes and higher operational stresses, making them vulnerable to cyberattacks (Tam Pás et al., 2021). Impacts of these attacks are wide-ranging, not just causing massive financial losses but also compromising patient safety and disrupting crucial medical services. The chaotic environment that the pandemic created presented difficulties for healthcare organizations to enforce earnest cybersecurity practices, expressing a causal link between operational stress and ransomware susceptibility. Moreover, malware and ransomware threats extend further than current operational inconvenience; they involve broader national security matters. The global attention to vaccine development and delivery during the pandemic has exposed underlying vulnerabilities in security frameworks. Cyberthieves have exploited such vulnerabilities, targeting intellectual property of vaccine research and delivery supply chain logistics, thereby undermining public health and safety (Tamás Palicz et al., 2021). Such revelation of such potential vulnerabilities demands a multipronged cybersecurity approach encompassing not just endpoint security but also closer intersectoral collaboration. As businesses operate through this ever more treacherous cyber space, hardening defenses against ransomware and malware cannot be overstated, necessitating a cross-cutting security strategy that spans all phases of operations (Figure 8.5).

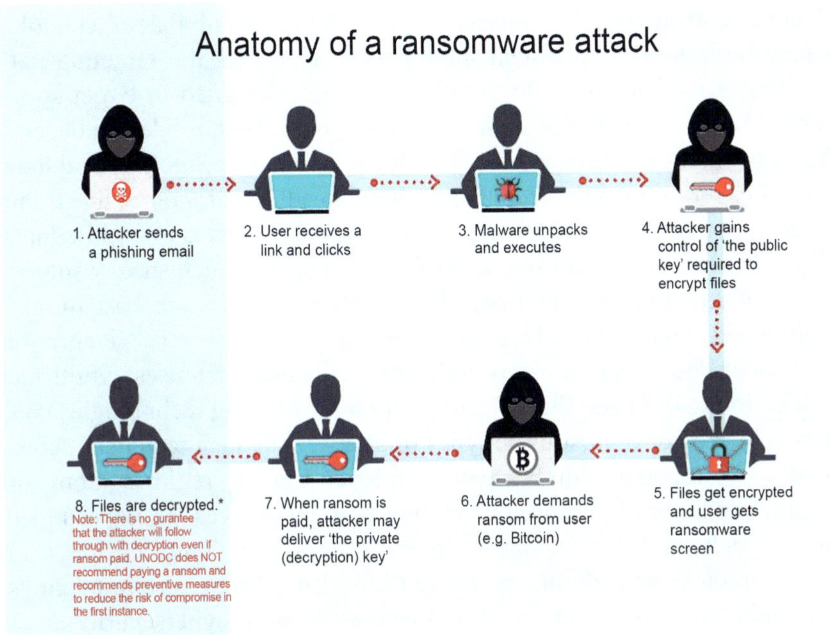

Figure 8.5. The rise of malware and ransomware attacks.

Source: https://www.unodc.org/roseap/uploads/images/2021/10/cybercrime/ransom_2.jpg.

8.2.2. Phishing and Social Engineering

Phishing and social engineering pose notable threats in the landscape of device and endpoint security, mainly because cybercriminals have significantly refined their skills in influencing human psychology. This has traditionally centered on targeting employees in organizations by playing on their trust to infiltrate sensitive information without authorization. The unprecedented surge in remote work, driven by the COVID-19 pandemic, has amplified these risks. In this evolving environment, traditional security measures may be inadequate; cyberattacks employ phishing messages and spoofed messages that mimic originals, taking advantage of the inadequacies existing in remote workspaces. The attacks compromise individual security as well as organizational integrity, and a multi-layered strategy towards endpoint protection with both technical as well as human factors for security is needed (Jena, 2025). On top of that, the development of advanced social engineering methods indicates the sophistication of modern-day cybercriminals. Techniques such as vishing, baiting, and pretexting are increasingly utilized to manipulate individuals into divulging sensitive details.

As elaborated in studies, their innovation has kept pace with that of technology, with new tools such as artificial intelligence enhancing the targeting nature of phishing. Attackers are able to tailor messages designed to target specific users based on monitoring of their own online behaviors, rendering deception all the more likely to be successful. This intersection of technology and human vulnerability serves to underscore the essential need for comprehensive training and awareness programs at the organizational level because a well-educated employee base is a powerful first line of defense against such sneaky strategies (Akeiber, 2025). Phishing and social engineering attacks are best mitigated through a proactive, multi-layered approach to security that incorporates technological defenses combined with human-focused defenses. Multi-factor authentication (MFA) and the utilization of email-filtering technologies can be used to help protect against these types of attacks, but such technical defenses must be supplemented with extensive employee training regimes. Companies must cultivate a security-conscious culture that encourages employees to flag phishing attacks and flag anomalies in real time. By so doing, businesses not only strengthen their endpoint protection but also create a strong talent pool that is equipped to fight persistent and next-generation cybersecurity attacks. The case of sophisticated attacks on even large companies is a pointer to the necessity of an integrated approach that marries state-of-the-art technology and psychological science to fortify defense against social engineering (Akeiber, 2025; Jena, 2025).

8.2.3. Insider Threats and Data Breaches

As companies increasingly rely on virtual platforms for their operations, the risk of insider threats and data breaches has grown as a universal tendency. In contrast to external threats, which may typically be sourced from hackers and cybercrime gangs, insider threats are sourced from members within the organization, including employees, contractors, or business partners. The threats can take several forms, such as deliberate data theft, negligence, or accidental security compromise. The specificity of insider threats arises from the fact that these individuals have legitimate access to confidential information, and thus it becomes complicated to detect malicious intent. As per continued research, it is essential to be aware of the human factor in order to mitigate these risks as vulnerabilities often lie not only in technology but also in user practices and organizational culture. Thus, the enforcement of robust endpoint and device security policies proves to be the most critical way of monitoring and managing such insider threats. Fighting against insider threats requires organisations to use a multi-layered security mechanism that relies on both technical controls and procedural controls. Endpoint security plays a prominent role in protecting sensitive information by locking down devices on which employees access organizational resources. This is done by implementing encryption, access control, and robust authentication mechanisms to secure valuable information. Furthermore, organizations need to incorporate ongoing monitoring and real-time analysis to identify unusual activity that may indicate a possible attack. The integration of high-security solutions, such as machine learning algorithms, can significantly enhance detection of anomalous user behavior. In cybersecurity standards, the definition of integrating these security functions includes regular training and awareness programs that allow employees to see the significance of their contribution towards data integrity.

Figure 8.6. Insider threats and data breaches.

Source: https://lh7-us.googleusercontent.com/RrPom7x4VvSHQl7aeyT4N2XcFciEh_1fb6R-0xkPksEJq2Wu7vZZGbrPQ97cNpcDlBx4vk4P6K5OZTf8wT2CP-T28YMA_n0r9qyLcCtKkUFgO5VRdhlOPwqsBsbRpqN499P0vCGIVTjdO-CYCkaOPdUk0.

Through emphasis on education alongside technology, organizations can build security culture that diminishes insider threat risks. Coordination among the stakeholders is vital in obtaining defenses against data breaches and insider threats. The primary stakeholders, including IT personnel, management, and third-party service providers, need to coordinate and assemble in order to develop inclusive security policies that govern access to sensitive information. Organizations must also have a layered security approach, referred to as Defense in Depth, to guarantee multiple security controls protect each layer of the network and endpoints. These initiatives not just de-enhance the risk of insider danger but also firm up the security position of the organization as an entity. The prognosis, as per current studies, is that proactive engagement by all parties can help lead to improved knowledge and increased vigilance against data breaches. The common alignment of responsibilities and interests across the hierarchy of the organization supports the sense of shared responsibility, ultimately for a stronger framework that can address the complexities of modern-day cybersecurity threats, as demarcated in the evolving debates regarding insider threats (Beliaev et al., 2023) (Figure 8.6).

8.3. SECURITY MEASURES AND BEST PRACTICES

Accurate security controls play a key role in the protection of endpoints and devices within an organization since they represent potential avenues for cybercriminals. Among the main methods of strengthening security is through the implementation of robust access controls. This involves the employment of multi-factor authentication (MFA) to ensure that several forms of authentication must be offered by the users prior to accessing sensitive data. In addition, the application of draconian permissions based on least privilege ensures that access is limited to just those resources that a user needs for his position. Frequent access audits and ongoing monitoring of permissions enable organizations to quickly identify outliers or misuse. This pre-emptive move lessens the danger of insider threats and encourages the culture of security awareness among employees, knowing well that they need to adhere to guidelines in the utilization of organizational resources. The incorporation of technologies such as encryption and endpoint detection and response (EDR) products also enhances the security position of organizations. Encryption plays a significant role in protecting data in transit and at rest such that even in the case of data breaches, data cannot be read unless decrypted using decryption keys. Further, EDR solutions scan endpoints in real time and continuously identify unusual behavior and respond to threats in near real time. Such vigilance allows organizations to combat sophisticated attacks that can bypass traditional security solutions. The use of these technologies, together with regular software updates and patching, minimizes vulnerabilities that might be exploited by cyber-attacks. While the cyber threats change, organizations must adapt their security systems by taking advantage of such technologies and developing a resilience response against a perpetually changing threat environment. Employee awareness and education programs are a fundamental pillar of any sound security plan, addressing the human factor most frequently exploited in breaches. Regular training sessions can instruct employees on recognizing phishing attempts, why secure password protocols are essential, and the proper procedures for reporting security incidents. These ought to be periodic rather than sporadic events, as a way of staying abreast with the dynamic nature of security threats and solutions. Empowering employees through simulations and practical exercises can prepare them better for probable attacks. By establishing an awareness-based security culture, organizations empower their workers to be the initial line of defense.

Organizations should also have adequate telecommuting guidelines because the increase in mobile and home-based workforces requires tailored security procedures for safeguarding data integrity, further highlighting the relevance of a holistic approach to best security practices.

8.3.1. Antivirus and Anti-Malware Solutions

In the world of cybersecurity, anti-malware and antivirus software are a foundation of protection against an increasingly sophisticated set of threats to endpoints and devices. Designed to detect, contain, and eliminate malicious code, they give users an essential line of protection against viruses, worms, Trojans, and ransomware. The effectiveness of these solutions relies on their capacity to detect not just known threats by signature-based detection but also unknown threats by heuristic analysis and behavioral monitoring. With the evolving nature of cybercrime, it is no longer sufficient to count on traditional methods alone. A holistic security strategy employs multiple layers of protection in order to effectively counter risks in conjunction with strategies like Defense in Depth. This multi-layered security system is instrumental in determining resilience against various attack vectors that keep emerging in the cyber environment. The speed at which cloud computing has progressed also altered the mode of delivery and efficacy of antivirus and anti-malware software. The cloud model has a number of advantages, including real-time updates and central management, that are instrumental in confronting the constantly changing threat landscape. Since solutions are kept in the cloud, companies can leverage shared intelligence and resources, which means detection and response to threats become easier. Additionally, this move towards the cloud eliminates the problem of scalability, and companies can shift their security position as they change. But as more companies turn to cloud solutions, they need to be cautious with backsliding on security, ensuring service providers possess solid security controls. This need is added to by the knowledge that security cannot be an add-on; it must be integrated into cloud services design itself to develop trust among users and facilitate wide usage (Getov et al., 2012). Besides, the various stakeholders within the cybersecurity landscape must play their roles in making antivirus and anti-malware effective. End users, the weakest link in the chain as is often assumed, are in charge of their own cyber hygiene. They must be practicing well-informed behaviors such as detecting phishing attacks and best practices for applying software updates. At the same time, developers must design their products robust enough to withstand evolving threats while producing user-friendly interfaces that simplify the process for end users. Moreover, Internet Service Providers, device manufacturers, and cloud providers must contribute towards creating an environment where the use of the internet is safe. Finally, establishing a collaborative way for all stakeholders is crucial for the establishment of a comprehensive framework of security that not just facilitates individual users but also influences the collective security position of the digital world (Beliaev et al., 2023) (Figure 8.7).

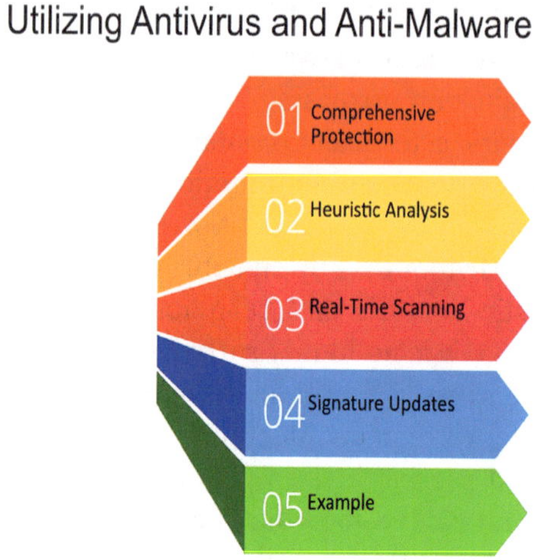

Figure 8.7. An antivirus and anti-malware solutions.

Source: https://fastercapital.co/i/Malware-Mitigation--Countering-the-1-51-Attack-Threat--Utilizing-Antivirus-and-Anti-Malware-Solutions.webp.

8.3.2. Firewalls and Intrusion Detection Systems (IDS)

The role of firewalls and IDS in endpoint and device security cannot be overlooked. Firewalls are critical barriers that prevent incoming and outgoing traffic based on pre-defined security rules. They create a controlled environment for network safety, discriminating between trusted internal resources and untrusted external threats. Modern firewalls have evolved a lot since primitive packet filters and are now advanced, multi-dimensional systems capable of performing deep packet inspection. This is needed, especially as organizations shift to more dynamic infrastructures such as cloud environments, where classical perimeter-focused security models are less applicable. The adoption of firewalls in cloud environments, such as the development of the BradStack platform on OpenStack, calls for embedding security controls deep into the operational infrastructure (Al-Shaidy et al., 2017). Therefore, the effectiveness of firewalls directly relates to the overall security posture of endpoints across different computing environments. IDS complement the defensive roles of firewalls with mechanisms for monitoring network traffic and detecting suspicious behavior. An IDS can be divided into two general categories: network-based and host-based systems, each having different roles in the security context of an organization. Network-based IDS monitor traffic at the network level, and host-based systems target individual computers. The data that the systems gather allows organizations to identify anomalies that would indicate a violation of security and respond accordingly in a timely manner. Further, recent advances in artificial intelligence and machine learning have further empowered IDS to be more effective by allowing predictive analysis of the likely threats. With organizations experiencing prolonged de-perimeterization trends, as classic security perimeters are disintegrating, the importance of IDS shines brighter in managing the advanced threat landscape that accompanies increased connectivity (Cleeff et al., 2008).

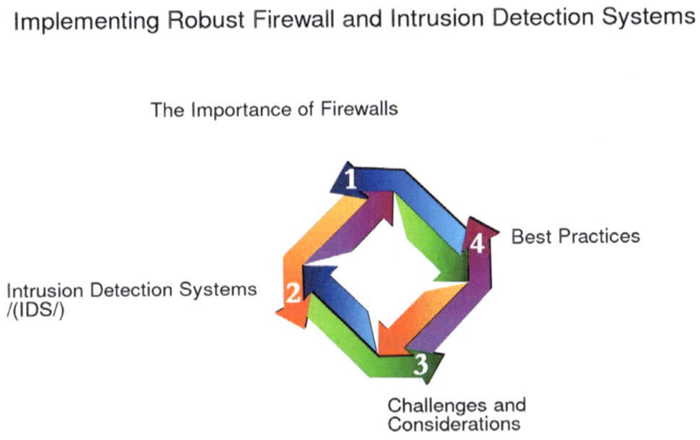

Figure 8.8. The role of firewalls and intrusion detection systems (IDS).

Source: https://fastercapital.co/i/E-commerce-security--How-to-Secure-Your-E-commerce-Website-and-Protect-Your-Customers--Data--Implementing-Robust-Firewall-and-Intrusion-Detection-Systems.webp.

Having both firewalls and intrusion detection systems integrated into an overall endpoint security strategy builds a robust defense strategy. These systems should never be isolated; rather, their deployment facilitates improved incident response and threat intelligence. For example, when an attack is denied by a firewall, an IDS will analyze the event in a way that provides information on the intrusion attempts to enable a more dynamic response.

Integration is particularly significant in cloud environments since traditional security perimeters are lost there. Besides, firewalls and IDS also require continuous updates and configuration to remain in pace with evolving threats. By systematically aligning these systems with an organization's security policy and tuning them to meet the needs of an ever-changing technological landscape, organizations can do much to mitigate risks associated with endpoint and device vulnerabilities (Al-Shaidy et al., 2017) (Figure 8.8).

8.3.3. Regular Software Updates and Patch Management

The secret to robust endpoint and device security lies in regular software updates and robust patch management. With evolving cyber threats, software updating becomes necessary for individuals as well as organizations. Security patches are typically released along with software updates that rectify known vulnerabilities exploited by attackers. Patching or isolating these vulnerabilities in timely updates can be quite helpful in countering unauthorized access and data breaches. Without regular updates, devices remain susceptible to malware and exploit kits that have been designed exclusively for attacking outdated software. Thus, making the strict update routine the highest priority allows organizations to build a more robust defense framework and offers an active defense against emerging threats, thus enhancing the overall integrity of the digital environment in which they operate. The inclusion of an ordered patch management procedure is an addition to regular software patches in that it allows for vulnerabilities to be resolved in sequence in a timely manner. Organizations tend to use tools that roll out patches automatically, thereby offering an additional layer of security. For instance, the utilization of Windows Server Update Services (WSUS) for controlling delivery of patches within Windows environments. This tool not only integrates the update process but also enables organizations to pre-screen patches before deployment to guarantee compatibility and reduce disruptions (Henderson et al., 2018). In WSUS, there are constraints that exist, particularly the availability of updates due to its metadata synchronization issues. Thus, the agencies must remain vigilant and explore alternative methods like the Custom Patching Manager, which can achieve higher efficiency in updating management tailored to the particular firm's requirements. Also, organizations must cultivate a culture of alertness toward the cybersecurity significance of regular updates and devastating consequences of neglecting patch management. Employee training is a vital task in creating end-users who are aware of their role in cybersecurity. For example, educating the users on the danger that comes with the use of out-of-date software will cause them to adopt processes that are in line with the organizations security policies. Other stakeholders such as device manufacturers and software developers bear the responsibility of making their products secure from inception, which comes with a promise to be updating and supporting them even after release (Beliaev et al., 2023). By encouraging a unified approach to cybersecurity, businesses can build a stronger infrastructure, thereby eliminating potential vulnerabilities and enhancing security in all endpoints of their networks.

8.4. ROLE OF ENDPOINT DETECTION AND RESPONSE (EDR)

In endpoint and device security, EDR is a critical defense solution that is designed to identify, analyze, and respond to sophisticated cyber threats targeting endpoints. Through the ongoing observation of endpoint behavior, EDR solutions provide real-time insights into potential threats and suspicious activity, allowing security professionals to respond rapidly to incidents before they escalate into bigger problems. Compared to traditional antivirus solutions, which are largely based on signature-based detection, EDR utilizes advanced techniques such as behavioral monitoring and machine learning models to recognize and identify threats based on patterns and anomaly. This proactive approach significantly enhances an organization's capability to neuter risk associated with advanced persistent threats and zero-day exploits. One of the central functions of EDR is incident response, which is imperative in today's more complex IT environments. Having the ability to not only detect but contain and remediate threats has made EDR a crucial component of security operations. As soon as it identifies suspicious behavior, an EDR system can trigger automated response, such as isolating infected hosts or blocking malicious processes, thereby preventing lateral movement within the network. This capability is crucial since it significantly reduces response time and mitigates potential damage from cyberattacks. Furthermore, the integration of EDR with more robust security models, such as the Zero Trust model, enhances its effectiveness, underlining the principle that trust must be repeatedly verified, even in endpoints themselves. Further, the evolving cyber threat environment demands more comprehensive detection and response systems, in which EDR plays a transformative role. Organizations increasingly realize that endpoint security cannot operate in isolation and hence give rise to cooperation amongst various security solutions such as Security Information and Event Management (SIEM) systems and Extended Detection and Response (XDR) solutions. By way of synergies from combined technologies, organizations can develop a more robust security posture able to address existing and emerging threats. By converging EDR with these complementary technologies, a unified understanding of security incidents and enhanced overall threat sense exists.

As automated, intelligent, and swift response capabilities are increasingly demanded, the role of EDR will further be vital in defining endpoint and device security's future (Cohen et al., 2024; Erdıvan et al., 2024) (Figure 8.9).

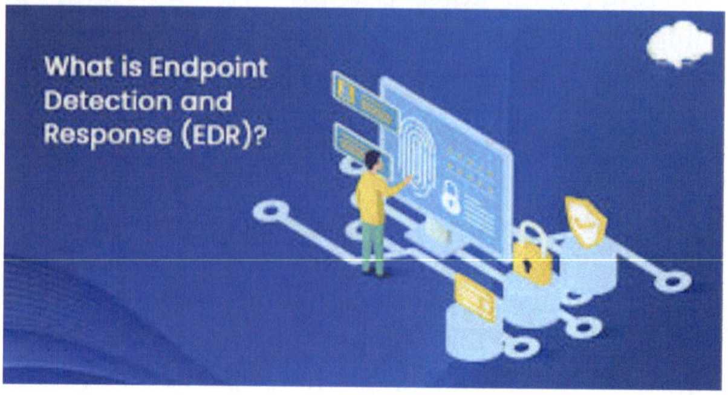

Figure 8.9. Role of endpoint detection and response (EDR).

Source: https://encrypted-tbn0.gstatic.com/images?q=tbn:ANd9GcQETrWOS6Cb7ZDgFP16WO_lbVXZrBo203wslg&s.

8.4.1. Definition and Functionality of Endpoint Detection and Response (EDR)

EDR has emerged as a critical component in the endpoint and device security arena, providing organizations with the means to detect, investigate, and respond to sophisticated cyber threats. Unlike legacy antivirus solutions that primarily operate on the principles of signature-based detection, EDR integrates advanced analytics and real-time monitoring. This proactive method enables EDR systems to scan behavior patterns and identify anomalies that might be indicative of potential security breaches. By collecting data from endpoint devices, EDR platforms allow security teams to better see network activity and therefore better develop strategic responses to incidents. The adaptability of EDR solutions allows them to be tailored to the individual needs of various organizational environments, rendering them an indispensable element in today's high-intensity threat environment. Functionally, EDR is rooted in a data gathering platform and real-time surveillance, handling a myriad of activities being executed on endpoints. This includes tracking file changes, network connections, and user behavior. The focus on real-time analysis allows for instant identification of malicious actions, which is critical given the dynamic character of threats in the cyber environment. In addition, EDR solutions typically incorporate machine learning engines to enhance detection and classification of anomalies and thereby eliminate the limitations associated with known indicators of compromise-based legacy security solutions. As organizations are faced more and more with sophisticated cyber-attacks, the ability of EDR to provide a model of layered defense is crucial, highlighting its status as a prevent and responsive tool within endpoint security management (Lewis et al., 2023). As much as EDR provides many advantages to users, there are problems encountered in its application and efficiency in actual situations. Typical problems include the need for expert professionals to interpret complex data and excessive reliance on mature configurations that cannot easily adapt to new attack approaches. In addition, an over-reliance on EDR solutions without a careful consideration of their limitations is found. For effective posture in security, organizations need to understand that EDR is not singular but one of the numerous contributory components of a dynamic security system subject to continuous updates and assessments. It is for this reason that understanding the entire scope of EDR capabilities as well as how it integrates into organizational security systems is essential to allow organizations to correct vulnerabilities while effectively managing newly arising cyber threats (Ikuesan et al., 2020).

8.4.2. Benefits of Implementing Endpoint Detection and Response (EDR) Solutions

Cyber threats are also constantly evolving with demands on robust security solutions, most notably for endpoints and devices that are the source of organizational operations. EDR solutions possess a significant advantage with the delivery of real-time monitoring and response capabilities to detect, analyze, and eradicate threats in real-time. Unlike traditional antivirus approaches, which rely primarily on signature-based techniques, EDR takes advantage of advanced analytics and behavioral detection to identify malicious activity that can indicate potential security breaches. With this innovative approach, an organization is far better positioned to respond rapidly to threats, effectively reducing dwell time and minimizing the risk of devastating data breaches or ransomware incidents. Thus, the implementation of EDR solutions becomes not just helpful, but necessary in maintaining a secure endpoint infrastructure that is able to keep pace with the dynamic nature of cyber threats, as evidenced by the urgent need for advanced security infrastructures in today's organizations (Seymour et al., 2023). The second main advantage of EDR solutions is their capacity for deep forensic analysis following an attack. Organizations can get to the root of security attacks by utilizing the wealth of information accumulated during EDR monitoring. With this, security experts can trace the attack chain, giving them insightful information on how an attack progressed, what vulnerabilities were being exploited, and whose endpoints were hit the most. With this knowledge, businesses are able to develop targeted defenses and remediation techniques to further enhance their security posture against threat vectors. Besides, EDR solutions typically integrate well with existing security infrastructure and provide centralized control and visibility. The integration enhances overall security in organizations and conforms to the layered security approach known as Defense in Depth and focuses on system-level accountability from all parties that are involved in cybersecurity (Beliaev et al., 2023). The cost-effectiveness of EDR solutions also highlights their strategic importance in endpoint and device security. Through automated incident response processes and faster recovery from compromise, EDR can considerably reduce the cost impact of cyber-attacks. Organisations are able to minimize business disruption and protect their valuable digital assets without spending enormous amounts of money on various, discrete security products. Secondly, EDR solutions help organizations address regulatory requirements by enabling organisations to effectively monitor and report security incidents. This is a critical skill in today's environment with the stricter data protection regulations. While firms persist in grappling with the complexities of digital security, deploying EDR solutions becomes the critical investment into a future defined by sophisticated cyberattacks and a relentless pursuit of strong cybersecurity behaviors (Seymour et al., 2023) (Figure 8.10).

Figure 8.10. Benefits of implementing endpoint detection and response (EDR) solutions.

Source: https://www.xcitium.com/blog/wp-content/uploads/2024/01/advantages-of-edr-solutions.jpg.

8.4.3. Challenges in Deploying Endpoint Detection and Response (EDR) Systems

The deployment of the EDR system presents intricate challenges to be overcome by organizations in an effective manner to ensure robust endpoint security. The greatest challenge is the large volume of security data generated by EDR technology. As cyber threats become increasingly advanced in terms of sophistication, parsing, analyzing, and responding to the data is increasingly critical but becoming cumbersome for security analysts. Companies typically have resource constraints that could limit their capacity to deal with and process intricate data streams effectively. Without a proper response process, businesses could not maintain a proactive security position, thus leaving them vulnerable to prospective vulnerabilities that would enable attackers to take advantage of them. This is a microcosm of the broader problem of cyber operations, where the nuanced interplay of human capital and technological capacity must be coaxed into alignment to minimize threats and maximize overall resilience against cyber-attacks (Beliaev et al., 2023). Besides issues of data management, integrating EDR systems within current IT infrastructures is another challenge. Most organizations possess a patchwork of security tools and practices. Merging new EDR solutions into this mosaic of existing solutions is a process that must be considered with great planning and budgeting, typically resulting in additional expenses and longer implementation timelines. Additionally, compatibility problems can exist, causing gaps that could invalidate the intended security safeguards of EDR systems. Seamless integration is paramount, as fragmented systems can create security blind spots that complicate incident response. Additionally, end-user training is required, considering that EDR system effectiveness relies on proper use and knowledge among employees. Failure to build an adequate knowledge base will de-weaken such systems, and individuals will turn to outdated methods to handle existing cyber threats (Ayittey et al., 2025). A second challenge in utilizing EDR systems is the dynamic nature of threats and the need to continuously adapt. Cybersecurity is far from resting; while organizations are trying to install EDR solutions, they also have to contend with round-the-clock updating of threat intelligence and shifting tactics used by cybercriminals at the speed of lightning. Round-the-clock development calls for round-the-clock monitoring and updating of EDR system configurations and threat models, which will stretch financial and human resources. Organizations must invest in continuous professional education for their cybersecurity experts to keep them current with the latest threats and advancements in EDR solutions. Additionally, as EDR products leverage artificial intelligence and machine learning to identify threats, how appropriate these systems' training data are in terms of relevance and quality is crucial to their effectiveness. Lastly, failure to address these changing challenges will leave organizations vulnerable to exploitation by ever more sophisticated attackers, undermining their endpoint security efforts (Beliaev et al., 2023).

8.5. COMPLIANCE AND REGULATORY CONSIDERATIONS

In a more digitalized ecosystem, compliance, and regulatory issues are central to endpoint and device security. Companies must operate in an advanced regulatory landscape with acts such as the Health Insurance Portability and Accountability Act (HIPAA) in the healthcare setting, necessitating secure data protection procedures. Effective endpoint security solutions can guarantee private health information is safeguarded and ensure compliance and secure against breaches that can lead to sizeable fines. Second, compliance-driven strategies encourage adopting best practices for managing data privacy and security throughout device ecosystems. By facilitating secure communication and secure access protocols, organizations not only reinforce their regulatory position but also build confidence amongst the stakeholders, including patients, healthcare providers, and regulatory agencies (Koneru, 2025). In addition to specific regulations, organizations need to be aware of broad frameworks that guide data protection across industries. The evolving landscape around digital therapeutics demands the adoption of compliance measures that harmonize regulatory imperatives with technological developments. For instance, endpoint security controls are required to adhere to regulations stipulated by the Food and Drug Administration (FDA) for software as a medical device, which calls for evidence generation via clinical trials. Moreover, integrating digital health applications with traditional healthcare modalities aims to enhance patient outcomes while safety and efficacy levels are met. Organizations are therefore encouraged to adopt holistic governance frameworks that incorporate regulatory compliance, operational efficiency, and technological innovation, but promote a responsive response to future regulatory challenges (Nalam, 2025). Finally, an active regulatory compliance approach enhances defenses against potential cyber threats to endpoints and devices. Cybersecurity standards such as the National Institute of Standards and Technology (NIST) Cybersecurity Framework provide companies with directives that enable them to possess a robust security stance, with endpoint security measures being both thorough and in compliance with the law. Cybersecurity standards foster ongoing monitoring, risk analysis, as well as incident response features that are essentials in the age of cyber. By implementing AI and machine learning technologies, as discussed in high-end endpoint security architectures, organizations can stay one step ahead of potential breaches while remaining compliant. Compliance to such regulatory models, therefore, not only mitigates risks associated with data breaches but also offers the foundation for a long-term operating model within the ever-evolving landscape of endpoint security (Koneru, 2025) (Figure 8.11).

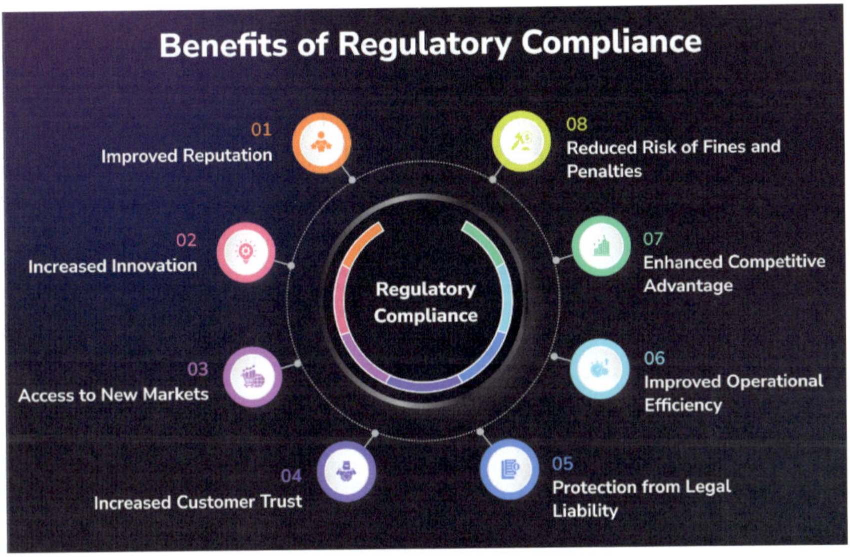

Figure 8.11. Compliance and regulatory considerations.

Source: https://cdn.prod.website-files.com/65e95c7fd0901fbe1a7733e0/66618ca714c9fb7f817b7dac_Benefits%20of%20Regulatory%20Compliance.png.

8.5.1. Overview of Relevant Regulations (e.g., General Data Protection Regulation (GDPR) and Health Insurance Portability and Accountability Act (HIPAA))

Regulations play a crucial role in establishing the endpoint and device security environment, particularly as organizations become increasingly engaged in sensitive data. The GDPR is an excellent example of an umbrella model of data protection through transparency, accountability, and consent by the user. Implemented in 2018, GDPR necessitates that organizations implement adequate security measures to safeguard personal data, thereby affecting the endpoint and device security measures in corporate networks. This legislation is an attempt at building trust with users through the assurance that their data will be processed securely and responsibly and also making organizations accountable for breaches and data protection failures. Hence, institutions, including institutions of higher learning, must adopt frameworks and regulations that align with GDPR compliance, particularly during a time marked by record levels of cyber-attacks, as highlighted by recent vulnerabilities in higher learning institutions' cybersecurity position (Kapatral et al., 2024). Under healthcare, the HIPAA provides a specialized framework that dictates guidelines for the protection of sensitive patient information. HIPAA mandates that healthcare providers, insurers, and related entities implement sound security controls for safeguarding electronic protected health information (ePHI) against unauthorized access, theft, and loss. As the use of mobile devices and remote access systems grows within the healthcare sector, HIPAA compliance is essential to endpoint protection.

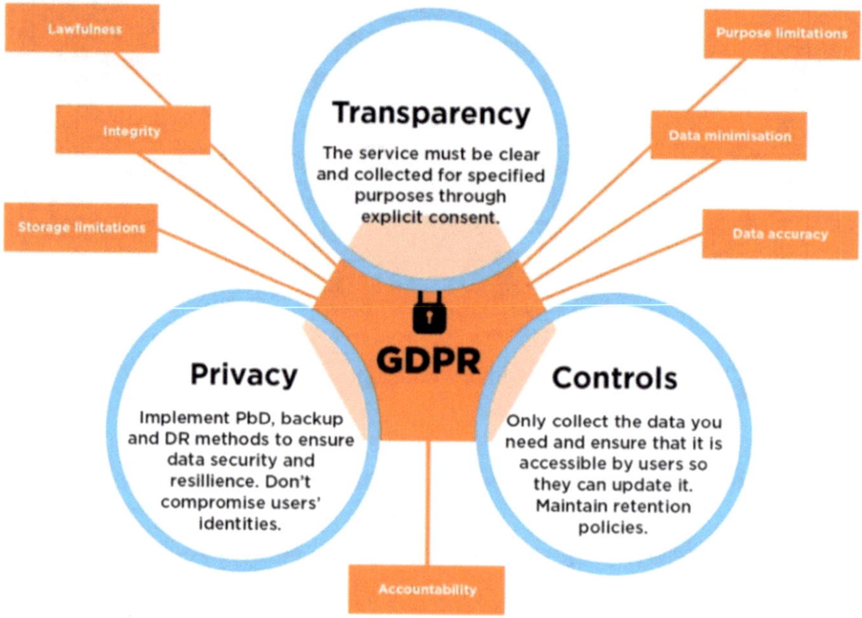

Figure 8.12. Overview of relevant regulations (e.g., General Data Protection Regulation (GDPR) and Health Insurance Portability and Accountability Act (HIPAA)).

Source: https://cdn-cjido.nitrocdn.com/OWBpPGhVFQSOSTFJpSSOzqWMqr-FnKYbA/assets/images/optimized/rev-9cf142c/www.ateamsoftsolutions.com/wp-content/uploads/2024/05/How-to-make-web-applications-GDPR-compliant-1024x690.png.

This rule dictates specific administrative, physical, and technical controls to be adopted by organizations in their operations, including the use of encryption and access controls in an effort to reduce exposure due to unauthorized access. Additionally, healthcare organizations have to conduct ongoing security audits to bring out vulnerabilities, in effect ensuring a proactive approach to endpoint security within a virtual setup riddled with exposure to potential cyber threats (Enberg et al., 2024). In addition, the ever-changing aspects of cyberattacks necessitate complete understanding of regulatory landscapes outside HIPAA and GDPR. New regulations such as the California Consumer Privacy Act (CCPA) also reflect the need for data protection and consumer rights. CCPA mandates organizations to disclose their data collection and processing activities, allowing consumers greater control over their data. Since organizations invest in enforcing robust endpoint security controls, it is imperative to follow these regulations, not only to avoid fines but also to earn consumer trust. Hence, organizations must embrace a multi-faceted approach embracing all these regulatory requirements. This demands end-to-end knowledge of the regulation environment, combining compliance with security planning to construct effective endpoint security stances with the capacity to counter the sophistication of the contemporary cyber environment (Enberg et al., 2024; Kapatral et al., 2024) (Figure 8.12).

8.5.2. Importance of Compliance for Organizations

Increasing maturity of cybersecurity threats has put a spotlight on compliance by organizations, particularly endpoint and device security. Compliance with regulatory needs and sector standards is not just a law-abiding step but also enhances an organization's security stance. Conforming to guidelines helps companies steer clear of data breach risks, hence preserving sensitive information and maintaining stakeholder trust. Further, specific compliance programs require organizations to implement strict security controls such as firewalls and encryption required for the safeguarding of devices and endpoints from unauthorized access. The proactive measure not only protects organizational resources but is also aligned with best practices in corporate governance, instilling a culture of accountability among management and staff. A further notable aspect of compliance is its ability to enhance better operational efficiency within an organization. Effective compliance programs can necessitate the use of standard processes and technology, streamlining workflows and reducing redundancies. This can be particularly relevant when considering endpoint security, where a systematic approach can minimize exposure to vulnerability in devices. The adoption of wide security frameworks, such as the Cloud Computing Adoption Framework (CCAF), can provide organizations with multi-layered security solutions. Such frameworks have been seen to possess high rates of effectiveness in threat detection, which is essential to guarantee business continuity amidst the challenges of escalating cyber threats. This intersection of compliance and business efficiency not only enhances security but also guarantees the company's responsiveness in fighting challenges in a dynamic regulatory landscape. Moreover, compliance is a key driver of trust among stakeholders and organizations, including customers, partners, and regulatory bodies. Demonstration of compliance provides a sense of credibility and trustworthiness, which is critical in today's data-driven economy. Organizations that emphasize compliance are generally well-positioned to acquire and retain customers who prioritize security and privacy of information. This is particularly relevant to businesses in highly regulated industries, where openness and control of risk are essential. Compliance not only shields institutions from potential legal ramifications and financial penalties but also enhances their competitive stature by evidence of commitment to responsible stewardship of data. Therefore, the importance of compliance to organizations is not merely about adhering to laws but a strategic imperative that bolsters wider business goals, like customer loyalty and sustained expansion in the context of evolving cybersecurity threats.

8.5.3. Consequences of Non-Compliance

Non-conformity with agreed security controls in endpoint and device security can be extremely damaging to organizations. Such non-conformity can lead to unauthorized access to confidential data, ultimately resulting in data breaches and the consequent loss of stakeholder trust. For example, noncompliance with requirements such as the Health Insurance Portability and Accountability Act (HIPAA) or the General Data Protection Regulation (GDPR) can attract substantial fines and penalties. Besides, companies can be sued by parties involved at law, further straining their finances. These legal penalties underscore the significance of robust compliance processes, as non-compliance not only breaches data integrity but also compromises corporate reputation and stakeholder confidence. Beyond monetary and image loss, non-compliance can also impact an organization's operational efficacy. In the event that security breaches occur due to relaxed measures, organizations have to invest additional resources in mitigating the breach. This diversion of focus can disrupt day-to-day business and result in a decline in productivity as the staff becomes preoccupied with coping with the consequences. In addition, imposing corrective measures or adjustments in existing processes can produce long-term issues in operation. As demanded by guidelines for best practices in the digital medicine environment, stringent adherence to compliance regimes can bring about a culture where not only is secure data management guaranteed but so are operational stability and efficiency. This cycle of compliance non-compliance serves to emphasize further the necessity of an active compliance strategy. The impact of non-compliance also carries over into technological advancement and innovation. Non-compliant organizations may be restricted in their ability to adopt new technologies or make improvements that require regulatory sanction.

One illustration is the continuous evolution of Digital Medicine, which presents opportunities for improvement but whose players have to be able to navigate complex compliance environments in order to be able to fully exploit these opportunities. Organizations risk losing beneficial technologies or suffering delays in adopting them due to non-compliance. This can ultimately place non-compliant organizations at a business disadvantage relative to those acting in their regulatory environments. As outlined in the proposed evaluation framework for Biometric Monitoring Technologies (BioMeTs), collaboration and proper knowledge of what is compliant are important to people working in the digital medicine industry, outlining that non-compliance can smother opportunities for growth and limit returns on technology advancement (Figure 8.13).

Figure 8.13. Consequences of non-compliance.

Source: https://thecsrdcompass.com/wp-content/uploads/2023/06/Legal.png.

8.6. FUTURE TRENDS IN ENDPOINT SECURITY

The evolution of endpoint security is part of a broader movement away from conventional models of security and towards more contextual and adaptive ones. Historically, organizations have had static security perimeters, relying on the belief that the threats arise externally from their communities. However, with evolving business and technology models, it has emerged as a need for organizations to reinvent their security models. Such a trend knows that assets are not contained within one boundary; therefore, organizations must implement security that honors the fluidity of information and user access, particularly with the increased prevalence of remote work. With this trend taking hold, organizations will have to better know their networks, which will result in a requirement for advanced monitoring equipment and policies capable of safely dynamically changing responses to a changing risk environment and user patterns (Cleeff et al., 2008). IoT presents unprecedented challenges and opportunities for endpoint security, and its future directions will be shaped in this respect. As sensors and connected devices grow, it is difficult for organizations to integrate such systems into their security infrastructure. The rapid expansion of IoT needs scalable architecture capable of integrating multiple communication standards while ensuring effective security against potential attacks.

When merging platforms such as Arm microcontrollers with cloud services such as Amazon Web Services, security elements should always be given priority, particularly with regard to data authenticity and integrity. As evident in instances of smart home application, the combination of real-time data transmission and cloud computing complicates security solutions further. Organizations must prepare for these threats by creating end-to-end security policies that are capable of shielding heterogeneous groups of interconnected endpoints (Cochran et al., 2018). Furthermore, new artificial intelligence (AI) and machine learning (ML) technologies are poised to transform endpoint security in ways not previously seen. These technologies enable auto-detection and response systems that are able to sift through huge volumes of data to identify suspicious patterns that point to security breaches. AI-driven security products not only react to attacks but may even predict potential threats in advance by tracking previous patterns, thereby maximizing proactive security. As organizations deploy these technologies, it also becomes more crucial to have skilled cybersecurity professionals able to analyze AI-generated inputs and integrate them into actionable security strategies. This reliance on technology-driven security solutions must be supplemented by effective governance and policy frameworks to balance the vulnerabilities of excessive reliance on automation. The endpoint security sector, as such, is set to become increasingly sophisticated, with synergy between human ingenuity and technological advancements in battling next-generation threats.

8.6.1. Rise of Artificial Intelligence (AI) and Machine Learning (ML)

The advent of ML and AI represents an evolutionary stage in endpoint and device security, transforming the way companies defend themselves against cyber threats. Traditional methods of security remain ineffective in the aftermath of changing attack vectors dynamically, and hence the necessity of adopting advanced AI-based techniques. These techniques employ enormous amounts of data to detect anomalies and predict possible security threats prior to their occurrence. Using machine learning algorithms, organizations are now better able to recognize patterns of activity, both malicious and benign, and act on threats beforehand. Not only is this more effective for security systems, it also lessens reliance on the input of human beings, which can be slow and error-prone in high-stakes situations. However, in order to ensure good security outcomes, the foundation AI systems must be well-trained and constantly updated to keep pace with the evolving threats and attack patterns. Second, with endpoints becoming increasingly pervasive, particularly with the rise of the Internet of Things (IoT), security complexity increases significantly. Each connected device is a possible entry point for cybercriminals, and hence organizations must embrace strong AI-based solutions that can deal with this enormous landscape of vulnerabilities.

The problem is not merely the enormous number of devices, but also their heterogeneous operating conditions and security features they employ. A successful solution must bring in multidisciplinary talent that spans cybersecurity, network architecture, and device management. With the enforcement of rigid security demands, organizations are able to make the IoT environment more resilient to known and unknown threats. As highlighted in recent studies, this can only be done by keeping an earnest interest in learning the unique security requirements of IoT architecture, with AI and ML remedy being tailored and updated to be effective against future threats to security (Tariq et al., 2023). While AI and ML hold tremendous promise to enhance endpoint security, they also introduce new threats and ethical dilemmas. The cutting-edge quality of AI can be reverse-engineered against an organisation by attackers using adversarial machine learning techniques that enable them to evade detection and amplify attacks. Moreover, usage of these automated systems raises concerns related to accountability and transparency.

Organizations need to navigate cautiously when creating AI systems to be used for security, where strong governance structures are to be established to manage their deployment and use. A multi-stakeholder approach needs to be encouraged that builds on inputs from a range of stakeholders, including security specialists, technology developers, and policy-makers. With the digital world evolving further, embracing the benefits of AI and knowing its limitations and unwanted aspects will be crucial to maintaining safe devices and endpoints. Ultimately, a harmonious strategy that promotes innovation as well as right use must be implemented to address the complexities of AI in the background of endpoint and device security (Muteeb et al., 2023) (Figure 8.14).

Figure 8.14. Rise of artificial intelligence (AI) and machine learning (ML).

Source: https://mytechdecisions.com/wp-content/uploads/2018/10/artificial_intelligence_robot_thinking.jpg.

8.6.2. Increased Focus on Zero Trust Security Models

Security has changed dramatically with the implementation of zero trust security models, particularly endpoint and device security. The process is based on the fact that no device or user, either inside or outside the enterprise network, will be trusted blindly. It encourages a rigorous validation process for every access request, raising the barriers of protection around valuable assets. With rising cyber threats, such models are becoming crucial in mitigating risk, particularly with consideration to the vulnerabilities created by multiple devices. With technologies such as the USB protocol growing in popularity, their built-in weaknesses are constantly exploited by cybercriminals through technologies such as BadUSB, and therefore there is a need to turn to a zero trust model that calls for constant authentication of users and devices, thus deeply enhancing endpoint protection (Butler et al., 2017). Secondly, the integration of electronic identification (eID) systems into this model is an indicator of the enhanced sophistication of security solutions. As eID becomes more common in a number of industries, it underlines the requirements for secure endpoints that are effective at authenticating people without divulging delicate information. The transition from single tokens to integrated solutions, such as those built into mobile phones, promotes a watershed shift in security and usability. Trusted hardware evolution, particularly through specifications like Trusted Platform Module (TPM) 2.0, facilitates more secure processing and storage of confidential data, which harmonizes perfectly with zero trust principles. By having every attempt at access properly authenticate and authorize, organizations can proactively retain control over threats emanating from numerous endpoint devices, further enhancing their security position (Asokan et al., 2014). As more businesses utilize zero trust security models, the impact on endpoint and device security grows stronger. Not only does it shift the manner in which access is granted, but it also impacts the means of monitoring and managing devices within a network. Continuous examination of device integrity and activity is essential, calling for sophisticated tools and frameworks to enable effective endpoint management. For instance, new technologies such as FirmUSB focus on scanning the firmware of USB devices for possible threats in advance before they are exploited, aligning with zero trust principles. This pre-emptive measure assists organizations in safeguarding themselves against threats caused by existing and new vulnerabilities in a rapidly evolving digital environment. As the threat environment evolves to become more advanced, zero trust frameworks give a robust structure to guarantee that every device on an enterprise network will be trusted only after they have undergone tightly enforced verification, further complementing overall cybersecurity efforts (Butler et al., 2017).

8.6.3. Evolution of Remote Work Security Strategies

Remote work security strategy development has been influenced significantly by the shift to a more flexible and distributed workforce. With companies beginning to adopt remote work, endpoint and device security became a priority due to the vulnerabilities in remote access. Most companies employed basic virtual private networks (VPNs) and firewalls to protect sensitive communications and data in the early stages. These conventional security measures proved to be unsatisfactory as cyberattacks developed with increased sophistication. So, organizations realized that they required end-to-end security architectures that would address not only network security but also endpoint security, by making sure all the devices connecting to corporate resources are well protected from emerging threats. As the nature of remote work expanded, so did the endpoint and device securing measures. Implementation of bring your own device (BYOD) policies necessitated stronger security to safeguard corporate data accessed through individual devices. This change necessitated an analytical approach to endpoint security, pairing advanced technologies such as encryption, endpoint detection and response (EDR), and multi-factor authentication (MFA). These solutions enable real-time visibility and rapid response to potential breaches, significantly lowering risks associated with remote work. Moreover, the proliferation of Internet of Things (IoT) devices makes remote work security even more complex to address, necessitating solutions that consider a multitude of devices which may not be similarly secured, and therefore presenting an even stronger call for holistic and dynamic security solutions (Demeester et al., 2015). The ongoing evolution of remote work security solutions has also been driven by the imperative for organizations to adopt a proactive stance towards cybersecurity. With advances in sophisticated cyberattacks, including ransomware and phishing attacks on remote employees, corporations have increasingly incorporated employee training and awareness as a central component of their security solutions. This shift is further enhanced by the realization that endpoint security is not necessarily an independent role of IT organizations but must be a collaborative effort within the organization. Organizations are making significant investments in robust security controls while fostering a culture of cybersecurity knowledge to allow workers to acknowledge and respond to potential threats. Finally, remote work security measures are increasing in response to a broader awareness of the complex interplay between technology, human behavior, and the necessity for strong endpoint and device security frameworks (Brunstrom et al., 2016).

SUMMARY

In conclusion, offering comprehensive endpoint and device security is now a critical need in today's digital environment. With advancements in the technological landscapes, organizations must embrace robust strategies that promote safeguarding against every threat type, from malware and unauthorized access to data breaches. This entails a collaborative approach involving not just high-end software solutions but also business policies that create a security-first culture among employees. By establishing this kind of environment, companies are able to eliminate threats and enhance their security posture. The dialogue between human factors and technological controls brings to the forefront the importance of continuous education and awareness in ensuring secure systems. In the end, the balance between technology and human factors forms the basis of a sound endpoint security program. Additionally, the rapid growth of cloud computing further complicates endpoint security. Increased reliance on platforms like OpenStack requires deploying secure processes that go beyond conventional methods. The illustrative example presented by Al-Shaidy et al. (2017) is that deployment of a secured cloud infrastructure must take into account legacy applications and growing blending of various applications. Notably, incorporating adaptable, open-source technologies allows organizations to adapt their security processes based on unique requirements and achieve compliance with industry standards. But as cloud environments continue to expand in scale, the threats also become dynamic in character as they demand the administrators to carry out penetration testing and vulnerability scanning frequently. These are proactive measures that assume a key role in ensuring the effectiveness of controls implemented, thereby building confidence in cloud solutions. In addition, endpoint security cannot remain restricted to traditional IT devices but must extend to new technologies such as continuous glucose monitoring (CGM) devices. As Bergenstal et al. (2017) observes, although CGM technology has been slowly adopted, there are several hindrances that discourage their use across the board. Breaking through these is essential in improving the safety and efficacy of these devices, especially as they become increasingly ingrained in healthcare infrastructure. The consequences are not only health; they reveal the necessity for a strategic solution focused on security in developing and implementing all devices that are networked. As the Internet of Things (IoT) expands further, endpoint security controls will need to be bolstered so sensitive data is kept secure and to foster user confidence in these emerging technologies. By a diversified approach that entails the learning of both cloud computing and healthcare technology, organizations can construct a robust security infrastructure that is able to mitigate future challenges.

REFERENCES

1. Akeiber, H. J. (2025). The evolution of social engineering attacks: A cybersecurity engineering perspective. Al Rafidain Journal of Engineering Sciences. https://www.semanticscholar.org/paper/8bfcc827f2378b43f51ed3de9caa3cb7e27b1cad (accessed on 13th September 2025).
2. Al-Shaidy, A. N. M. K., Kamala, M. A., Kinteh, S., Kiran, et al. (2017). Technical report on deploying a highly secured OpenStack cloud infrastructure using BradStack as a case study. http://arxiv.org/abs/1712.09152 (accessed on 13th September 2025).
3. Asokan, N., Ekberg, J. E., & Nyman, T. (2014). Citizen electronic identities using TPM 2.0. http://arxiv.org/abs/1409.1023 (accessed on 13th September 2025).
4. Ayittey, G., & Ayittey, G. (2025). A security operations and analytics framework: Continuous detection and response. https://core.ac.uk/download/655889301.pdf (accessed on 13th September 2025).
5. Aziz, S. M. (2016). Ransomware in high-risk environments. https://core.ac.uk/download/144553121.pdf (accessed on 13th September 2025).
6. Bakker, J., Bent, B., Coravos, A., et al. (2020). Verification, analytical validation, and clinical validation (V3): The foundation of determining fit-for-purpose for biometric monitoring technologies (BioMeTs). https://core.ac.uk/download/305121271.pdf (accessed on 13th September 2025).
7. Beliaev, S. (2023). Personal cyber security. https://core.ac.uk/download/657104651.pdf (accessed on 13th September 2025).
8. Bergenstal, R. M., Fleming, G. A., Heinemann, L., et al. (2017). Improving the clinical value and utility of CGM systems: Issues and recommendations: A joint statement of the European Association for the Study of Diabetes and the American Diabetes Association Diabetes Technology Working Group. https://core.ac.uk/download/132211400.pdf (accessed on 13th September 2025).
9. Bronfman, J. (2016). Weathering the Nest: Privacy implications of home monitoring for the aging American population. https://core.ac.uk/download/62570457.pdf (accessed on 13th September 2025).
10. Brunstrom, A., Damjanovic, D., Fairhurst, G., et al. (2016). De-ossifying the internet transport layer: A survey and future perspectives. https://core.ac.uk/download/80685935.pdf (accessed on 13th September 2025).
11. Butler, K. R. B., Fowze, F., Hernandez, G., et al. (2017). FirmUSB: Vetting USB device firmware using domain informed symbolic execution. http://arxiv.org/abs/1708.09114 (accessed on 13th September 2025).

12. Chang, V., Kuo, Y. H., & Ramachandran, M. (2016). Cloud computing adoption framework: A security framework for business clouds. https://core.ac.uk/download/46523058.pdf (accessed on 13th September 2025).
13. Cleeff, A. van, & Wieringa, R. (2008). De-parameterization as a cycle: Tearing down and rebuilding security perimeters. https://core.ac.uk/download/pdf/11468848.pdf (accessed on 13th September 2025).
14. Cochran, S., Heidari, H., Martin, S., et al. (2018). Arm Mbed–AWS IoT system integration [Open access]. https://core.ac.uk/download/151170100.pdf (accessed on 13th September 2025).
15. Cohen, K., & Weinberg, A. I. (2024). Zero trust implementation in the emerging technologies era: Survey. http://arxiv.org/abs/2401.09575 (accessed on 13th September 2025).
16. Demaiter, L., Qureshi, A., & Rapajic, P. (2016). Modern human machine interface in industrial wireless networks. https://core.ac.uk/download/74244059.pdf (accessed on 13th September 2025).
17. Demeester, P., Hoebeke, J., Moerman, I., et al. (2014). Fine-grained management of CoAP interactions with constrained IoT devices. https://core.ac.uk/download/55781199.pdf (accessed on 13th September 2025).
18. Demeester, P., Hoebeke, J., Moerman, I., et al. (2015). Integration of heterogeneous devices and communication models via the cloud in the constrained Internet of Things. https://core.ac.uk/download/55746971.pdf (accessed on 13th September 2025).
19. Duncan, B., Jaatun, M. G., Whittington, M., et al. (2017). Could the outsourcing of incident response management provide a blueprint for managing other cloud security requirements? https://core.ac.uk/download/80686095.pdf (accessed on 13th September 2025).
20. Enberg, P. (2024). Behavior analytics in cyber security. https://core.ac.uk/download/638718226.pdf (accessed on 13th September 2025).
21. Erdıvan, C. (2024). Process, technology and human aspects of a security operations center. https://core.ac.uk/download/620607518.pdf (accessed on 13th September 2025).
22. Getov, V. (2012). Security as a service in smart clouds: Opportunities and concerns. https://core.ac.uk/download/161121742.pdf (accessed on 13th September 2025).
23. Ghazala, M., Rehman, M. T., Shahwan, M., & Aatif, M. (2023). Origin of antibiotics and antibiotic resistance, and their impacts on drug development: A narrative review. Pharmaceuticals, 16, 1615–1615. https://doi.org/10.3390/ph16111615 (accessed on 13th September 2025).

24. Hallnäs, L., & Redström, J. (2006). Interaction design: Foundations, experiments. https://core.ac.uk/download/11433511.pdf (accessed on 13th September 2025).
25. Henderson, B. M. (2018). Custom Windows patching methodology – Comparative analysis. https://core.ac.uk/download/229302429.pdf (accessed on 13th September 2025).
26. Ikuesan, R. A., Karie, N. M., & Kebande, V. R. (2020). Real-time monitoring as a supplementary security component of vigilantism in modern network environments. https://core.ac.uk/download/387849778.pdf (accessed on 13th September 2025).
27. Jena, J. (2025). Adapting to remote work: Emerging cyber risks and how to safeguard your organization. Turkish Journal of Computer and Mathematics Education (TURCOMAT). https://www.semanticscholar.org/paper/c2971f8f70d6b3cd7344c1a5a592672c69591c44 (accessed on 13th September 2025).
28. Kapatral, S., & Kumar, P. (2024). Develop comprehensive cyber security recommendations for future development at Haaga-Helia UAS. https://core.ac.uk/download/638719936.pdf (accessed on 13th September 2025).
29. Koneru, S. H. (2025). AI-driven endpoint automation for patient monitoring: Transforming healthcare infrastructure. World Journal of Advanced Engineering Technology and Sciences. https://www.semanticscholar.org/paper/1d319761f7cac96897551e788e52d6344ab89a4b (accessed on 13th September 2025).
30. Lewis, T. M. (2023). Endpoint Defense as Code (EDAC): Configurable contextual Analysis of Process Behaviors from Kernel/User Event Tracing. https://core.ac.uk/download/567836940.pdf (accessed on 13th September 2025).
31. Nalam, S. (2025). Hybrid therapeutic modalities: Scalable data infrastructure for converging digital and pharmacological treatments. Journal of Computer Science and Technology Studies. https://www.semanticscholar.org/paper/9957b6c310250ee56c960907f65f8fe2219ad246 (accessed on 13th September 2025).
32. Palicz, T., Bencsik, B., & Szócska, M. (2021). Kiberbiztonság a koronavírus idején–a COVID–19 nemzetbiztonsági aspektusai. Scientia et Securitas. https://www.semanticscholar.org/paper/9a3c2634c1df0f5f46771c4ce198f160393a6a86 (accessed on 13th September 2025).
33. Seymour, N. L. (2023). Zero trust architectures: A comprehensive analysis and implementation guide. https://core.ac.uk/download/599656584.pdf (accessed on 13th September 2025).

CHAPTER 9

CYBERSECURITY IN INDUSTRIAL IOT AND CLOUD INTEGRATION

Contents

INTRODUCTION

The advent of the Industrial Internet of Things (IIoT) has revolutionized the industrial landscape through the provision of unprecedented connectivity and data exchange among devices. The revolution is associated with many prospects of increased efficiency, productivity, and reduction in costs but is accompanied by an unprecedented explosion of cyber threats. As IIoT devices proliferate, they pose an increased attack surface that can be attacked by attackers. Integration of these devices with cloud infrastructures also increases the complexity of cybersecurity, as the cloud infrastructure comes with its own set of security intricacies, including data confidentiality, system vulnerabilities, and requirement of robust access controls. Therefore, it is important to understand the convergence of cybersecurity with IIoT and cloud integration so that organizations can not only tap into technological advancements but also defend themselves against evolving threats. Critical analyses of existing cybersecurity frameworks demonstrate a critical gap in the domain of organizations' capacity for self-evaluation of their IoT cybersecurity stance. Existing research calls for a successful framework in examining and facilitating the enhancement of the adopted cybersecurity practices in IIoT systems. Organizations must, as recent research advises, institute a process of self-assessment measuring their susceptibility to IoT-specific cyber-attacks. The evaluation of many cyber risk assessment approaches, as highlighted in recent empirical research, draws attention to the need for achieving a target state of cyber risk that meets organizational security requirements. With the identified threats from evolving cyber threats, an objective-oriented dependency model has proven to be a highly valuable tool for assessing complex IoT environments and thereby enhancing overall cybersecurity readiness (Ani et al., 2020). Besides, the general objective of industrial cybersecurity should be pursued through holistic measures across all dimensions of cyber protection. Theoretical discussions concur that training, risk management, and heightened awareness are essential components of a proactive organizational cybersecurity culture. The integration of IIoT into current operational frameworks requires an extensive security strategy involving the identification of key technologies to be safeguarded and implementation of arrangements that address prospective vulnerabilities. Ongoing research and development of security controls are crucial in establishing cyber threat resilience. By placing horizontal initiatives and community efforts aside for cyber defense, organizations are likely to be better prepared to deal with the complexities associated with IIoT and cloud convergence (Anglano et al., 2018). Through this comprehensive step, the industrial sector can best leverage the benefits of technological advancement without jeopardizing its infrastructures.

Definition of Industrial Internet of Things (IIoT)

The arrival of the IIoT is a revolutionary shift in industrial and manufacturing processes through networked devices, data analysis, and cloud computing. Basically, IIoT combines physical equipment with digital technologies to track and control industrial systems in real time. Combining physical equipment with digital technology provides added efficiency, predictive maintenance, and operational flexibility via machines that communicate with each other in networks and exchange data that inform decision-making. The combined operation among various components, such as sensors, devices, and analytics platforms, makes it possible for industries to optimize operations in a way that prevents wastage. Consequently, this shift not only accommodates cost optimization but also allows innovation through the provision of a foundation for advanced applications, such as artificial intelligence and machine learning, that further enhance capability. As IIoT continues revolutionizing industrial frontiers, its necessity triggers the importance of efficient cybersecurity capabilities. It combines physical and virtual systems to generate novel vulnerabilities, to which industrial environments are vulnerable as a result of cyber-attacks that can slow processes and compromise confidential data. Protecting such networks relies on extensive awareness of potential attack vectors, as well as the commitment to implementing defense in depth security configurations to both the devices themselves and the cloud infrastructure upon which they depend. As discovered by recent research, continuous innovation in technologies and infrastructures for cyber protection must match pace with IIoT advancements to protect critical infrastructure and provide operational integrity (Anglano et al., 2018). Addressing these vulnerabilities requires ongoing investment in security practices, employee training, and risk management methods, maintaining industries' resistance to the deluge of new cyber-attacks. Aside from cybersecurity, the use of IIoT calls for re-examination of existing governance frameworks and sustainability processes. For small and medium-sized European towns, the integration of IIoT technologies can be an innovation driver and economic growth catalyst. The merging of devices, data, and processes can significantly improve operation effectiveness and resource management in these towns, driving sustainable development policies. There has been some evidence demonstrating the governance and sustainability roles as primary enablers of innovation in small and medium-scale businesses (SMSTEs), with technology integration and environmentalism intersecting (Mercadé Melé et al., 2020). IIoT solution industries must navigate how well-governance can be synergistic with cybersecurity strengths together with fostering an innovative culture responsive to sustainability aims. This engagement thus emphasizes the importance of establishing strong strategies that address both technological advancement and responsible management in the industrial sector.

Importance of Cloud Integration

Integration of cloud computing with the Internet of Things (IoT) has been a milestone towards developing more effective cybersecurity strategies in industry. This alignment allows organizations to integrate massive amounts of data being generated by IoT devices, making them better equipped to analyze and act on this data. With more use of IoT technology in the business for operational effectiveness, the risks associated with the use of these devices become imperative to grasp. It calls for a robust cybersecurity approach, and hence organizations have to foresee, deter, and respond to cyber-attacks. The cyber threat fluidity inherent within IoT devices brings into the spotlight the analysis and the need to enhance cybersecurity systems that govern cloud integration. To cater to this involves new models of risk analysis to quantify IoT cyber risk posture effectively, synchronize security interventions with industrial IoT ecosystems' operational goals, as the recent empirical research on such vulnerabilities (Ani et al., 2020) attest. Furthermore, industrial IoT cloud integration helps ensure that security solutions are scalable and flexible. With cloud services, organizations can deploy end-to-end security solutions that have the capacity to react to an evolving threat landscape. This capacity for responsiveness is especially necessary considering that the volume of data IoT devices will increase in the future. Cloud environments enable the deployment of next-generation security solutions, including machine learning-based algorithms that enhance threat detection and response. These solutions can analyze data access and usage patterns to derive deviation from the norm that can be an indicator of imminent cyber threats. As highlighted in current discussions on cybersecurity frameworks, there is a requirement for organizations to implement overall security frameworks that are supplemented by such cloud features and drive resilience toward sophisticated cyberattacks (Anglano et al., 2018). This incorporation ultimately leads to improved operational continuity, reducing downtime and potential monetary losses incurred due to security breaches. Apart from flexibility and scalability, cloud integration allows for the implementation of centralized security administration, a key in addressing the fragmentation of industrial IoT environments. With aggregation of security practices under a single framework within the cloud, organizations can ease their cybersecurity operations, with the promise of even application across all the IoT devices. This centralization also facilitates real-time monitoring and management, enabling better detection of threats and coordination of responses to security breaches. In addition, cloud-based solutions facilitate full logging and tracking of user activity, assist with compliance with regulatory requirements, and foster a culture of accountability. The integration facilitates a proactive security stance, in that organizations can utilize learning from centralized data analytics to further enhance their immunity to cyber-attacks. Therefore, putting cloud integration on top not only fortifies security measures but also provides organizations with mechanisms to navigate the complexities of industrial IoT ecosystems of cybersecurity, thus resulting in secure business processes (Ani et al., 2020) (Figure 9.1).

Figure 9.1. Cloud integration.

Source: https://boomi.com/wp-content/uploads/blog-seowhat-is-cloud-integration.png.

Overview of Cybersecurity Challenges

With more and more industries leveraging Industrial Internet of Things (IIoT) technology, the environment is exposed to a wide range of cybersecurity risks that undermine operational integrity as well as data privacy. Interconnectivity of IIoT devices increases vulnerabilities, exposing them to cyberattacks. These attacks are capable of exploiting weak device authentication protocols, weak encryption practices, and weak update mechanisms. Additionally, the volume of data generated by such devices is even harder to secure since organizations cannot keep up with monitoring and tracking information in real-time. It requires an in-depth knowledge of potential threats and laying down defenses which address every angle of the IIoT environment in order to address these problems. With no adequate measures, industries risk not only losing significant amounts of money but also reputations that can undermine stakeholder trust and customer confidence. In addition to the inherent risks associated with IIoT devices, the application of cloud computing introduces an additional dimension of difficulty in the aspect of cybersecurity. Smooth storage and processing of data are facilitated by cloud infrastructure, but so is an avenue for exposure to unauthorized access and data breach. Organizations must make sure that robust encryption methods are employed, both in transit and at rest, to safeguard sensitive data. Moreover, third-party cloud reliance calls for rigorous vendor assessment and constant checks. According to Anglano et al. (2018), building concrete infrastructure and controls is an important aspect of crafting successful cyber defense. Training and awareness programs must be implemented to sensitize employees to potential exploits, ultimately contributing towards an awareness-based cybersecurity culture. By addressing these complexities, organizations will better protect their cloud-integrated IIoT systems from future risks. The merging of IIoT and cloud technologies holds promise for unprecedented industry breakthroughs, but it also demands an urgent response to fight cybersecurity issues. As has been discussed in the arguments surrounding Blockchain technology, though offering new solutions that can make it a more secure option, its implementation must be done with care so as not to bring new complexities such as increased computational loads and bandwidth demands (Abdolee et al., 2019). Sectors will have to spend on developing a multi-layered security architecture that caters not only to technological controls but also organizational procedures and response plans. Furthermore, regulatory matters and data privacy concerns need to be resolved to avoid potential legal repercussions. As companies conduct business in these security environments, it will be essential that the development of technology is balanced against stringent cybersecurity practices to ensure the integrity of their IIoT and cloud infrastructure.

9.1. THE RISE OF INDUSTRIAL INTERNET OF THINGS (IIOT)

The advent of IIoT is a revolutionary time in manufacturing and operational efficiency across sectors. By seamless incorporation of smart sensors and devices into industrial systems, IIoT facilitates unmatched connectivity and automation. By doing this, businesses are empowered to reap and analyze vast levels of data in real-time, providing unmatched levels of productivity and decision-making. With this technology, however, as industries embrace it, they also get a multitude of challenges, most significantly on the aspect of cybersecurity. Protection of sensitive operational data against cyberattacks and data breaches is crucial in the security of these networked devices. As underscored, the biggest challenges facing the achievement of widespread IoT functionality—ranging from cybersecurity, data privacy, and scalability—impose the use of robust protective measures in this rapidly evolving landscape (Abdolee et al., 2019). Cloud integration is the other key aspect of the IIoT revolution, which provides access to data and collaboration between various industrial players. Cloud platforms enable IIoT devices to share data with ease, and thus organizations are able to unleash the full capability of analytics and machine learning. Cloud integration also provides remote monitoring and management, further enhancing operational efficiencies. But the larger attack surface of cloud deployment raises the cybersecurity threats of IIoT. Organizations then need to implement robust security controls in order to shield against potential vulnerabilities that could be exploited by cyber attackers. Strides taken towards establishing critical infrastructure and effective cyber defense mechanisms are core steps in building resilient systems. As emphasized, it is important for organizations that want to enhance cyber resilience against this paradigm shift in the digital world (Anglano et al., 2018).

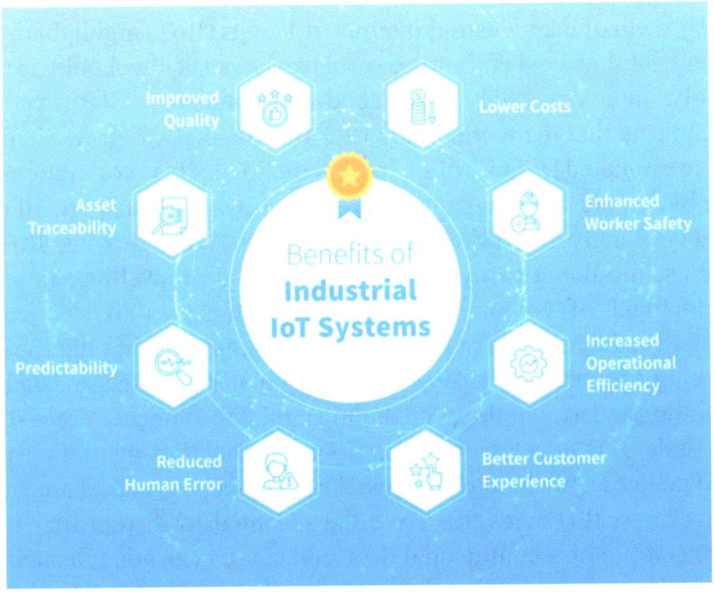

Figure 9.2. The advent of industrial internet of things (IIoT).

Source: https://cdn.prod.website-files.com/63dea6cb95e58c42d8b98c97/658688505d10862e8257591c_fyZ19fmCydJ0DJFUzOCZ8GcVNvQOW0NMxWWkEN-DOmI-d8sATAiUbImUNy-1Yao-GI0GgyD6qeY8XmZTfzCwOHh_Qa8LcKFqwJOOt_9HQwfYjLih9X7iqxFDtB0jh311Gq2fXlEbuooyMNXYwWewH3A.png.

Moreover, the intersection of IIoT and blockchain technology presents a potential solution for reducing some of the cybersecurity challenges of this industrial revolution. Blockchain's decentralized form is one where it can bring integrity as well as security to data, with the additional security of protection against unauthorized access and data manipulation. With blockchain protocols, industries can make communication channels between IIoT devices more secure, where data would remain confidential and authenticate origination. However, organizations must consider the computational overhead and potential delays while executing blockchain, balancing the benefit against such disadvantages. The blend of these advanced technologies necessitates a nuanced approach to implementation that takes into account operational capacity and security requirements. Addressing these multifaceted challenges is key to realizing the full promise of IIoT while also safeguarding critical infrastructure against emerging cyber threats (Abdolee et al., 2019) (Figure 9.2).

9.1.1. Growth Statistics and Trends

The rapid growth of the Industrial Internet of Things (IIoT) highlights a pressing requirement for close cybersecurity, particularly as such technologies become increasingly interconnected with cloud environments. Statistics released recently indicate that the world market for IIoT is likely to grow significantly, reaching an estimated USD 1 trillion by the year 2026. This acceleration is driven primarily by sectors such as manufacturing, energy, and transport, all of which become increasingly networked with each other. With growing IIoT device deployments, the threat to cybersecurity produced is swelling, necessitating the development of comprehensive security strategies to safeguard such ecosystems. The intersection of IIoT with cloud computing presents the potential for enhancing operation efficiency and risking matters related to data integrity, confidentiality, and availability, demanding the establishment of robust security practices under this development direction. As industries increasingly rely on connected devices and cloud platforms, the cybersecurity stakes are enormous. Cyber-attacks on IIoT networks have the potential to disrupt manufacturing processes, compromise confidential data, and cause enormous financial losses. In various cybersecurity studies, the number and sophistication of cyber-attacks on IIoT systems have been rising exponentially. These findings highlight the necessity for organizations to adopt advanced security models that include threat detection, incident response, and constant monitoring. Furthermore, research reveals that integrating cybersecurity into the design process of IIoT products can enhance resilience against possible intrusions. Against this backdrop, it is clear that long-term sustainability and success of the IIoT ecosystem hinge on proactive cybersecurity strategies that keep pace with the evolving momentum of technology advancements. The future of cybersecurity in the face of IIoT and cloud computing will depend on the establishment of education paradigms and worker preparedness. As recent studies have highlighted, more and more of a demand is being made to cultivate a talent pool possessing the needed expertise to manage the complexities of emerging technologies. Institutions are finally conforming to the demand by designing curricula focusing on technical knowledge as well as interdisciplinary solutions for cybersecurity to empower students with the ability to address future problems in an evolving environment. For example, education programs for professionals in the field of risk management, creation of awareness, and innovation of key technologies are central to the development of an educated workforce. The heightened focus on education across this industry is in alignment with the industry's need for innovative solutions and, ultimately, contributes to the creation of a secure and resilient IIoT environment in line with the broader narrative offered by cybersecurity studies (Anglano et al., 2018) (Alrumaih et al., 2015).

9.1.2. Key Industries Utilizing Internet of Things (IoT)

The integration of IoT is changing spearheading industries, particularly the manufacturing industry, which is at the forefront of this innovation. IoT solutions streamline processes by optimizing in-the-moment monitoring and predictive maintenance through machine interconnectivity. These innovations contribute to the ideas of Industry 4.0, which include the smart manufacturing systems that optimize productivity as well as efficiency. As companies deploy IoT solutions, they receive insights which enable them to reduce wastage and downtime, thus facilitating cost savings and business efficiency. As this increased reliance on connected devices becomes more of a reality, it presents severe cybersecurity challenges since if the manufacturing infrastructure contains vulnerabilities within it, it can lead to catastrophic failures or breaches if not controlled. Hence, solving these cybersecurity issues is the need of the hour to realize the complete potential of IoT in the manufacturing sector so that there is a safe yet innovative atmosphere for future growth. The health sector is another major area where IoT is taking revolutionary leaps. Integrated health monitoring solutions and medical devices enable constant care of patients, remotely monitoring vital signs, and facilitating healthcare professionals to act pre-emptively towards medical requirements. IoT technologies facilitate more informed decision-making, improve patient outcomes, and facilitate the collection of enormous amounts of data for analytics and research. The increased connectivity that comes with such development brings in more risks related to data privacy and system security. Cyber-attacks can compromise sensitive patient information, which can have major legal and ethical impacts. Therefore, balancing the application of IoT for improved patient care against protection from potential cyber-attacks is crucial to health industry trust and making sure that technology is performing as expected. Lastly, the agricultural industry displays the functionality of IoT by integrating clever farming technology to optimize crop care and resource distribution. Internet of things-enabled and sensor devices monitor plant health, weather, and soil health to facilitate data-driven decision-making that increases yields and reduces wastage. It is part of a larger trend of using technological advancements to respond to major industry challenges, such as sustainability and food security. But like with any business, the application of IoT for agriculture comes with attendant cybersecurity risks that can render operations nonfunctional or sensitive data exposed. Agricultural infrastructure and data must be secured against cyber risks so that it is possible to guarantee that IoT technologies are able to fulfill their role and contribute to the development of the industry. Universal solutions to such cybersecurity problems will be vital as the IoT becomes more of a central function in major industries, particularly with cloud integration (Figure 9.3).

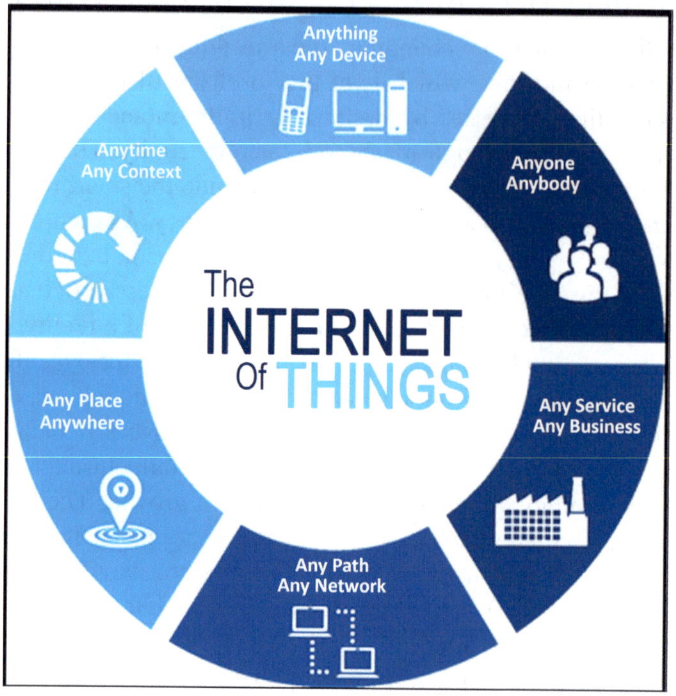

Figure 9.3. The integration of the Internet of Things (IoT).

Source: https://www.ntrixit.com/assets/img/services/iot.png.

9.1.3. Benefits of Internet of Things (IoT) in Industrial Applications

The use of the IoT in industrial applications has brought significant advancements in terms of operational efficiency and productivity. By utilizing networked devices and sensors, businesses are able to monitor their processes in real-time, which makes them able to make immediate interventions and adjustments. Such capability results in increased responsiveness to inefficiencies or anomalies, ultimately reducing downtime and operational costs. Moreover, the creation of vast amounts of data from these IoT devices enables predictive maintenance methodologies, eventually extending equipment lifespan and reducing unplanned maintenance costs. Such operational enhancements are crucial, given that they not only enhance profitability but also enhance overall production quality and reliability in competitive markets, an overwhelming shift in the manner industries operate and expand. Moreover, the benefits of IoT spill over into superior decision-making processes within industrial settings. The data analytic aspects of IoT technologies offer decision-makers well-researched analyses based on actual, up-to-date information. By the application of advanced analytical tools, organizations can better understand operational patterns and customer behavior and thus plan strategically with more knowledge. Such a level of evidence-based intelligence allows for responsive action in response to changes in the market and drives innovation, since companies can try new approaches backed by facts. While corporations struggle to change in reaction to regulatory pressure and fluid customer need, IoT integration irreversibly reshapes the landscape of operational architecture, triggering forward-looking practices that form the foundation for sustainable competitive leadership. The transition to massive-scale IoT integration is not, however, without peril, particularly in terms of cybersecurity. As there is growing sophistication in IoT systems with more and more devices connected, the threat of cybersecurity threats and vulnerabilities is greatly on the increase. Effective safeguarding of such systems is of prime importance, ranging from the deployment of cutting-edge cybersecurity measures to the development of a culture of awareness and risk management among employees. This is aligned with the requirements disclosed in Italy's national research agenda on cybersecurity, which highlights the necessity of integrated plans to defend strategic industrial infrastructures from evolving cyberattacks (Abdolee et al., 2019; Anglano et al., 2018). Ultimately, while the benefits of IoT are enormous, they must be tied to the same strict cybersecurity regulations to provide a secure industrial framework.

9.2. CLOUD INTEGRATION IN INDUSTRIAL SETTINGS

The application of cloud integration in industrial settings is at the forefront of the performance and efficiency of manufacturing processes. With the application of cloud-based technology, industries can be able to provide enhanced operational agility by utilizing optimized management of resources as well as real-time information sharing. This technology enables the optimization of production lines using data analytics, predictive maintenance, and remote monitoring. With these benefits, however, comes a matching increase in cyber exposures. As industries embrace the Internet of Things (IoT) connecting gadgets and systems, the attack surface for potential cyber threats widens, putting sensitive operating data and infrastructure at risk. Hence, it is essential that industries implement robust cybersecurity measures along with cloud integration in order to safeguard their information and develop resiliency against cyberattacks and hence enable a secure industrial ecosystem with innovation and efficiency. In order to be able to utilize the potential of cloud integration to its best, the stakeholders must consider the specific cybersecurity risks arising in the industrial Internet of Things (IIoT). As pointed out in current discussion around cybersecurity, there is a need to expose technological architectures to continuous analysis to prevent risks of interdependent systems (Anglano et al., 2018). This transcends threat detection and also involves measures that should be put in place to preserve the integrity of data throughout.

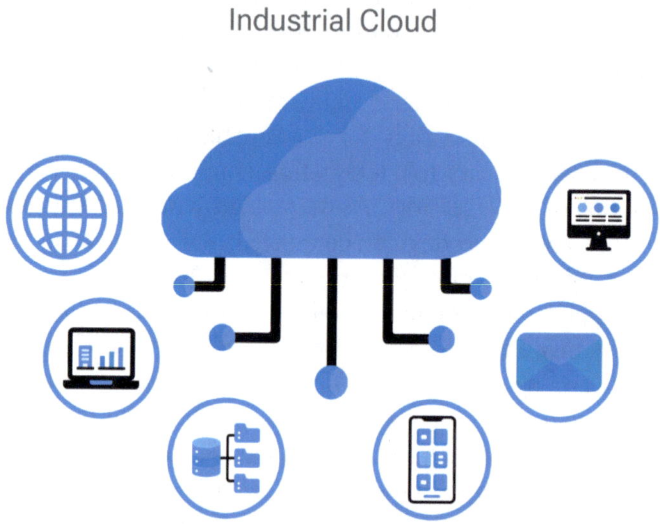

Figure 9.4. Cloud integration in industrial settings.

Source: https://www.fogwing.io/wp-content/uploads/2022/08/Industrial-Cloud.png.

The industrial configuration often includes legacy systems that were not planned to integrate with connectivity, thus requiring integration of cyber defense elements within the framework of cloud systems. Companies need to invest in advanced monitoring features and regular security audits to check for compliance and protection against evolving cyber-attacks. It is thus crucial to implement a culture of cybersecurity awareness through technology innovation coupled with effective risk management features to further secure industrial network defenses. Besides, the implementation of Industry 4.0 values implies the greater necessity of cloud integration in the industrial sector, particularly in such sectors as the food industry, and it implies the stronger economic effects (Córdoba-Roldán et al., 2017). While industries move toward smart manufacturing, which is marked with increased flexibility and customization, they ought not to forget the cybersecurity implications to these advancements. For instance, process and data streams interdependence demands stringent security protocols for preventing data leakage and unauthorized access. Integration of cloud technologies must therefore also feature safeguarding measures tailored to each industry's specific proprietary operation frameworks. This forward-looking strategy not only fosters competitiveness but also decreases the risk of interruptions caused by cyber events, ultimately producing a stronger industrial setting that can adapt to emerging challenges without sacrificing secure operational practices (Figure 9.4).

9.2.1. Advantages of Cloud Computing

Cloud computing provides a whole range of advantages that significantly enhance the efficacy and efficiency of cybersecurity solutions in industrial IoT settings. One of the main strengths is its scalability. With increasing numbers of IoT devices being incorporated into business processes, cloud computing helps them scale resources in real time. This adaptability not only provides space to expanding data requirements that arise with IoT but also helps to distribute the processing needs evenly, which is crucial as far as providing effective cybersecurity is concerned. As a result, companies are able to implement their cybersecurity measures over a scalable platform, so that when new threats arise, the defenses can be increased or modified without considerable downtime or resource utilization. In the end, the marriage of scalable cloud technologies and IoT infrastructure makes organizations better security postures possible, allowing organizations to react nimbly to changing threats. Another significant advantage of cloud computing is increased data analytics that plays a vital role in bolstering cybersecurity in industrial settings. Cloud infrastructures often have advanced analytical software that utilizes machine learning and artificial intelligence, enabling organizations to analyze huge volumes of data generated by IoT devices. Such software can identify patterns and abnormalities indicating potential security breaches or equipment failure. Accessing these insights allows proactive security and quick responses to potential threats, thus minimizing the risk-exposure level. In an environment characterized by Industry 4.0, where factory processes become increasingly connected utilizing IoT networks, the ability to draw on data analytics helps in making wiser decisions. As quoted, there is a need to adopt the Industry 4.0 model, and cloud computing assists in doing so by augmenting the precision and reliability of cybersecurity actions, as emphasized in industries such as the food industry (Córdoba-Roldán et al., 2017). Moreover, cloud computing supports the intensification and dissemination of collaboration across different industrial industries, which is crucial in countering cybersecurity challenges in the integration of IoT. With the availability of centralized data storage and management, cloud services enable various stakeholders to share common resources and knowledge easily. The shared strategy not only enables easy sharing of cybersecurity news and best practices but also allows organizations to learn from each other's experiences and failures as a collective group. This kind of collaboration becomes even more important with the interdependent nature of modern manufacturing processes, where a miscalculation in one stage can greatly influence another. Noted in discussions surrounding the Internet of Things, successful integration hinges on overcoming significant barriers like cybersecurity, and greater collaboration through cloud computing becomes an unavoidable method of addressing these problems (Abdolee et al., 2019). With these collaborative structures, organizations can build a more secure cybersecurity stance, pushing past the complexity of an increasingly connected industrial world.

9.2.2. Types of Cloud Services Used

The integration of cloud services into the Industrial Internet of Things (IIoT) provides various forms of cloud models to enable operational efficiency and responsiveness. Among them, Infrastructure as a Service (IaaS) is significant for its ability to provide necessary computational resources, including virtual machines, storage, and network services. IaaS makes it possible for industries to expand infrastructure according to need, decreasing costs significantly while improving flexibility. Adoption of IaaS becomes particularly vital in scenarios where rapid data processing and real-time analysis are of critical importance. While industries capitalize on such assets, they also must be prepared with robust cybersecurity measures to counter potential threats that could arise from cloud weaknesses. Dependence on IaaS is a quintessential example of how cloud can empower IIoT and yet indicate the unavoidable need for an across-the-board security framework in protecting sensitive operational data. In conjunction with IaaS, Platform as a Service (PaaS) offers invaluable support in creating and deploying applications in the IIoT landscape. PaaS provides a sophisticated collection of development tools and services that enable organizations to create scalable applications without having to manage the infrastructure. This cloud service is particularly beneficial for innovation and speed-of-novation application and for companies that are innovation and speed-of-novation application centric along with rapid application iteration required to process and analyze IIoT device data. Further, PaaS usage might involve additional security mechanisms to shield applications from potential exploits during the development stage. As can be seen in recent research, these platforms must be bolstered with strong coding standards and high-grade testing to minimize vulnerabilities, which accords with general debate around sound cybersecurity procedures required in the assurance of technological innovation in IIoT environments (Anglano et al., 2018). Lastly, Software as a Service (SaaS) is a flexible option for organizations in the IIoT environment, with pre-developed software ready for use through the internet. This type of cloud service delivers optimized processes such as data management, analytics, and communication across devices and platforms. Organisations can obtain sophisticated functionalities through SaaS without significant initial investments or complicated installations. There are, however, inbuilt cybersecurity concerns associated with SaaS, including data privacy and regulatory compliance. With more organizations looking for operational efficacy via SaaS, there needs to be strong access controls and encryption methods available for protecting key information. Recent studies highlight the importance of such controls in enhancing the cybersecurity position of organizations against cloud usage and IIoT complexities while also highlighting an increasing need for a uniform cybersecurity strategy in this evolving ecosystem (Fernández-Caramés et al., 2019) (Figure 9.5).

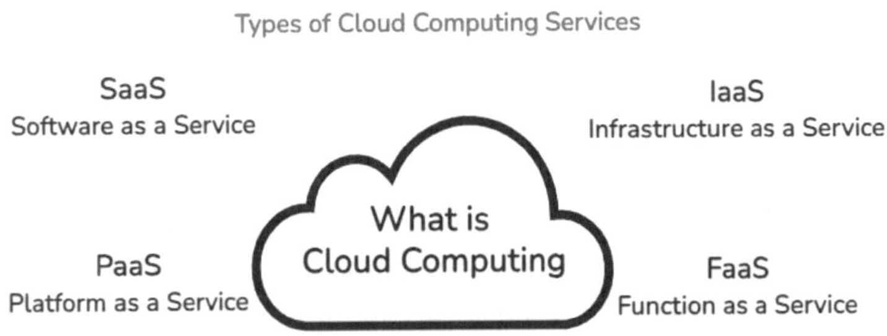

Figure 9.5. Types of cloud services used.

Source: https://media.geeksforgeeks.org/wp-content/uploads/20240429111954/Types-of-Cloud-Computing-Services-(1).webp.

9.2.3. Role of Cloud in Data Management

The future of data management is significantly influenced by cloud computing, particularly when it comes to industrial IoT. The cloud makes it possible for organizations to store and process amounts of data generated by connected devices in real-time and analyze them. This capability is a requirement for industries that critically rely on IoT solutions to optimize operational efficiency.

With the scalability and agility of cloud resources, organizations are able to manage data from different sources to have actionable intelligence whenever needed. Cloud architectures also facilitate enhanced interdepartmental collaboration in a way to have an improved integrated usage of data, which is needed for the improved optimization of IoT systems within industry. But integrating cloud services into data management creates significant cybersecurity concerns, particularly industrial ones. As IoT devices become more common, the attack space for possible cybersecurity threats widens, and a robust mechanism is required to assess and respond to vulnerability. Unlike traditional IT systems, IoT communities possess aspects that are unlike one another and require specially designed risk management strategies. To this end, an empirical assessment of existing cyber risk assessment practices reveals the lack of methodologies focused on IoT environments. By concluding a goal-oriented dependency model to be the most applicable, organizations can measure their IoT cyber risk position numerically, improving their predictive ability to predict, deter, and counter cyber security threats (Ani et al., 2020). Further, cloud computing in data handling intersects with end-to-end security solutions to protect infrastructure and sensitive information. Risk assessment lessons from the ground on which efforts are taken to protect cloud resources from emerging new cyber threats all the time. Cybersecurity solutions will be required to address technology-oriented solutions as well as organizational steps like training and awareness creation. By fostering a culture of cybersecurity among the workforce, organizations can tremendously expand their capacity to resist potential cyber threats. As security framework forums have identified, thorough understanding of both the infrastructure required to provide cyber protection and the actions required to effectively manage risk is critical for protecting the increasingly interdependent industrial IoT landscape (Anglano et al., 2018).

9.3. CYBERSECURITY THREATS IN INDUSTRIAL INTERNET OF THINGS (IOT)

With industries increasingly embedded with IoT technologies, they are increasingly vulnerable to a variety of cybersecurity threats. The merging of IoT devices—most commonly described as possessing low processing powers and infrequent updates—creates an arena for offenders. The loopholes inherent in these devices present criminals with avenues to exploit vulnerabilities, typically leading to disastrous hacks. These hacks can have extensive impacts, including operations disruption, monetary loss, and a tainted reputation. These threats necessitate a robust cybersecurity solution to counter potential threats and safeguard vital systems from incessant cyberattacks that change and modify at the same rate as changing technology. It is essential to comprehend these threats in an effort to develop a robust security plan capable of handling the unique challenges presented by the dynamic nature of industrial IoT. Further, the advanced nature of cyber threats in industrial IoT settings highlights the urgent need for thorough risk assessment methods. As is mentioned, there is a lacking area at this time as far as self-assessment methods to measure an organization's cybersecurity posture across IoT go. This deficiency poses difficulties in robustly predicting, preventing, and reacting to cyber threats. With the environment continuing to evolve, organizations need to focus on adaptive solutions encompassing ongoing vulnerability and risk assessment of devices in constant communication. Empirical research and comparison of various cyber risk assessment methods can guide industrial operations to develop more robust systems.

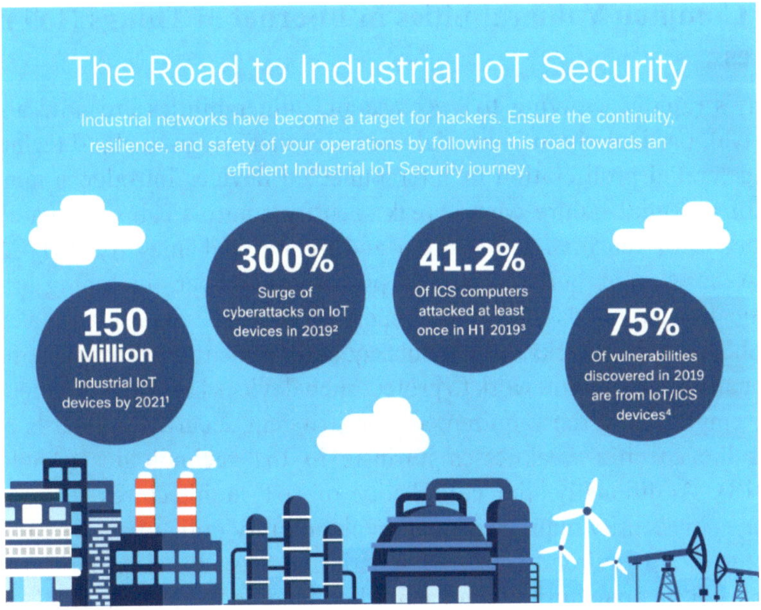

Figure 9.6. Cybersecurity threats in industrial Internet of Things (IoT).

Source: https://cloudtweaks.com/wp-content/uploads/2020/10/IoT-Industrial-Security.png.

This anticipatory stance not only minimizes exposure to potential vulnerabilities but also prepares the ground for continuous enhancement of security measures enforced in the industrial IoT setup, ultimately protecting critical infrastructures from catastrophic disruptions. Training and awareness programs for the personnel operating within industrial IoT settings are required to further enhance resilience through instilling a security-conscious culture. The incorporation of successful security interventions should also involve identification and deployment of horizontal interventions like extensive training, creation of awareness, and cautious risk management as indicated in cybersecurity literature (Anglano et al., 2018). Additionally, the goal-oriented dependency model is a helpful tool for quantitative self-assessment of cyber risk stance, in which organizations can transition to a target status that significantly minimizes uncontrollable risks represented by complex IoT systems (Ani et al., 2020). By maintaining a clear roadmap for continuous improvement and training, organizations can successfully keep up with the sudden changes in technology, thus enhancing their cybersecurity position and ultimately making their defenses more robust against potential attacks (Figure 9.6).

9.3.1. Common Vulnerabilities in Internet of Things (IoT) Devices

In today's rapidly evolving IoT ecosystem, vulnerabilities are now a matter of survival, particularly in industrial use cases that integrate cloud technology. The accelerated proliferation of interconnected devices introduces numerous points of potential failure where weak security controls can culminate to the compromise of sensitive information and operational integrity. IoT devices, in most cases, are built with low processing power, including insecure authentication protocols and limited encryption. These exposures make them vulnerable to cyberattacks, which subsequently penetrate entire systems. The need for an effective framework to protect such devices has become exceedingly urgent, emphasizing the requirement for stringent security standards able to address the specific weaknesses intrinsic in IoT technologies (Anglano et al., 2018). Additionally, the complexity of IoT architectures poses severe challenges in security management and vulnerability mitigation. When devices connect to cloud environments, they inherit diversity in their security postures that may not be compatible with each other, leading to what is widely known as an "implementation gap." This incompatibility highlights the difficulty of applying best practices in all layers of an IoT ecosystem. For instance, a number of organizations may overlook configuring and keeping connected devices, leaving backdoors for attackers in the process. Furthermore, security's time-related aspects—i.e., when to update and handle incidents—add another layer to complexity. As industries embrace smart technologies, detection and development of how to handle these vulnerabilities are essential to sustainable operational resilience (Accenture Technology et al., 2018). Besides technical vulnerabilities, regulatory landscapes also define the security environment for IoT devices. New legislative reforms, such as the EUs General Data Protection

Regulation (GDPR) and Network and Information Security (NIS) Directive, have compelled organizations to redesign their cybersecurity strategy. These laws not only attempt to protect data but also emphasize the protection of operational technology from emerging threats. Although cybersecurity violations grow and include DDoS attacks on critical infrastructure, stakeholders have to adopt proactive risk management. The issue lies in developing the culture of cybersecurity awareness beyond compliance, and this calls upon industry leaders to acknowledge the collective responsibility for ensuring the integrity of the industrial IoT environment (Accenture Technology et al., 2018).

9.3.2. Types of Cyber Attacks on Industrial Systems

The cyber threat landscape to industrial systems has changed tremendously, necessitating immediate discussions on how best to secure industrial systems from cyber threats. Cyber-attacks on industrial systems can take different forms, and each of these forms has distinct challenges and vulnerabilities. For example, Distributed Denial-of-Service (DDoS) attacks are so menacing in that they spam a network with traffic, making essential services inaccessible. These intrusions not only halt manufacturing operations but also jeopardize personnel safety and sensitive data integrity. The effects of these attacks are not only confined to direct monetary losses but may also extend to bleed over and touch national infrastructure, particularly in sectors like energy and manufacturing. As industrial equipment becomes networked and cloud-integrated, strong defenses need to be established that can repel these attacks in a proper way and ensure the resiliency of industrial operations. Phishing attacks are another significant industrial threat, primarily serving as an entry point for cyber attackers. Phishing attacks exploit human vulnerabilities by duping users into divulging sensitive credentials or opening malicious links. In an industrial setting, the consequences can be dire, with unauthorized entry potentially leading to data breaches, process interference, and other ill-intentioned activities. The unique challenges inherent in human factor vulnerabilities demand comprehensive training programs that not only enhance technical defenses but also instill an organizational culture of cybersecurity awareness among employees. Through the emphasis on cybersecurity training, companies will be able to equip their employees with a better ability to identify and counter such threats, overlaying a human sense of awareness over their cyber security. Such a multi-dimensional strategy will be critical to ensuring the integrity of industrial systems in the face of changing cyber threats. Ransomware attacks against industrial systems have become more common, highlighting the importance of robust cybersecurity frameworks. Such attacks typically involve encrypting vital data and demanding a ransom to obtain decryption keys, effectively grinding operations to a halt.

The strategic significance of industrial facilities makes them desirable targets for cyber criminals and, as such, necessitates multi-pronged defensive strategies with technology, procedure, and individuals. As discussed in recent studies, firms must create an elaborate awareness of their cybersecurity position to anticipate, avoid, and respond to prospective ransomware attacks. Combining findings of empirical cyber risk assessment method studies can inform organizations of their vulnerabilities and recommend proactive measures to strengthen their defenses. Developing a goal-oriented dependency model for the assessment of cyber risks in Industrial IoT systems is necessary in navigating this complex terrain of threats and ensuring operation continuity in spite of disruptive cyber-attacks (Anglano et al., 2018; Ani et al., 2020) (Figure 9.7).

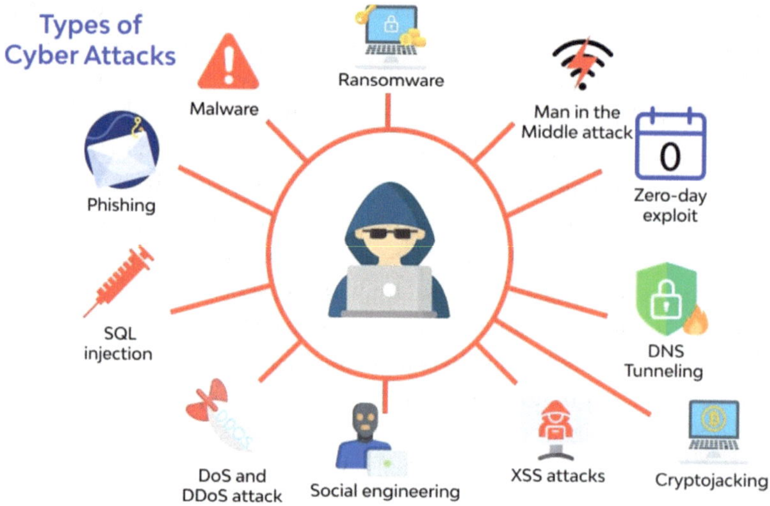

Figure 9.7. Types of cyber-attacks on industrial systems.

Source: https://cdn.prod.website-files.com/5ff66329429d880392f6cba2/66ceeb4ca81c6263e6aaa8e2_622b1df712adfa4ebff027a5_types%2520of%2520cyber%2520attack.jpeg.

9.3.3. Case Studies of Notable Breaches

The increasing merging of industrial Internet of Things (IoT) systems and cloud computing has exposed a multitude of vulnerabilities, well-illustrated in major cybersecurity incidents. One notable case study was when a large manufacturer suffered a breach due to poor security on its IoT devices, exposing unauthorized access to information. This episode illustrated how traditional security models are not designed to anticipate the particular challenges of networks of interrelated devices. Amidst widespread concern over such susceptibilities, experts are certain that the IoT will continue to expand in the future, incessantly linking devices and optimizing productivity. Although this growth has many advantages, it simultaneously increases the potential for breaches, highlighting the imperative necessity of strong security measures to counter these threats effectively (Anderson et al., 2017). There is another relevant case study that came in the healthcare industry, as seen in a major data breach on a leading hospital network. This breach, the byproduct of combining old cybersecurity architectures with newer IoT technologies, put patient information at risk and compromised essential healthcare services. The attack highlighted an unpleasant truth: modern-day cybersecurity architectures are often too vague to address the specific challenges in modern healthcare organizations. As our recent research in this field indicates, there tend to be substantial differences in cybersecurity maturity across industries. Especially, companies that finance their cybersecurity defenses not only protect sensitive data but also enhance their attractiveness to foreign investors, who nowadays put strong emphasis on robust security as an investment decision-making factor (Iyer et al., 2025). Moreover, analyzing these case studies reveals an evident connection between cybersecurity maturity and resistance of industrial IoT systems to attacks. The pattern indicates that organizations with aggressive cybersecurity practices tend to endure fewer and less damaging breaches. By further bolstering defenses, institutions are able to efficiently safeguard their operations and customers' data, foster trust and business continuity. This is particularly applicable to industries relying on sensitive data and operational integrity, such as manufacturing and healthcare. The message is clear: embedding strong security measures crafted to address the unique realities of IoT and cloud infrastructures is essential for risk mitigation and towards a more secure digital society, thus promoting economic sustainability and public trust (Anderson et al., 2017; Iyer et al., 2025).

9.4. SECURITY FRAMEWORKS AND STANDARDS

In the evolving landscape of Cybersecurity for Industrial IoT and Cloud Integration, security standards and frameworks are key components that provide structured methods towards identifying, mitigating, and managing cyber threats. These frameworks facilitate an integrated form of security thinking in resolving the unique challenges with IoT-enabled environments. Particularly, organizations must conduct self-assessment of their IoT cyber security position in order to identify vulnerabilities inherent in their systems. Such assessments offer insights into current and potential dangers and guide the implementation of security solutions that meet the unique needs of an organization. The integration of a goal-based dependency model can significantly enhance the ability of an organization to foresee, prevent, and contain emerging risks that come with the extensive use of IoT, hence offering an efficient security platform for industrial operations (Ani et al., 2020). Furthermore, the effectiveness of best practices in sound cybersecurity for industrial IoT relies on international-standard frameworks and standards. These frameworks offer interoperability that is crucial where multiple devices and applications converge in a cloud setup. As long as organizations comply with established standards, they can assure that security is comprehensive as well as compatible with existing technology. Frameworks such as the NIST Cybersecurity Framework enable organizations to have clearly documented guidelines and best practices necessary for enhancing their overall security posture. Furthermore, security frameworks offer a roadmap to continual improvement, training, and awareness programs that are vital to embedding a culture of cybersecurity in organizations. Such security pledge is important in protecting confidential data and upholding the integrity of industrial processes associated with cloud technology (Anglano et al., 2018). In summary, embracing entrenched security patterns and norms is a cornerstone in reducing the security threats in Cybersecurity in Industrial IoT and Cloud Integration. Organizations must routinely update these frameworks to meet new threats and the dynamic nature of technology. By taking a proactive stance in cybersecurity—through intermittent examination of risk, extensive self-auditing, and best practices implementation—organizations can significantly enhance their defensive position. Lastly, the use of robust security standards strategically not only enables regulatory compliance but also evokes stakeholder trust, thereby facilitating sustainable industrial development in an increasingly interwoven industrial age (Anglano et al., 2018; Ani et al., 2020).

9.4.1. Overview of Existing Cybersecurity Frameworks

Internet of Things (IoT) technology and cloud computing have necessitated the use of robust cybersecurity frameworks to secure industrial environments. These frameworks serve as official standards to guide organizations to comply with and mitigate risk from cyber-attacks. Examples of popular frameworks include NIST Cybersecurity Framework, which emphasizes an integrated approach to risk management, and the ISO/IEC 27001 standard, which emphasizes establishing, implementing, maintaining, and continuously enhancing an information security management system. Organizations use these models with particular sectoral requirements in mind, ensuring that the actions they are taking prove to be effective as much as relevant to the particular problem caused by the interplay of IoT and cloud technologies. With an ever-evolving threat landscape, there is the necessity for constant updates and incremental improvements to these frameworks to keep pace with emerging cyber adversaries.

In the industrial IoT context, convergence of cybersecurity frameworks is even more critical. Interconnected systems and devices can pose vulnerabilities that will be used by malicious individuals if proper safeguards are not in place. For instance, industry group-created frameworks, such as those from the Industrial Internet Consortium, provide industry-specific guidance aimed at protecting interconnected, complex systems. The significance of these frameworks is that they provide solutions for specific issues that impact industries, from manufacturing to food processing. One exemplary case is the food industry, which is also of significant value to local economies, so it is essential to apply cybersecurity best practices, as dictated in available frameworks. Córdoba-Roldán et al. (2017) emphasizes the industry flexibility in applying these guidelines towards establishing a secure operational environment, pointing out that the compliance with cybersecurity frameworks opens the door to innovation and resilience. Moreover, existing cybersecurity frameworks facilitate a culture of awareness and readiness amongst employees and stakeholders. Companies with formal training and awareness schemes, such as outlined in models such as that referenced in Anglano et al. (2018), can address very effectively the risk of human error—a major weak spot in cybersecurity. Technical measures alone are not sufficient to be effective in combating cyber-attacks, but the human element is also crucial; educating employees and running awareness campaigns ensure that the staff remains familiar with potential vulnerabilities and how to spot them. By establishing an educated culture, organizations reinforce their defenses and stimulate an active posture toward cybersecurity. With organizations ramping up cloud technologies with IoT capabilities, adherence to defined frameworks will always be imperative in safeguarding precious information and offering business continuity in industrial environments (Figure 9.8).

Figure 9.8. Overview of existing cybersecurity frameworks.

Source: https://img.helpnetsecurity.com/wp-content/uploads/2023/12/16180902/cybersecurity_frameworks-1400.jpg.

9.4.2. Importance of Compliance and Regulations

In case of Cybersecurity for Industrial IoT and Cloud Integration, regulations and compliance are paramount. As more IoT devices become integrated, organizations are becoming under mounting cyber-attacks that have extensive implications for their business and the wider public. Compliance with current cybersecurity legislation ensures that organizations adopt required measures to secure sensitive data and maintain operational integrity. Furthermore, adherence to such regulations fosters a culture of accountability, where firms are compelled to regularly assess their cybersecurity position to avoid possible vulnerabilities. The study reaffirms that to date, there are no normalized methods of measuring IoT cyber risk posture, and therefore, full compliance frameworks that not only meet regulatory but also allow for self-measurement during the course of outlining risk areas are vital (Ani et al., 2020). Besides, compliance and regulations' relevance extends beyond internal control to include external influences of stakeholders and consumers demanding transparency regarding data management and cybersecurity activities. With the proliferation of IoT technology, organizations must navigate a web of compliance requirements, from data privacy legislations to industry-specific legislations.

Violations of such legislations can result in expensive financial sanctions, reputation damage, and consumer trust loss. Therefore, it is essential for any organization to build a robust compliance strategy in order to thrive in the IoT and cloud services competitive landscape. A good regulatory framework not only provides compliance with best practices but also spurs innovation since it conveys to stakeholders a commitment to ethical cybersecurity practices and risk management (Abdolee et al., 2019). Furthermore, the balance between compliance and cybersecurity regulations guarantees the higher reliability of IoT systems in actual practice. Data privacy and energy efficiency concerns continue to be the primary issues that should be addressed in order to maximize the potential offered by IoT technology.

Compliance frameworks can provide guidelines to effectively counter such issues while guaranteeing secure and optimal operations. With decentralized IoT technology and upcoming technologies like Blockchain, regulatory compliance must be employed as a guide for responsible implementation by organizations. The use of this regulatory compliance assists in maintaining a balanced innovation practice with a provision to venture out for innovative solutions while maintaining tight security and ethics. Thus, the intersection of compliance and risk management is not only a matter of law but also one of strategy to achieve sustainable growth in the rapidly evolving space of Industrial IoT (Ani et al., 2020).

9.4.3. Best Practices for Implementing Security Standards

In the area of cybersecurity, particularly in the context of Industrial IoT and cloud connectivity, the enforcement of robust security standards is critical to safeguard confidential data and maintain operational integrity. A root best practice is the development of a general cybersecurity policy that explicitly covers risks involved in interconnecting systems. The policy must contain guidelines for assessing risks, developing incident response plans, and establishing access controls to keep unauthorized individuals from entering the system. Groups also have to constantly revise their policies according to evolving cyber threats and ensure all those involved are trained and aware of risk areas. By developing a solid foundation based on risk assessment and response planning, organizations can create the cornerstone for a solid cybersecurity position in accordance with industry standards and regulatory requirements. Regular self-evaluation of the cybersecurity position of an organization is another key best practice for implementing security standards within Industrial IoT environments. With the additions of new IoT devices and services, there is a need to assess the cybersecurity strength of an organization to predict, prevent, and respond to potential threats. An IoT cyber risk posture measuring gap has been realized in existing practice. Empirical research comparing various cyber risk assessment approaches has demonstrated the necessity of goal-oriented dependency modeling. This model enables organizations to understand and assess uncontrollable risk states in advanced IoT systems, hence enabling a quantitative self-assessment process. Through the adoption of such methods, organizations are capable of synchronizing their security policies with the ever-evolving threat landscape and enhancing the manner in which they handle vulnerabilities (Ani et al., 2020). Furthermore, introducing a proper cybersecurity training and awareness program is critical to educating all employees on their specific roles in maintaining security levels. Continuous professional learning on cybersecurity can empower employees to recognize and take countermeasures against threats beforehand. All these activities should include training workshops on phishing attempts recognition, learning about data breach consequences, and adopting best practices in secure data handling. Additionally, incorporating risk management strategies and awareness into organizational culture will help cultivate an environment where security is prioritized at all levels. To further bolster protective measures, organizations should enhance their technical defenses alongside employee training. This multidisciplinary framework strengthens not only individual competence sets but also overall organizational cyber resilience, thus making integration of IoT and cloud technology more secure (Anglano et al., 2018) (Figure 9.9).

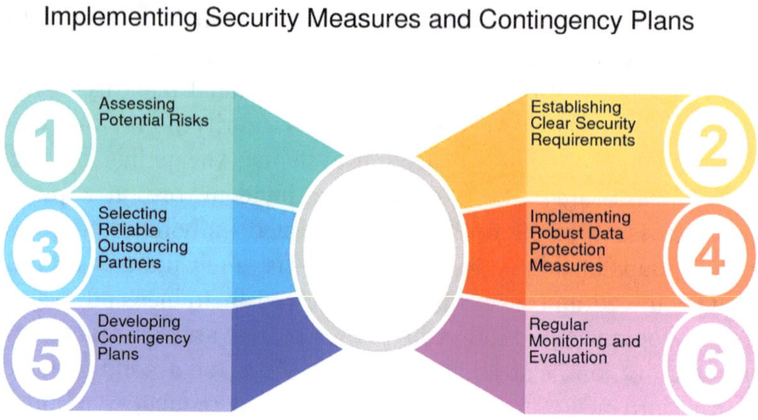

Figure 9.9. Best practices for implementing security standards.

Source: https://fastercapital.co/i/Outsourcing-best-practices--How-to-follow-the-best-practices-and-standards-of-outsourcing-your-tasks--Implementing-Security-Measures-and-Contingency-Plans.webp.

9.5. STRATEGIES FOR ENHANCING CYBERSECURITY

Enhancing cybersecurity in Industrial IoT (IIoT) and cloud integration involves an interdisciplinary approach that includes technological fixes as well as systematic risk assessments. An effective method is the development of a sound security posture, which is possible through structured self-assessments of an organization's IoT cybersecurity features. Through assessing the sufficiency of their cybersecurity framework, organizations can establish weaknesses and develop specific plans to reduce cyber-attacks. Specifically, adoption of a novel goal-based dependency model can facilitate quantitative self-analysis of non-controllable risk conditions in complex IoT setups, thereby making it possible for organizations to perform quantitative self-analyses of their cyber risk stance, as propounded in the empirical study of cyber risk assessment approaches presented by Ani et al. (2020). Such proactive measures not only enable organizations to anticipate and prevent cyber-attacks more effectively but also enhance their overall resilience to emerging threats. Along with self-assessment methodologies, promoting a cybersecurity awareness and training culture within employees is an imperative strategy. In the majority of cases, the security chain's weakest link is human error. Therefore, by having routine training sessions dedicated to the identification of possible cybersecurity threats—e.g., social engineering attempts and phishing attacks—a company can reduce vulnerabilities significantly. Moreover, according to the general literature on cybersecurity practice, having adequate procedures and guidelines can enable employees to react appropriately to security breaches when they occur. By integrating educational courses with experiential simulations, organizations can build an employee base that is not only educated on cybersecurity best practices but also skilled at behaving responsibly in safeguarding sensitive industrial data as well as cloud assets, as noted in Anglano et al. (2018). Surprisingly, collaboration between multiple stakeholders—ranging from industry moguls to academia—seems to be another strong weapon in enhancing cybersecurity in IIoT and cloud environments. By sharing information and tactics in protecting technological infrastructures, stakeholders are able to collectively address changing threats and develop innovative solutions for protection against cyber-attacks. This can be achieved through membership of cybersecurity consortiums or regional collectives for identification of critical technologies to be safeguarded. Additionally, membership in research initiatives can assist in the construction of infrastructure and controls that are the pillars of effective cyber defense. Such collaborative models not only facilitate cooperative cybersecurity efforts but also offer avenues for constant improvement in risk management procedures, thereby building the resilience of organizations undertaking the increasingly interconnected regime of Industrial IoT and cloud integration, as quoted in Anglano et al. (2018).

9.5.1. Risk Assessment and Management

Within industrial IoT and cloud security, effective risk analysis and mitigation are the prime foundations to ensure system resilience against ongoing, dynamic threats. Risk analysis involves a broad evaluation of IoT and cloud-based architecture vulnerabilities and the effectiveness of current security controls. A careful analysis detects possible threats, aurus, evaluates the effect of security compromises, aurus, and identifies the probability of such occurrences. With the ever-widening scope of the IoT environment, enterprises are forced to embrace a proactive approach with a focus on self-assessment of their IoT cybersecurity stance. This self-assessment is critical in predicting, preventing, and responding to cyber-attacks as it allows organizations to ground their security strategy both in operational requirements and inherent risk to design good cybersecurity architectures that can withstand anticipated threats (Ani et al., 2020). Moreover, bridging the wide resource gap in current approaches to measuring IoT cyber risk posture calls for use of novel models for risk assessment. The study of goal-directed dependency modeling is a key advancement in this field. Through focus on dynamic interdependencies between system components, this framework enables more realistic understanding of uncontrollable states of risk within complex IoT environments. The model can be applied by organizations for quantitative self-assessments, ultimately enabling them to realize vulnerabilities within their cyber defense schemes. This empirical analysis complements the efficacy of risk management practices to facilitate informed, effective interventions for strategic improvement and an improved position of security. The development of such comprehensive evaluation methods underscores the imperative for continuous monitoring in fast-changing threat landscapes (Ani et al., 2020). In addition to good risk assessment techniques, the application of a proper risk management approach is key to securing industrial IoT and cloud infrastructures.

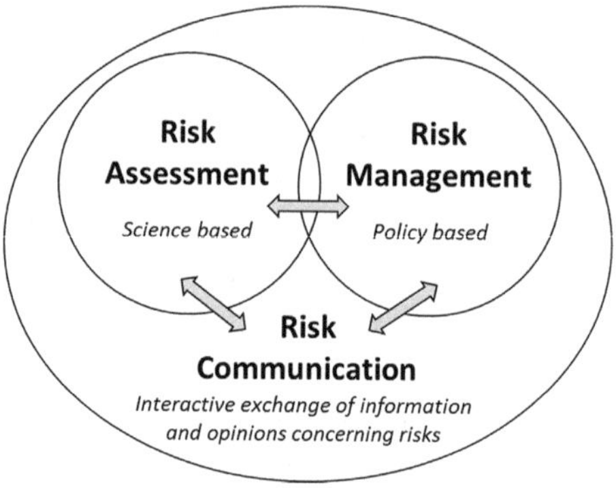

Figure 9.10. The effective risk assessment and management.

Source: https://www.researchgate.net/profile/Maarten-Nauta-2/publication/324681978/figure/fig1/AS:619936116379666@1524815696649/The-risk-analysis-framework-with-the-elements-risk-assessment-risk-management-and-risk.png.

This approach entails not just the technical aspect of cybersecurity but organizational policies and training frameworks to create awareness culture among workers. Greater awareness enables detection of possible security loopholes and allows personnel to respond accordingly. The use of horizontal steps, such as training and awareness programs, is part of the implementation that allows an active security culture. With empowerment of employees with knowledge to recognize and neutralize cybersecurity threats, organizations are able to drastically reduce the exposure to cyberattacks. These comprehensive approaches of risk management are best practices in cybersecurity and lay the basis for a strong industrial IoT network infrastructure (Figure 9.10).

9.5.2. Employee Training and Awareness Programs

Employee training and awareness programs are most vital in the scenario of cybersecurity, especially in the Industrial IoT and cloud integration environments. The rapid pace of technology evolution in these areas introduces a high number of vulnerabilities that can be exploited by malicious actors. To mitigate these threats, companies must prioritize as a top goal the extensive training programs that equip employees with probable cyber-attacks and how to protect confidential information. Such training enhances the chances of avoiding security breaches due to human error, which remains a top issue in the reporting of security breaches. Besides, these programs must be updated every now and then to include the dynamic nature of cyber threats so that employees are better informed with the most up-to-date knowledge and skills they need to deal with an increasingly complicated technology setup, thereby creating a more efficient organizational culture against cyber threats. More significantly, deploying employee training programs is not just a key defense strategy but an organizational development strategy. As noted in recent scholarly perspectives, organizations are called upon to design models of training that focus on the important nexus between technological underpinnings and human factors (Anglano et al., 2018).

Through this approach, workers know how their individual behavior impacts overall cybersecurity, which instills accountability and vigilance. By tapping the workers through interactive education, organizations can provide a sense of ownership and responsibility regarding cybersecurity protocols. Lastly, successful training programs transform employees into productive custodians of their firms' digital properties, strengthening the corporation's defense post in times of crisis. Integrating cybersecurity consciousness as an integral aspect of the organizational culture requires a coordinated push towards aligning that employees realize their own vital role in safeguarding digital properties. The Industry 4.0 challenges demand renewed focus on the participation of workers and a commitment to ongoing professional development in the field of cybersecurity, as noted in current research on asset management systems (Happonen et al., 2023).

These types of systems rely on human control and technology-based defense systems working together seamlessly. Through incorporating cybersecurity practices into standard processes and decision-making, organizations can attain a culture in which consciousness of cybersecurity is second nature. This type of cultural integration not only mitigates threats but also positions the firm to be able to maximally leverage technology advancements while having robust security that would protect it from possible vulnerabilities associated with cloud integration and IoT technology.

9.5.3. Advanced Technologies for Threat Detection

In the evolving security world of today, particularly in Industrial IoT and cloud integration, new threat detection technologies are an indispensable ingredient for effective defense systems. Machine learning and artificial intelligence are now focal points in identifying anomaly and potential threats in large networks. The technologies analyze system activity and user behavior patterns to enable anticipatory action against suspicious behavior. The addition of these intelligent systems not only enhances the speed and accuracy of detection for threats but also reduces the burden for cybersecurity professionals to sift through huge amounts of data. As interconnected systems gain more usage across organizations, these advanced threats require detection technologies not only as a valuable but also as a necessary measure to maintain operational integrity and secure confidential data, as highlighted in existing literature on cybersecurity models (Anglano et al., 2018).

Figure 9.11. Advanced technologies for threat detection.

Source: https://dbac8a2e962120c65098-4d6abce208e5e17c2085b466b98c2083.ssl.cf1.rackcdn.com/advanced-threat-detection-empowered-by-ai-logo-5-w-4758.JPG.

The rise in complex cyber-attacks necessitates the use of specific advanced technologies designed for the unique challenges posed by the Industrial IoT. For instance, behavioral analytics is significant in monitoring users' behavior and detecting anomalies that might indicate compromise. This kind of approach is most appropriate for applications where traditional perimeter defense is made obsolete by ubiquitous cloud connectivity and continually growing attack surface. Secondly, the interplay between regulatory settings and evolving technologies sets firms ahead of compliance requirements, as witnessed by the impacts of the EUs NIS Directive and GDPR. By aligning threat detection practices with adhering to these evolving regulations, organisations enhance their security standing while also evading the possible expense of fines for non-adherence (Accenture Technology et al., 2018). Further, the increasing complexity of industrial environments calls for end-to-end threat detection solutions that can simply extend to both legacy and new infrastructure. Combining threat intelligence from multiple sources provides a fuller picture of cyber threats, and automated response capability offers prompt action against detected threats. As the virtual world expands, the reliance on the technologies that enable continuous monitoring and rapid response will be crucial. Businesses will need to assign top priority to making investments in next-generation threat detection technologies as a part of their overall cybersecurity strategy to safeguard key assets and maintain operational resiliency during a fluid threat environment (Anglano et al., 2018) (Figure 9.11).

9.6. FUTURE TRENDS IN CYBERSECURITY FOR IOT AND CLOUD

As cybersecurity evolves, future trends in the Internet of Things (IoT) and cloud technologies will play a key role in industrial system protection. One of the emerging trends is the evolution of self-assessment tools that enable organizations to quantify their IoT cyber risk posture. Introduction of new IoT devices provides new vulnerabilities that necessitate a future-focused risk management approach. To date, there is a significant shortage of methodologies for effective assessment of these risks. Goal-oriented dependency modeling has been newly recognized by research works as a dominant technique in the analysis of IoT cybersecurity models, offering organizations with clear paths to achieve robust security levels. By adhering to a more formalized and quantifiable process of evaluation, business organizations will be in a position to predict, prevent, and respond effectively to emerging cyber threats, a safer infrastructure overall (Ani et al., 2020). Another prominent trend shaping the future of cybersecurity for IoT and cloud integration involves blockchain technology integration. The decentralized nature of blockchain can address certain of the fundamental issues common to IoT systems, namely scalability and privacy of data. Applying blockchain protocols could facilitate the ease of achieving more stringent security controls using unalterable records of transactions, greater transparency, and facilitation of trust between devices. Implementing such technologies, however, must be undertaken sensibly considering their computational needs and bandwidth overhead, which can introduce new vulnerabilities if not suitably addressed. Through the balance of leveraging blockchain for its security benefits and mitigating its limitations, organizations can develop a more integrated cybersecurity plan that bridges IoT and cloud technologies' loopholes (Abdolee et al., 2019). Lastly, artificial intelligence (AI) and cybersecurity synergy will drastically impact future trends in IoT and cloud environments. AIs capability to process enormous amounts of information and adapt to emerging patterns of threats offers unprecedented potential for adaptive threat detection and response. Machine learning models can identify anomalies in real-time, thereby enhancing an organization's capabilities to lock down threats before they escalate into major breaches. As AI systems become mainstream, however, they also offer new attack surfaces for threat actors to exploit. Organisations will need to put in place robust governance frameworks around AI and machine learning technologies, with ethical parameters and security best practices at the front. Engaging AI in use in cybersecurity strategy will eventually become a cornerstone for securing industrial IoT and cloud infrastructure against increasingly sophisticated cyber-attacks (Ani et al., 2020).

9.6.1. Emerging Technologies and Innovations

The advent of new technologies in the context of Industrial Internet of Things (IIoT) and integration with the cloud introduces revolutionizing possibilities in numerous industries, most notably manufacturing and food processing. In seeking greater operational efficiency, the deployment of Industry 4.0 concepts has become unavoidable. The innovation, characterized by connectivity among machines and processes via smart devices, offers enhanced data analysis and decision-making ability. The food industry, in particular, stands to benefit a lot from this shift. Through the use of tailor-made production processes and immediate feedback gathering, sectors are not only made more productive but also ensured compliance with stringent safety standards. The fusion of smart technologies transforms production environments, as per the overall push for sustainability and efficiency that Córdoba-Roldán et al. (2017) identifies. Thus, innovation through the aid of IIoT is a quantum jump for industries that have traditionally relied on outdated methods. At the same time, increased reliance on cloud computing services to handle data raises the stakes for secure cybersecurity measures. With industrial systems increasingly interfacing with cloud services, there is a vulnerability that is introduced, which necessitates advanced mechanisms to secure sensitive information. The interconnectivity between IIoT devices also provides the opportunity for one attack to sabotage an entire network, hence the importance of cybersecurity in this setup. In addition, planning for actions and technology to boost defenses against cyber threats is now an imperative area of focus. As observed in pioneering works pertaining to national cyber defense plans, anticipation through awareness, risk management, and stakeholder engagement remains critical (Anglano et al., 2018). Consequently, safeguarding industrial systems against potential weaknesses requires an interdisciplinary approach, integrating technology innovations into cybersecurity. Where emerging technologies and cybersecurity overlap within IIoT, it requires continuous updating of practice and procedure. As industries apply fresh technology advancements, they also need to evaluate and fortify their security measures in parallel to beat more sophisticated cyberattacks. This dual focus not only engenders resilience but also generates a culture of security within organizations, which allows employees to recognize potential vulnerabilities. The agri and food processing sectors are just two examples of industries where the growing imperative of bridging these two worlds comes sharply into relief, with the demand that progress in cloud computing and IIoT is paralleled by sound cybersecurity protocols. Such an integrative approach is in consonance with a forward-looking vision needed to steer sustained growth and competitiveness in an increasingly digitized world. To thrive in such an intricate environment, industries must be dynamic and evolve continuously to respond to technological developments as well as the ensuing security concerns they raise, as per Anglano et al. (2018) and Córdoba-Roldán et al. (2017) (Figure 9.12).

Figure 9.12. Emerging technologies and innovations.

Source: https://investingnews.com/media-library/robot-hand-touching-human-hand-with-a-spark-in-between.jpg?id=33553684&width=1200&height=800&quality=80&coordinates=0%2C0%2C0%2C1.

9.6.2. Predictions for Cybersecurity Developments

With Industrial Internet of Things (IIoT) and cloud solutions continuing to transform industrial operations, cybersecurity future projections suggest more efforts toward adaptive security. Businesses should increasingly prioritize constructing robust cyber defenses to protect sensitive data, particularly in the light of rising threats such as ransomware and advanced persistent threats (APTs). The ongoing evolution of cyber threats calls for an evolving strategy that continues to evolve at the same rate as the technologies it is supposed to safeguard. Sophisticated frameworks with the integration of machine learning and artificial intelligence are likely to play key roles in pre-empting and responding to would-be threats. As argued in the ongoing debate among security researchers, these adaptive approaches will be essential in shaping the future of cybersecurity to ensure that industries can effectively counter the increasing complexity and sophistication of cybersecurity threats (Anglano et al., 2018). In the context of the IIoT and cloud convergence, the role of big data analytics is changing as a cornerstone in predicting and countering cybersecurity threats. The retention of high amounts of data from various devices connected offers unique opportunities for organizations to identify vulnerabilities and intrusions in advance. By utilizing big data analysis, stakeholders can perform thorough investigations of trends and anomalies within network activity that may represent security incidents. Being proactive in this way enables quicker reaction and risk management. As articulated by the recent academic controversies, applications of big data analytics not only enhance the identification of threats in real time but also enhance the general security stance of industrial systems, providing an end-to-end solution against cyber threats (Gharehpetian et al., 2019). As the future unravels, coordination in the cybersecurity ecosystem is likely to become increasingly important. Industry, academia, and governments are expected to come together in partnerships with the aim of sharing knowledge, resources, and innovation related to cybersecurity. This platform of collaboration will help to advance a better understanding of the emerging threats and development of standardized incident response and risk reduction measures. With organizations increasingly adopting cloud technology, cloud security compliance will become increasingly important, driving organizations to adapt to best practices and regulatory compliance. These collective measures will not only enhance individual organizational safeguards but also contribute to a safer global cyber space, ultimately enhancing the resilience of the IIoT and cloud integration infrastructure (Anglano et al., 2018).

9.6.3. The Role of Artificial Intelligence (AI) in Security

With industries increasingly integrating Internet of Things (IoT) devices into cloud infrastructure, the application of AI to enhance cybersecurity is at the forefront. The volume of data generated by such devices is phenomenal, and traditional security controls lag behind the speed and intricacy of IoT networks. AI provides a solution in the form of improved analytics and machine learning algorithms, which can process massive data sets. By leveraging big data analytics, organizations can identify unusual patterns indicative of incipient cyber threats in real-time, hence enabling prompt responses that are critical in risk prevention. The ability of AI systems to learn and adapt from past incidents also enhances their predictive capabilities, offering a stronger defense against complex cyber threats that are likely to characterize industrial IoT environments. In this, AI is a support pillar for improving security operations and safeguarding sensitive infrastructure. Furthermore, AI technologies are affecting the development of more advanced security solutions required for the proper management of industrial IoT environments. The development of IoT-enabled devices makes it a certainty to shift from reactive to preventive security arrangements, and this is achievable with the use of AI-based tools. For instance, AI algorithms can dynamically adjust security configurations based on real-time behavior monitoring of the network, thus minimizing the exposure of vulnerabilities. In this manner, the integration of AI and cybersecurity not only detects threats but also boosts the overall resilience of the system through continuous learning and adaptation. As recently brought to the forefront of cybersecurity debates, collaboration between big data and AI is vital in addressing the issue of connected devices and platforms in industry (Gharehpetian et al., 2019). Moreover, inclusion of AI within cybersecurity extends beyond detection; it impacts incident response and recovery as well. Machine systems have the capability of executing pre-programmed routines upon detection of a breach, thereby significantly reducing response times and minimizing potential damage. Such capabilities are especially critical in industrial settings where wasted time equates to significant economic loss. Furthermore, the dynamic nature of cybersecurity threats renders it essential to continuously train AI systems to detect new channels of threats, as stressed by existing literature on comprehensive cyber defense measures (Anglano et al., 2018). As companies continue to harness the synergy of AI and IoT, their ability to harness these technologies will determine the effectiveness of their cybersecurity frameworks, underlining the crucial role of encompassing next-generation tech solutions in an industrial environment.

SUMMARY

Lastly, the increasing deployment of IIoT with cloud functions unveils unprecedented opportunities and translates into serious challenges from a cybersecurity perspective. The integration of these technologies provides easier access to increased operational effectiveness, data analysis, and real-time monitoring that allow organizations to streamline their processes and innovate quickly. However, with more connected devices and cloud infrastructure, so does the risk of increased cyber-attacks and vulnerabilities. As industries rely more on digital twins for real-time assets management and simulation, it is critical that their deployment consideration addresses the cybersecurity issues involved to protect sensitive data against potential attacks or breaches, as highlighted in the forthcoming debate around (Botín-Sanabria et al., 2022). In addition, auditing and forging strong cybersecurity structures are the call of the hour for organizations to safeguard themselves against the intricacy of cyber threats in IIoT systems. Leverage the power of artificial intelligence (AI) to enhance the security position of companies by actively detecting threats and automating response actions. AI-driven models possess the capability to adapt with the changing landscape of cybersecurity threats, showcasing their worth in maintaining integrity and security in cloud-integrated IIoT systems. Institutions therefore need to prioritize investments in smart security solutions, understanding that the evolving threats require similarly evolving countermeasures, as highlighted by the experiences of AI-focused methods outlined in Sarker (2022). Lastly, it is evident that the convergence of IIoT and cloud computing will only develop further, with stakeholders being required to stay watchful and proactive in their cybersecurity efforts. The necessity to implement whole-security measures cannot be overstated, because failure to do so could have negative implications for individual companies and entire industries. Future research must focus on developing standards and best practices that particularly address the unique security needs of IIoT and integration with clouds. As we continue to step forward towards an increasingly connected future, establishing a culture of cybersecurity awareness and resilience will be central to long-term organizational success and security in an interconnected world.

In the evolving landscape of cybersecurity in IIoT and cloud integration, there are a number of issues of overriding concern that must be closely scrutinized. The rapid proliferation of IoT devices created a bigger attack surface for cyber-attacks, and the intricate process of introducing cloud technologies only increases these weaknesses. With IoT infrastructure providing monitoring and control over most industrial processes, the need for robust cybersecurity is self-evident. According to Abdolee et al. (2019), these kinds of systems possess serious issues, including data protection and cybersecurity, which must be addressed in order to enable their total functional efficiency. Without any strategic initiatives in the direction of enhancing the security environment, businesses will expose their confidential data and essential infrastructure to hostile intentions. To prevent such vulnerabilities, application of cutting-edge technologies, e.g., blockchain, presents a promising future. Decentralized architecture of blockchains offers a potential solution to all the major issues of most of the IIoT systems, particularly data integrity as well as transparency. Abdolee et al.(2019), nonetheless, notes that using blockchain protocols must be approached with caution since their computation has the potential to pose additional problems like delays as well as bandwidth overhead. This duality necessitates a comprehensive assessment of how blockchain must be strategically integrated within IIoT systems in a way that enhances security without burdening it. Through the development of a symbiotic ecosystem of cooperation among different technologies, organizations can collaborate towards developing innovative solutions that enhance the security position of IoT and cloud infrastructures. A converged and multi-faceted cybersecurity strategy is essential to secure industrial applications in IoT and cloud environments. As quoted in, it is important for effective cyber defense to understand fully the infrastructure and controls necessary. This does not only constitute the identification of key technologies to be protected, but horizontal strategies such as training and risk management. As awareness is enhanced and continued investment made in devising new security protocols, organizations can develop strong buildings that keep pace with the dynamic nature of threats. With growing IoT embedding within infrastructure, constant evaluation of cybersecurity processes will be critical in ensuring that current as well as emerging concerns are managed effectively.

REFERENCES

1. Abdolee, R., Maroufi, M., & Tazekand, B. M. (2019). On the convergence of blockchain and Internet of Things (IoT) technologies. http://arxiv.org/abs/1904.01936 (accessed on 13th September 2025).
2. Accenture Technology, Adams, A., Alperovitch, A., Anderson, B., et al. (2018). Avoiding the internet of insecure industrial things. https://core.ac.uk/download/141198917.pdf (accessed on 13th September 2025).
3. Alrumaih, H., Byers, B., Gudoniene, D., et al. (2015). Multinational perspectives on information technology from academia and industry. https://core.ac.uk/download/86427274.pdf (accessed on 13th September 2025).
4. Anderson, J., & Rainie, L. (2017). The Internet of Things connectivity binge: What are the implications? https://core.ac.uk/download/89472988.pdf (accessed on 13th September 2025).
5. Anglano, C., Aniello, L., Antinori, A., et al. (2018). The future of cybersecurity in Italy: Strategic focus area. https://core.ac.uk/download/188830442.pdf (accessed on 13th September 2025).
6. Ani, U., Anthi, E., Burnap, P., et al. (2020). Dynamic real-time risk analytics of uncontrollable states in complex Internet of Things systems, cyber risk at the edge. https://core.ac.uk/download/491119431.pdf (accessed on 13th September 2025).
7. Botín-Sanabria, D. M., Mihăiţă, A. S., Peimbert-García, R. E., Ramírez-Moreno, M. A., Ramírez-Mendoza, R. A., & de Lozoya-Santos, J. J. (2022). Digital twin technology challenges and applications: A comprehensive review. Remote Sensing, 14, 1335–1335. https://doi.org/10.3390/rs14061335 (accessed on 13th September 2025).
8. Córdoba-Roldán, A., de las Heras García de Vinuesa, A., Luque Sendra, A., et al. (2017). State of the Industry 4.0 in the Andalusian food sector. https://core.ac.uk/download/185608711.pdf (accessed on 13th September 2025).
9. Dehghanghadikolaei, A., Emamian, S. S., Fotovvati, B., et al. (2019). The potential of additive manufacturing in the smart factory Industrial 4.0: A review. https://core.ac.uk/download/231748826.pdf (accessed on 13th September 2025).
10. Fernández-Caramés, T. M., & Fraga-Lamas, P. (2019). Towards next generation teaching, learning, and context-aware applications for higher education: A review on blockchain, IoT, fog and edge computing enabled smart campuses and universities. https://core.ac.uk/download/250404797.pdf (accessed on 13th September 2025).

11. Gharehpetian, G. B., Marzband, M., Moradi, J., et al. (2019). Attributes of big data analytics for data-driven decision making in cyber-physical power systems. https://core.ac.uk/download/305121394.pdf (accessed on 13th September 2025).
12. Happonen, A., Usmani, U. A., & Watada, J. (2023). Advancements in Industry 4.0 asset management: Interoperability and cyber security challenges and opportunities. https://core.ac.uk/download/630297890.pdf (accessed on 13th September 2025).
13. Iyer, V. R., Babu, K., & Guruswamy, V. R. (2025). Evaluating cyber assurance maturity levels and their impact on foreign direct investment in the Indian healthcare sector. https://core.ac.uk/download/647975930.pdf (accessed on 13th September 2025).
14. Mercadé Melé, P., Molina Gómez, J., & Sousa, M. J. (2020). Technology, governance, and a sustainability model for small and medium-sized towns in Europe. https://core.ac.uk/download/322918717.pdf (accessed on 13th September 2025).
15. Sarker, I. H. (2022). AI-based modeling: Techniques, applications and research issues towards automation, intelligent and smart systems. SN Computer Science, 3. https://doi.org/10.1007/s42979-022-01043-x (accessed on 13th September 2025).
16. Schmitt, M. (2023). Securing the digital world: Protecting smart infrastructures and digital industries with artificial intelligence (AI)-enabled malware and intrusion detection. https://core.ac.uk/download/620949718.pdf (accessed on 13th September 2025).

CHAPTER 10

INCIDENT RESPONSE, FORENSICS, AND FUTURE TRENDS

Contents

INTRODUCTION

With the rapidly evolving world of cybersecurity, the importance of incident response and forensic analysis has become particularly significant to both professional and academic circles. With increasing reliance on digital infrastructure by more organizations, cyber incidents are increasingly frequent and sophisticated. This increase in threat environment highlights the importance of efficient incident response mechanisms and forensic methods for identifying, containing, and drawing meaningful conclusions from cyber incidents. Against this background of exigency, it is critical to understand the scientific basis of cybersecurity practice. Despite critics who argue that dominant approaches are devoid of a scientific basis, recent scholarship suggests otherwise, citing the intricate interactions among structured observation, evidence judgment, and expert domains like forensics as cornerstones for constructing trustworthy cybersecurity science (Moore et al., 2017). Systematic research approaches integration is needed not only to respond to incidents but even to understand the broader implications of cybersecurity controls on organizational resilience. As technology continues to evolve, cybersecurity education and accompanying guidelines must modify suitably as well. Pedagogy is increasingly getting closer to filling the gap between theory learning and industry-scale application. In meeting with industry partners, the academic community has developed models like IT2017 that try to align curriculums with existing requirements of the information technology sector. Experience based on Latin American universities is that some attempt is being made to align education in academia with practical utility, but obstacles are considerable (Angles et al., 2016). On the horizon are some trends that promise to redefine incident response and forensics in profound ways. The growth of artificial intelligence and machine learning technologies offers the promise of greatly enhancing threat detection and response capabilities, enabling organizations to act swiftly on massive amounts of data. Meanwhile, the increased technical complexity of cyber-attacks makes a forward-looking approach to cybersecurity more crucial than ever before. This paradigm will require continuous evolution and innovation of incident response methodologies to ensure that they are not only useful but also robust against evolving threats. Moreover, understanding the philosophical underpinnings of cybersecurity, like the recognition of the incapacities of logical empiricism, may equip the discipline with the body of knowledge required to refine methodology and augment the scientific standing of the discipline (Moore et al., 2017). By interdisciplinary teamwork and a focus on both theoretical and practical knowledge, the future of forensics and incident response promises revolutionary breakthroughs.

Definition of Incident Response

The term incident response refers to an organized approach used by organizations to plan, detect, handle, and recover from security incidents. In today's active cybersecurity landscape, with attack frequency and sophistication continually more dangerous, a solid incident response plan is imperative. The plan not only delineates responsibilities and roles but also escalation procedures for incidents according to severity. Proper incident response can greatly reduce the costs and impact of security breaches, furthering the mandate for businesses to prepare by training and practicing. Further, as companies adapt to accommodate the latest technology and security concerns, the function of repeatedly improving their incident response mechanisms becomes crucial in ensuring sustained resilience against cyber-attacks. Notably, incident response must be informed by empirical evidence, which sheds light on the fallacies underpinning available response mechanisms. For example, reports from the industry quote that the majority of organizations lose significant amounts of money due to poor responses to incidents, further highlighting the necessity of employing tried and tested standards for evaluating such plans.

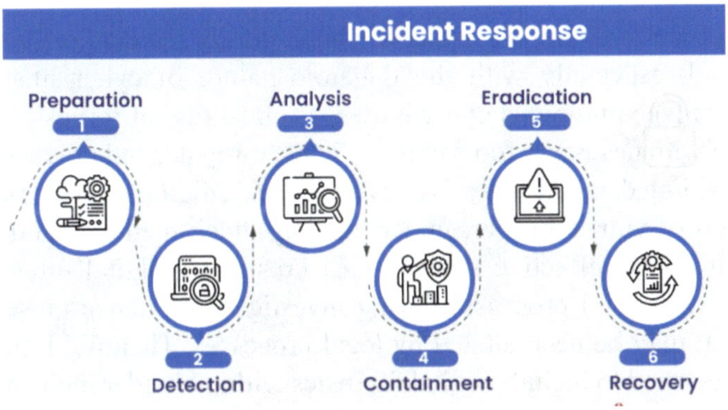

Figure 10.1. The concept of incident response.

Source: https://www.stealthlabs.com/wp-content/uploads/2020/10/the-six-steps-of-incident-response.jpg.

The Security Incident Response Criteria (SIRC) is a key guideline that can be employed by organizations to enhance their capabilities of handling incidents. Upon the foundation of rigorous research that involved interviews with a Global Fortune 500 firm, the standards listed below are derived from and play two key functions: they provide benchmarks for existing solutions and guide future improvement initiatives. With firms adopting SIRC, their ability to make streamlined responses to incoming threats will be bolstered, allowing for better prevention of likely breaches (Glisson et al., 2015). The field of incident response keeps evolving, and focus must be given at the intersection of incident management and forensics. By studying actual cases conducted by forensic investigators, organizations can develop actionable intelligence that enables them to enhance their response processes. Incident reports resulting from thorough investigation, and not self-reporting, provide a third-party perspective through which organizations can examine the effectiveness of their existing incident response processes. This critical examination aids the development of a better understanding of modern security breaches, and through this, organizations can identify systemic weaknesses and adjust their strategies in response. With forensics being integrated into incident response on a wider scale, organizations can reap the advantage of improved preparedness and effectiveness in countering advanced cyber-attacks and position themselves in a superior place in the ever-changing cybersecurity landscape (Maple et al., 2010) (Figure 10.1).

Importance of Digital Forensics

The significance of digital forensics in contemporary incident response cannot be overstated, especially with the dynamic nature of cyber-attacks. While organizations face numerous cyber issues, a robust digital forensics capability equips them with necessary tools to investigate, mitigate, and prevent incidents. Through enabling detailed analysis of cyber-attacks, digital forensics enables the identification of vulnerabilities, understanding attack methods, and providing insights that inform effective security measures. Also, digital forensics offers compliance with legal processes during investigations, hence preservation of evidence that may be necessitated by legal processes. This multi-purpose role highlights the need to include digital forensics within broader incident response plans in order to enhance overall organizational resilience against cyber-attacks. In the recent few years, new threats such as ransomware and cryptojacking have brought into sharp focus the need for sophisticated digital forensics processes, particularly for sectors such as finance. Ransomware attacks are now a regular occurrence, and ransomware actors continue to evolve as they target weaknesses in digital frameworks. The banking sector has particularly compelling challenges brought by both ransomware and cryptojacking that demand sophisticated forensics processes. Digital forensics also assists in the recovery of data and investigation of the exact methods utilized by the attackers in addition to assisting financial institutions in anticipating impending attacks. As discussed, a Digital Forensics and Incident Response (DFIR) plan would assist in enhancing threat hunting processes significantly in order to minimize the impact of such attacks on financial institutions and their clients (Kshetri et al., 2023). This focus on proactive forensics is important in limiting the impact of these sophisticated cyber threats. Further, with technology and tools constantly changing, the application of new-age tools in digital forensics provides an arduous landscape toward further enhancing incident response capacity. The advent of Generative AI, and more importantly, large language models (LLMs), provides fresh avenues for analysis of advanced data patterns associated with cyber incidents. The traditional method is likely to encounter problems with managing the amount of network traffic generated by sophisticated attacks, which acts to slow down responses. To counter this, the new model with open-source LLMs and vector databases holds potential to recover cyber incidents faster (Rahman et al., 2024). Such convergence does not only provide a better understanding of attacks but also calls for the reorientation of forensic capability to be able to address new forms of cyber threats. Thus, investing in advanced digital forensics will be high on the agenda for organizations that seek to be more secure as they navigate the complexities of an ever-changing digital world.

Overview of Future Trends in Cybersecurity

As we advance to a period of increasing digital connectivity, the future of cyber security is quickly becoming more complex and sophisticated. Among the prominent trends is greater reliance on predictive analytics and artificial intelligence (AI) for dealing with incidents and identifying threats. Organizations today employ sophisticated algorithms that analyze vast volumes of historical data to identify patterns and anticipate potential cyber-attacks before they happen. This proactive strategy, entailing the use of a combination of machine learning and sophisticated statistical models, plays an important role in increasing threat visibility and counteracting cyber vulnerabilities. Of special study is that using high-quality analyst-approved datasets could improve the efficacy of such predictions to allow cybersecurity providers to plan resource use and act on related threats accordingly (Barabasi et al., 2018). Thus, not just are organizations reacting to incidents but also preemptively enhancing defenses ahead of time through actionable insights derived from predictive modeling. In addition to predictive analytics, another key trend shaping the future of cybersecurity is collaboration between government agencies, private sectors, and academic circles. The requirements of cybersecurity have fostered an ecosystem where stakeholders come together to work together and exchange strategies. Current seminars, such as those held by cyber experts in Australia, emphasize inter-sectoral cooperation, such as banking, IT, defense, and broader business circles (Woodall et al., 2025). An approach aggregated in this manner not only encourages greater awareness of newly faced threats but also offers channels to exchange best practices and novel solutions for combatting cyber-related challenges. Through interdisciplinarity, cybersecurity specialists are able to draw upon various lines of thought and expertise, ultimately resulting in more effective defenses against a continuously evolving environment of cyber threats. Lastly, the future of cybersecurity will incorporate increased adoptions of automation as a key component of incident response as well as forensic analysis. Automation solutions such as robotic process automation (RPA) and automated incident response software enable easier detection, assessment, and mitigation of threats. Since cyber-attacks are growing in the aspects of volume and complexity, manual response is no longer adequate within response procedures. Automating the tiresome processes such as scanning logs and filtering alerts allows security teams to channel their focus to more complex issues that require expert judgment. In addition, research has highlighted the objectivity of automated systems in verifying cyber events, thereby minimizing false positives and providing actionable intelligence (Barabasi et al., 2018). As organizations look to the future, smooth integration of automated technologies into cybersecurity models will become central to enhancing overall resilience to cyber threats, an active not passive approach to defending digital infrastructure (Figure 10.2).

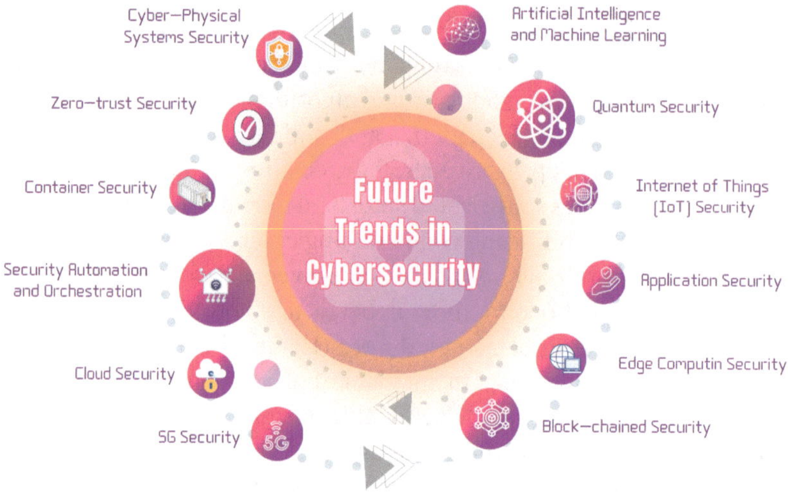

Figure 10.2. An overview of future trends in cybersecurity.

Source: https://www.zenarmor.com/docs/assets/images/future-trends-in-cybersecurity-bd7487de4467592d9237e001b544c91b.png.

10.1. THE INCIDENT RESPONSE PROCESS

Incident Response Process is a crucial process for companies that are to control and mitigate the impact of cyber incidents. Incidents usually happen through regular basic steps: preparation, detection and analysis, containment, eradication, recovery, and post-incident review. The steps follow consecutively with an emphasis on following a structured process to minimize damage and get back to normal operations. Under the preparation phase, organizations establish and enforce policies, procedures, and training exercises to get ready for potential incidents. The effectiveness of the overall incident response depends on this early foundation since it allows teams to react in the correct manner once genuine threats are realized. Detection and analysis comprise the second phase, where organizations identify alerts and potential breaches through various tools and methods. This phase comprises a comprehensive threat analysis from information collected from network infrastructures, honeypots, and intrusion detection systems. The utmost significance of accurate, analyst-verified data is also reflected in the incorporation of state-of-the-art forecasting techniques in cyber defense. For instance, data collected through time-series analysis of verified cyber incidents can significantly influence an organization's perception of threats and plans for resource allocation, allowing teams to optimally deploy responses. Such advanced information not only heightens situational knowledge but also helps with proactive measures, reducing the likelihood of future incidents (Barabasi et al., 2018). The incident response proceeds to containment, eradication, and recovery phases after detection. Containment is focused on restricting the spread of the incident to protect sensitive information and provide operational integrity. Elimination involves eliminating the root cause of the compromise so that already-compromised systems are purged of vulnerability. Finally, the recovery process involves restoring systems to operational state, supplemented by a thorough examination of what went wrong. The post-incident review is most critical to this process and is the learning exercise, where organizations look at their response and incorporate lessons learned into future operations. As technology advances in the area of cybersecurity, new methods acquired through such analyses will assist in enhancing the incident response process and anticipating emerging threats (Woodall et al., 2025).

10.1.1. Preparation and Planning

Preparation and planning are the cornerstones of an effective incident response approach, where organizations not only prepare to respond to probable threats but are also able to reduce the effect of security incidents. By developing comprehensive procedures and incorporating regular training exercises, organizations can improve their resistance to cyber-attacks. A well-written incident response plan establishes the procedures used to detect, contain, and remove threats, and recovery and post-incident assessment. This forward-looking approach establishes a security-conscious culture, with personnel informed of possible threats and their responsibilities in safeguarding sensitive information. Having stakeholders from all departments involved in the process also assists in ensuring that the plan reflects the operational context of the organization, bridging the gap between conceptual thinking and actual practice (Angles et al.,

2016). Further, integrating forensics in planning is fundamental to building a robust incident response capability. Forensic examination illuminates the nature of incidents, exposing weaknesses and attack channels through which future defenses may be formulated. By placing forensic preparedness at the forefront, organizations can ensure that evidence is collected and preserved in a manner compliant with legal and regulatory requirements. This does not just enhance the integrity of the response process but also enables organizations to make informed decisions when responding to security incidents. Proper forensics can also enable organizations to gain clearer insight into the broader threat landscape and drive future education and training programs within the organization. Prioritizing forensics in advanced planning is important to simplify the learning process from incidents and enhance readiness for the future (Augustinos et al., 2016).

Figure 10.3. Preparation and planning serve as the backbone of an effective incident response strategy.

Source: https://fastercapital.co/i/Block-Policy-and-Incident-Response--A-Coordinated-Defense-Strategy--Creating-an-Incident-Response-Plan.webp.

Moreover, as new technology emerges, adaptive planning and preparation become increasingly important. Organizations must stay current with fresh developments in cybersecurity threats and technology that can shift vectors of attack. This entails not just updating incident response plans to reflect new threats but also leveraging next-generation tools that enhance detection and response capabilities. For instance, the emergence of machine learning and artificial intelligence can revolutionize threat detection and incident response strategy so that organizations have the ability to identify anomalies in real time. With adaptive thinking and taking on new technologies, organizations can strengthen their incident response frameworks and remain agile in a constantly evolving threat environment. In such situations, preparation and planning are essential to make responses not just reactive but strategically placed in line with future cybersecurity directions (Figure 10.3).

10.1.2. Detection and Analysis

Detection and analysis are the fundamental pillars of the incident response and forensics model, particularly within the context of the constantly evolving digital ecosystem. Efficient detection and analysis of incidents can determine the effectiveness of countermeasures and recovery protocols. Developments in detection software like machine learning capabilities and behavioral analytics enable rapid detection of anomalies that can lead to security breach or other incidents. Along with this, the integration of real-time monitoring facilities helps organizations to identify patterns of behavior that fall outside the norms, staging threats for mitigation before they become threats of a more worrying sort. Increased vigilance is paramount in the protection of sensitive information and the assurance of operational continuity in an increasingly interconnected world, opening the way for more effective incident response policies. As businesses adapt to the maturity of modern cyber-attacks, the role of artificial intelligence (AI) in detection and analysis has become increasingly central. While AI enhances abilities in monitoring and detection of potential incidents, it also raises a new level of concerns regarding transparency and explainability. The majority of AI systems are black boxes, providing very little regarding their decision-making. This lack of comprehensibility can hamper trustworthiness and suitability for cybersecurity operations, as stakeholders will not be willing to trust solutions that cannot explain their behavior. The recent work has pointed to the merit of Explainable Artificial Intelligence (XAI), which seeks to address these challenges by offering models that explain their actions and decisions. The balance between the use of AI for anticipatory detection and preventing the system from rendering itself unintelligible is a critical part of consideration for the future development of cybersecurity, as it further emphasizes the need for ongoing innovation in this regard. The future of incident response and forensics detection and analysis lies in accordance with the advent of the metaverse and associated technologies that vow to transform the digital world. Security concerns are primarily informed by the inherent character of this new world, since experiential immersion can scale invasions of privacy and security breaches exponentially. Stopping these from happening demands developing sophisticated tools for analysis that are able to perform within this new digital world, and integrating real-time detection capabilities aligned to operate in accordance with the metaverses complex, multi-faceted design. It entails not only fostering technology innovation but also forging inter-disciplinary scholarship in information technology, law, and ethics. With the organizations poised to respond to these advancements, there is a need to research new metrics and incident detection methods followed by analysis, en route to robust security policies that can keep up with the ever-evolving digital world, as described by Capuano et al. (2022) and Wang et al. (2022).

10.1.3. Containment, Eradication, and Recovery

Incident management in the field of information security relies on three key factors: containment, eradication, and recovery. Containment is the first action that aims to limit damage created by an incident to security. This action is crucial as it stops the incident from expanding and further penetrating systems or sensitive data. Containment methods may range from isolating affected systems to implementing temporary solutions that serve as a buffer against the spread of the threat. A successful containment process not only protects organizational assets but also lays the groundwork for the subsequent eradication and restoration procedure, emphasizing the necessity of having an ordered incident response framework to ensure that security incidents have little business impact. Following containment successfully, the second response to an incident is eradication, wherein the causative factor of the security breach is removed. Eradication is important because containing an incident but not eliminating the vulnerabilities may lead to it re-occurring and thus weaken the security posture of the organization. Eradication might entail advanced forensics to examine the threat and necessitate changes needed for security systems and procedures. This process typically includes the integration of technical fixes along with procedural testing to ensure that fault systems responsible for the breach are effectively corrected. Through the integration of data from these tests, organizations can bolster their defenses against future threat attempts, responding to the ongoing need for adaptive security measures that evolve in response to emerging threats, as reinforced by the Security Incident Response Criteria (SIRC) (Glisson et al., 2015). The recovery phase is the culmination of the incident response process, with focus shifting towards restoring systems to normal business operation while making sure that existing vulnerabilities are attended to. Good recovery is not only important for returning organizations to pre-incident levels but in minimizing long-term impacts on business operations. This stage includes restoring data from backup, re-enforcing security controls, and conducting rigorous testing to ensure that all systems are secure and functioning correctly. Furthermore, recovery has to include an evaluation of the incident response process itself to identify areas for improvement and inform future plans. Cooperative incident response actions, as emphasized in recent research, show that encouraging inter-organizational collaboration can improve improvement outcomes based on common knowledge and practices, thus enhancing trust and resilience to future incidents (Agbanusi et al., 2014) (Figure 10.4).

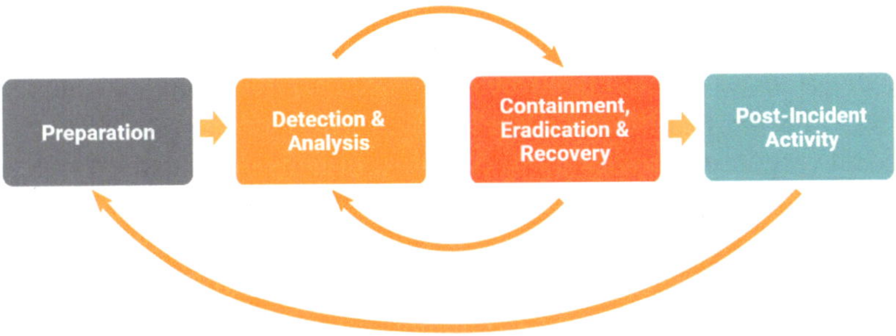

Figure 10.4. Containment, eradication, and recovery.

Source: https://docs.lumu.io/galleryDocuments/edbsn7a512fa0d8e2f00fae741f2e662f195a2c1156644adffef639e2e68cb1bfae1d21d519070cffdd9ba9f94efd95c2809f?inline=true.

10.2. DIGITAL FORENSICS FUNDAMENTALS

Digital forensics is a vital incident response sub-discipline that involves the systematic recovery, analysis, and presentation of data from electronic equipment. The technology is crucial not just for cyber threat detection but also for supporting legal proceedings in cases of digital crime. The fundamentals of digital forensics involve a sequence of processes that vary from data preservation to analysis and reporting. Specialized tools and techniques have to be used by practitioners to maintain the integrity of the data during the investigation process. Utilizing a systematic approach, digital forensic investigators can give organizations an understanding of the cause of cyber incidents, enabling them to enhance their cybersecurity posture and formulate effective incident response measures. The constantly speeding evolution of cyber technologies doubles the demand for effective forensic methods to keep up with fresh threats and complexities in cyberspace. As more organizations wake up to the importance of cybersecurity, the recruitment of competent practitioners in digital forensics takes center stage. The mounting need for professional qualifications doubles the significance of interdisciplinary competency frameworks unique to the field of digital forensics. Literature indicates that the most important skills for cybersecurity professionals, particularly incident response and management, are crucial to achieve maximum performance in real-time decision-making processes (Poe et al., 2018). Their identification not only helps to prepare appropriate training curricula and certification programs but also shapes the recruitment efforts of organizations to build effective incident response teams. By ensuring that practitioners are extremely capable in technical competence as well as situational awareness, organizations are able to enhance their capacity to deal with and handle cybersecurity attacks effectively. For the future, trends in digital forensics are likely to be influenced by technological innovation, regulatory change, and the evolving landscape of cyber threats. As the nature of incident response becomes more complex, organizations can choose to hire people with relevant certifications in digital forensics since they are perceived to bear critical competences that ensure a successful incident response process (Jarocki et al., 2021). Also, the interface between technical ability and soft skills such as communications and critical thinking will be critical in ensuring effective teamwork in incident response teams. Emphasizing these characteristics in recruitment and training processes will become critical to empower response teams not just to operate in the technical landscape well but to engage with stakeholders as well, laying the ground for a future-ready anticipatory cyber-defense.

10.2.1. Definition and Scope of Digital Forensics

The extent of digital forensics encompasses the systematic process of finding, preserving, examining, and reporting digital evidence from electronic devices. As technology has progressed, the extent of digital forensics is no longer confined to the conventional examination of computers but comprises a wide variety of digital sources such as mobile phones, cloud storage, and network traffic. This broadening is crucial for an effective response to incidents that happen in today's technologically advanced world. Digital forensics is also applied legally for criminal investigations but is also of critical relevance in cybersecurity because it helps organizations comprehend intrusions and assist in safeguarding against future intrusions. Forensic experts must have a thorough understanding of both the technical characteristics of these devices and legislatures regarding evidence gathering and analysis given the nature of digital devices. In the past couple of years, digital forensics has been a focal point after the high rate of cybercrime and security breaches. The need for accurate and reliable methods has instigated the emergence of new digital forensic analysis processes. For instance, the Case-Based Reasoning Forensic Triager (CBR-FT) method aims to enhance the effectiveness of evidence data identification by reuse of knowledge gained from past investigations. This approach addresses challenges posed by traditional triage methods that have been questioned in terms of their capacity to reveal useful information. By applying CBR-FT, investigators have a better chance of locating likely points of evidence within a suspect's online environment, thereby streamlining the investigative process and promoting results in general across forensic analysis (Horsman et al., 2014). Consequently, this evolution not only bears witness to the growing sophistication of digital forensics but serves to bring into focus the necessity of continuous adaptation to emerging technologies. Moreover, empirical research plays a critical role in shaping the practices and methodologies of digital forensics. A great case in point is a study conducted by the forensic investigation unit at 7Safe, who utilized breach investigation data from real cases in contrast to self-reported cases in order to bring accuracy and validity to findings. It transcends bias often introduced by subjective reporting, therefore offering more precise windows into issues of the modern security scene (Maple et al., 2010). Such rigorous analysis informs best practice and further development in the field so that forensic techniques remain abreast of rapid technology evolution. As digital forensics becomes more advanced, incorporation of evidence from legitimate investigations will be the key to making incident response plans more effective and the security posture of organizations wiser to respond to challenges of the digital age (Figure 10.5).

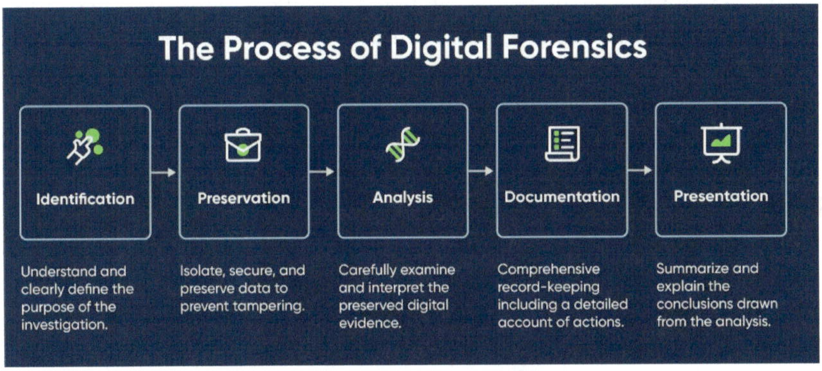

Figure 10.5. The digital forensics encompasses the systematic process.

Source: https://www.servicenow.com/content/dam/servicenow-assets/public/en-us/images/ds-what-is-pages/what-is-digital-forensics.png.

10.2.2. Types of Digital Evidence

For the digital forensics community, types of digital evidence exist and can impact the findings of investigations significantly. Digital evidence can be in various forms, starting from exchanges by email to web browsing history and social media activity, all of which can become critical to an incident being understood. The importance of these records is that they can recreate the timeline of actions leading to a cyber-attack or criminal activity. More significantly, when law enforcement agencies effectively utilize this electronic evidence, they increase their investigation outcomes.

For instance, studies have shown that an active community policing presence and collaborations with external agencies positively affect clearance rates for homicides, since proper evidence collection enables an improved chance of solving homicide cases (Carter et al., 2015). Such findings underscore the necessity of modern law enforcement integrating digital evidence into investigation methods. Additionally, data found within cloud services and databases is yet another form of digital evidence that requires specialized forensic approaches. With increasing organizational reliance on cloud computing and data storage systems, the expertise of retrieving and analyzing such data becomes an important factor. For example, database forensic analysis focuses on recognizing improper operations within such systems, both in the data and their metadata.

This emerging field of study is encouraging ongoing research to establish standard procedures, such as the Generic Database Forensic Investigation Process Model (DBFIPM), to explain primary procedures employed to analyze database forensic evidence effectively (Abd. Razak et al., 2016). Through the establishment of an agreed-upon framework, identification of digital evidence, collection, and preservation are more efficacious processes, ultimately resulting in enhanced quality and validity of forensic investigations. Finally, the scope of digital evidence is also widened to the ever-growing realm of Internet of Things (IoT) devices, which are finding their way more and more into daily life. Smart appliances for the home and wearable fitness monitors are all IoT devices, which store vast amounts of data that can prove to be extremely useful evidence in incident response scenarios.

The challenge for forensic examiners is presented by the amount of data produced and the complexity of the network environments where these devices are used. Acquiring proficiency with working with and analyzing data from IoT devices will become increasingly critical as their use spreads. Moreover, bringing these kinds of evidence into the overall context of digital investigations can improve incident response effectiveness and strategy. By adapting to such advances, forensic and law enforcement officers can make sure that they remain effective against future challenges introduced by technological developments, positioning them for success in the days ahead.

10.2.3. Legal Considerations in Digital Forensics

Digital forensics is highly grounded in legal considerations because any outcome can have a serious bearing on criminal prosecution and investigation results. Of course, the most critical is the preservation of the integrity of evidence collected digitally. Legal systems generally insist on digital forensics processes being followed in line with strict procedures to prevent them from being inadmissible in court. Ensuring a strong chain of custody is imperative because it documents who received possession, handled, and examined the evidence, thereby giving the evidence credibility. Compromised or tampered-with evidence will be dismissed, and this can disrepute legal proceedings seriously. The consequences of not meeting these legal standards highlight the importance of training and certification of experts in digital forensics to enable them to have the necessary skill and expertise to manage complex legal situations effectively. Further, mobile devices and the associated forensics procedures bring additional legal intricacies. The increasing usage of smartphones in both personal and professional lives generated the need for more sophisticated methods of collection and analysis of digital evidence. Volatile memory acquisition from cellular phones, for example, has to be undertaken with great caution since it can contain crucial, time-sensitive information essential to an investigation. The use of such tools as Linux Memory Extraction (LiME) is controversial based on their legality and ethics, especially in the context of users' privacy rights and device security (Heriyanto et al., 2014). Thus, forensic analysts must balance the technical needs of acquiring evidence with the law provisions for privacy and data security and ensure all their actions comply with applicable legislation and regulation throughout the investigation process. Legal considerations in the context of database forensics are also sophisticated. Database content and their metadata management is highly sensitive because databases will be containing sensitive and personal information. Designing a Generic Database Forensic Investigation Process Model (DBFIPM) is a significant measure toward developing standard procedures that can be utilized to aggregate techniques within various database environments (Abd. Razak et al., 2016). These models enable conformity to legal standards because they clearly define procedural steps such as identification, collection, maintenance, examination, and presentation of evidence. With further development of database forensics, it is critical that investigators embed legal considerations into their practices so that not only are their methods based on accurate findings, but also the legal integrity of the evidence gathered is maintained. Such convergence is essential to creating an authentic and reliable forensic practice that will stand scrutiny in court and deliver the objectives of justice (Figure 10.6).

Figure 10.6. Legal considerations in digital forensics.

Source: https://cellebrite.com/wp-content/uploads/2023/07/CLBT_Training_Why-Your-Digital-Forensics-Credibility-Matters-More-Than-Ever-scaled.jpg.

10.3. TOOLS AND TECHNOLOGIES IN INCIDENT RESPONSE

In today's cybersecurity environment, the effective management of incidents is largely contingent on a broad array of tools and technologies. These solutions are designed to simplify every step in the incident response process, from preparation to detection, containment, eradication, and recovery. Advanced threat detection platforms utilize machine learning-powered algorithms that monitor network traffic for unusual behavior so that instantaneous detection of potential threats is possible. Additionally, incident response platforms consist of automated workflows that automate communication among team members to simplify the entire process of managing incidents. These tools not only foster a future-oriented work culture concerning cybersecurity but also enable organizations to be better at workarounds in order to ward off threats prior to becoming significant breaches. Those with robust analytics and reporting functionality spot trends in security incidents, thus enabling subsequent defense planning and enabling continuous organizational improvement in resisting cyber risks. Last but not least, innovation in digital forensics is changing the character of investigations subsequent to an attack. The utilization of forensics tools in incident response ensures that any unforeseen activity is analyzed systematically, and potential evidence is preserved for further legal analysis. The techniques that apply Case-Based Reasoning Forensic Triager (CBR-FT) are excellent illustrations of this synthesis of tools and technologies in incident response. Through the application of past investigation as a model to project where evidentiary information would be found on a suspect's system, CBR-FT enhances the efficiency of forensic analysis. This innovative approach not only helps investigators make informed choices but also settles the somewhat controversial role of triage in digital forensics, giving it a sense of legitimacy in the application within the sector. Such advancements are a step towards more precise and reliable forensic techniques, hence improving the overall effectiveness of incident response programs. With organizations continuing to deal with evolving cyber-attacks, the implementation of advanced tools and technologies is the way forward. The arrival of cloud-based technology, for instance, brings with it versatile incident response functions, allowing organizations to quickly scale to fluctuating levels of threat intensity. Furthermore, the integration of threat intelligence platforms allows incident response groups to remain ahead of attackers by providing real-time indicators of novel vulnerability and attack techniques. Further, incident response automation can quite simply reduce response times, thus limiting the damage caused during a security incident. Through fostering a culture of continuous learning and adaptation, supported by robust tools and technologies, organizations not only respond to incidents promptly but also build a defensive posture that evolves with new trends in cybersecurity. The interplay of technological advancement and incident response policies is a fundamental pillar in enhancing organizational resilience to ever-changing cyber threats.

10.3.1. Software Tools for Incident Detection

The integration of advanced software tools to detect incidents is the basis of enhancing cybersecurity systems in various fields. With increasing data breaches and cyber-attacks, organizations have no option but to adopt specialized tools that facilitate real-time incident detection. Such tools employ algorithms that scan network activities in real time, making it possible to identify malicious activity or abnormalities instantaneously. By converging these systems with incident response frameworks, organizations can best minimize the effects of security intrusions and improve their investigative effectiveness. The anticipatory nature of these tools not only minimizes potential threats but also instills an alertness and readiness culture at organizations, bringing with it an effective security posture. In mobile forensics, with the emergence of professional software tools, the processing of data from mobile devices has been transformed, and such data are frequently instrumental in the detection of incidents. Linux Memory Extraction (LiME) and Volatility 2.3 are among the tools that are instrumental in obtaining and analyzing the volatile memory of devices, which could contain vital evidence during investigations. Effective deployment of such tools is dependent on the availability of root privileges to the examiner, highlighting the technical challenges faced by mobile forensics. The tools also provide memory dump analysis that reveals user actions and potential malicious activity performed on mobile devices. However, according to recent research, the inherent limitations of such applications have caused a need for continuous development to enhance their performance and attain comprehensive solutions in incident detection and forensic analysis (Heriyanto et al., 2014). In addition, the creation of mobile forensic tools highlights the necessity to continue evolving in light of evolving technology. A comparative overview of forensic tools developed for different mobile platforms, such as iOS, Android, and BlackBerry, illustrates divergent capabilities and limitations inherent in every platform. This highlights the importance of selecting appropriate forensic tools specific to environments for optimal data acquisition and preservation on discovery of incidents. Regular updates of tools, facilitated by advancements in mobile operating systems, are needed with a commitment to ongoing research and development from the digital forensics world. As technology keeps evolving, it is most likely to enhance data extraction, analysis, and preservation capacity, leading ultimately to more efficient incident detection approaches in the future (Martini et al., 2025) (Figure 10.7).

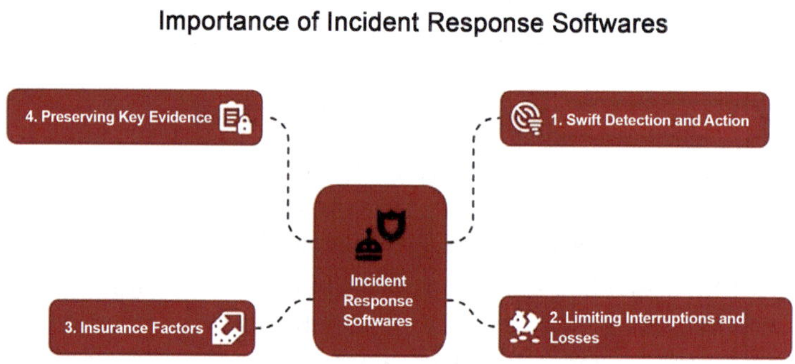

Figure 10.7. Software tools for incident detection.

Source: https://www.atlassystems.com/hs-fs/hubfs/Importance%20of%20incident%20response%20softwares%20.webp?width=630&height=357&name=Importance%20of%20incident%20response%20softwares%20.webp.

10.3.2. Hardware Tools for Data Recovery

Hardware tools for data recovery are of essence in incident response and digital forensics in minimizing the effects of data loss. Their importance is heightened in the current age where data breaches are extraordinarily common, especially among SMEs that do not have resources for inclusive incident response measures. In most cases, data breaches leave organizations scrambling to recover important information. Hardware solutions such as disk drives, data recovery appliances, and recovery cards can provide essential capabilities to help recover lost data. In addition to providing a means of recovering compromised information, these solutions also allow organizations to perform forensic analysis, which is crucial in understanding the nature of the breach that occurred as well as how to prevent such breaches in the future. Consequently, the inclusion of hardware data recovery tools in incident response planning is now essential for organizational resilience (Archibald et al., 2018). Furthermore, the creation of hardware tools for data recovery has brought about better methods that are more effective in enhancing incident response techniques.

The majority of modern tools are designed to make the recovery process easier through automated procedures that boost efficiency and accuracy. For example, network monitoring platforms can alert network anomalies that indicate possible data breaches. This functionality is important so that incident response teams are able to identify and mitigate security incidents rapidly with minimal downtime in operations. Moreover, visualization and resilience information offered through such solutions facilitate decision-making so that teams are able to develop proactive defense strategies against future threats. This focus on taking advantage of advanced hardware in recovery and monitoring is consistent with the ongoing requirements of the forensic domain, highlighting the importance of preparedness to handle diverse incident scenarios and articulating a strategic response model (Johnson et al., 2012). The future of incident response and digital forensics will more likely than not see a greater enhancement in the functionality of the data recovery hardware tools with the complicating nature of cybersecurity threats as its foundation. As data breaches evolve, so do the technologies employed to recover data effectively. Hardware developments may include faster execution of recovery processes, better interfaces to ease user involvement, and packaged systems to support seamless interaction between incident response groups and forensic investigators. In addition, the transition towards greater automation of the recovery process will increasingly reduce the time and resource burden on organizations. These advances will be transformative for organizations that wish to enhance their incident response platforms in light of the complex backdrop of new threats and technologies. As incident response practice matures, offering top-of-the-line hardware equipment to teams will help them guard important information and support overall organizational integrity and growth (Archibald et al., 2018) (Figure 10.8).

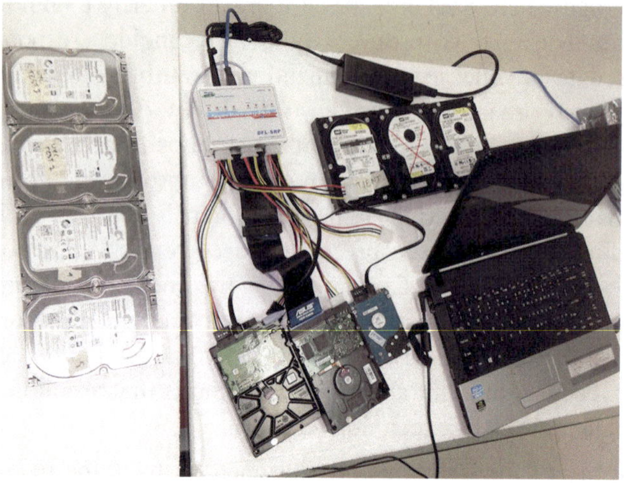

Figure 10.8. Hardware tools for data recovery.

Source: https://www.dolphindatalab.com/wp-content/uploads/2015/02/DFL-SRP-USB3.0-Data-Recovery-Equipment-2.jpg.

10.3.3. Emerging Technologies in Incident Response

The incident response arena is evolving rapidly due to the incorporation of emerging technologies that enhance efficiency as well as efficacy. The most emerging development is the use of artificial intelligence (AI) and machine learning (ML) algorithms, which are increasingly utilized to automate incident discovery as well as response. These technologies allow organizations to process vast volumes of data in real time and identify patterns that indicate potential security violations. As attacks are becoming more and more sophisticated, traditional response channels can fail; hence, AI and ML become the primary tools that provide organizations with the capability to respond in advance of the threat. These technologies allow rapid responses to evolving threat vectors, with incident response teams staying ahead in safeguarding their environments and enhancing overall security resilience, in accordance with evolving national security strategies outlined in planning frameworks like the Capability Plan for Australia (N/A, 2025). Further, the utilization of cloud computing platforms has also transformed incident response plans so as to allow scalable tools and resources to enhance responsiveness to security incidents. Organizations may adopt incident response solutions that are not just effective in different environments but also facilitate collaboration among different stakeholders who are dealing with incident management. In such a model, teams are able to leverage advanced forensic analysis capabilities facilitated by cloud technology to preserve evidence and analyze data more effectively. The cloud resources are versatile and easily deployable, enabling rapid deployment of resources in case of an attack, minimizing downtime and ensuring proactive as well as reactive measures are executed successfully. This enables a coordinated effort towards national security, enabling agencies to enhance coordination and intelligence sharing in the wake of new cyber threats (N/A, 2025). With the evolution of digital forensics, new technologies such as Case-Based Reasoning Forensic Triager (CBR-FT) are emerging to enhance the identification of evidence and effectiveness of investigations. The method utilizes previous case data to inform investigations at the current moment, enabling forensic experts to effectively prioritize their investigation. Such targeted approaches are vital in data-rich environments because they make evidence recovery efficient and increase the potential to find critical information in the shortest time possible. The success of CBR-FT would suggest a new direction of more intelligent, data-based decision-making in digital forensics as a significant evolution in incident response models. By adopting such new technologies and methodologies, incident response teams are not only better positioned to improve their operational response but also fall in line with broader trends in national security strategy, positioning the nation better to defend itself against new threats.

10.4. THE ROLE OF CYBERSECURITY TEAMS

Cybersecurity teams have a critical role to play in organizations, upholding the integrity and confidentiality of confidential information within an atmosphere replete with cyber-attacks. Their role does not end at defensive actions; they assume proactive functions in incident response, risk assessment, and compliance monitoring. Effective cybersecurity units also create culture of awareness and readiness by continued training and simulation exercises, both transmitting necessary skills to the staff and planting a forward-thinking strategy for the next wave of attacks. Engaging stakeholders from across industries, such as in workshops involving government and industry stakeholders, enhances the teams strategic thinking and cooperative capacity. By sharing perspectives through differing experiences, such workshops as the one held in Canberra prior to the Seoul Conference on Cyberspace enable cyber security professionals to develop creative solutions to changing issues in digital security (Woodall et al., 2025). Cyber security teams are increasingly shaped by the need to be adaptive and anticipatory in response to new technological advancements. With the frequency of cyber-attacks increasing and sophistication taking place, employing cutting-edge threat detection and response strategies becomes inevitable. That's what drives the adoption of analytics and machine learning in cybersecurity. By employing techniques such as time-series forecasting from authenticated sources of data, groups can turn indefinite and unorganized patterns of cyber activity into quantifiable forecasts. Knowledge obtained through data obtained over years is invaluable, as it gives the ability to identify systematic trends in cyber-attacks, ultimately helping cybersecurity teams to use resources more effectively and anticipate possible threats ahead of time (Barabasi et al., 2018). The tasks of cybersecurity teams will probably increase in the future as they introduce more new technologies and practices to their operations. Advancements such as artificial intelligence and computerized threat detection solutions not only enhance the teams competency but also necessitate cybersecurity professionals to stay sharp in critical thinking and ethical decision-making. The threat environment is constantly changing, which necessitates a commitment to ongoing education and evolving within the cybersecurity teams. As they navigate the intricacies of cyber forensics and incident response, these professionals are leaders in securing digital infrastructure to ensure not only the security of their own organizations but also the resiliency of the overall digital ecosystem. As cyber threats grow more sophisticated, it remains important that such teams draw on collective experience and emerging technologies to maintain pace with the demands of a more connected world.

10.4.1. Structure of Incident Response Teams

The incident response team structure is a critical component of the demonstration of organizations' capacity to effectively respond to and contain cybersecurity incidents. The teams typically have staff with diverse roles that complement one another to provide a comprehensive approach towards response. At the heart of the team is the incident response manager, who coordinates the team's operations and communication with stakeholders during the response process. Supporting this function are various specialists such as security analysts, forensic experts, and IT support personnel, each bringing some special skill that enhances the team's ability to respond effectively. The team environment in such teams is very important because divergent perspectives lead to improved problem-solving and innovation in response strategies, thus addressing the sophisticated nature of modern-day cybersecurity threats. In order to develop a strong incident response capacity, many organizations pay special attention to some qualifications and certifications when forming such teams. Recruitment processes typically favor candidates with recognized incident response certifications since the credentials not only express familiarity in the technical skills but also professional commitment to competency refreshment in a dynamic setting. Both managers and non-managers, according to research, regard candidates with certifications as being better suited to enhance the incident response process, affirming that both hard and soft skills are important in effective incident management. Soft competencies such as communication and critical thinking have been cited as crucial to build teamwork, given that technical knowledge alone could be insufficient to respond to intricate security breaches (Jarocki et al., 2021). Therefore, organizations must have a multimodal orientation in mind when structuring their incident response teams. As more organizations face heightened cyber threats, incident response teams are defined by the need for agility and responsiveness to fluid environments. This kind of development is reflected in the majority of IT curriculum programs that move more emphasis towards the intersection of academic preparation and applied use and linking education to industry requirements (Angles et al., 2016). Partnership with academia and the cybersecurity community alike ensures that incident response teams possess not just fundamental technical skills but also remain up to date with current trends and technologies. Training and professional development do a lot towards keeping teams prepared and effective as well. With the ongoing evolution in cybersecurity, the structure and roles of incident response teams must also evolve to make them effective for addressing a variety of possible threats and vulnerabilities (Figure 10.9).

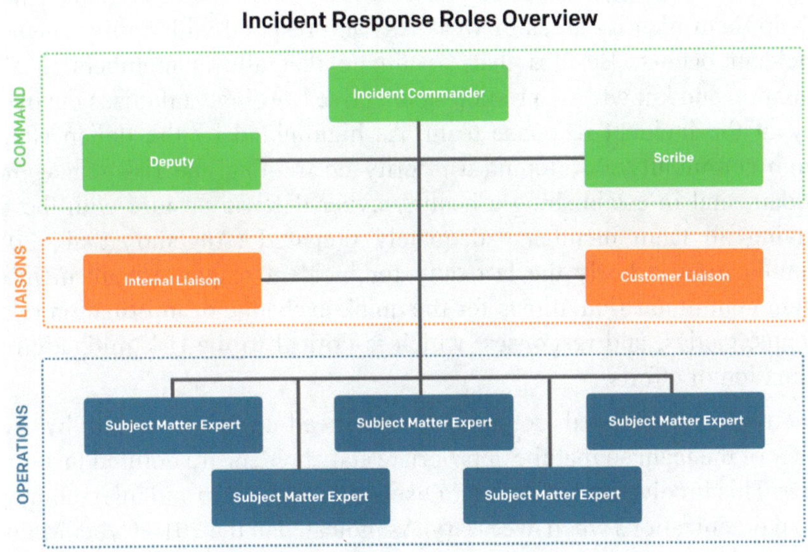

Figure 10.9. The structure of incident response teams.

Source: https://response.pagerduty.com/assets/img/misc/incident_response_roles.png.

10.4.2. Responsibilities of Team Members

In the incident response and forensics discipline, members' responsibilities are at the heart of successful management and mitigation of cyber incidents. Each member must know clearly what his/her contribution is to the incident response scheme, because this clarity generates teamwork and simplifies decision-making during emergencies. Responsibilities are highly varied and may involve activities such as threat assessment, forensic examination, communication, and recovery planning. By defining these roles, incident response teams are then able to function with effectiveness and precision, ultimately leading to improved results within cybersecurity incidents. Furthermore, a defined set of roles also creates a culture of responsibility where every member knows how crucial their role is to the success of the organization overall in fending off cyberattacks. The dynamic and changing nature of cyber threats mandates that team members continually update themselves on the latest trends and tools of cybersecurity.

Training and awareness programs prove to be helpful here, equipping team members with vital skills and knowledge required to react urgently in real time. Regular training can add to the technical skills base of members, which will help them identify areas of weakness and respond with agility whenever an incident occurs. Besides that, a climate that allows members to share information and knowledge promotes collective learning, and raises the overall ability of the incident response team. As highlighted by the Italian Nations research community, developing a priority on training and risk management is fundamental in establishing a healthy cyber defense posture with the goal of having all team members adequately prepared (Anglano et al., 2018). Communication is lastly the last duty for incident response team members. Efficient communication allows for the quick exchange of information related to threats, tactics, and responses, which is critical to the risk mitigation and coordination of efforts.

Internal and external communications need to be mastered by every member of the team so that the appropriate stakeholders are notified in a timely manner. This involves the updating of executive leadership and interacting with external organizations when necessary. As indicated in the 2016 Cyber Maturity report, understanding the governance structure of cyber policy enhances communication strategy and helps teams manage complicated environments successfully (Woodall et al., 2025). Summarily, incident response responsibility by team members is threefold, consisting of role definition, continued training, and successful communication, all of which are crucial to adapting to the ever-complicated cybersecurity environment.

10.4.3. Importance of Training and Simulations

In terms of incident response and forensics, training and simulation cannot be overstressed. These are essential for preparing groups to handle actual situations appropriately. Training itself enhances not only the technical skill of staff but also a culture of awareness and vigilance. Simulation allows responders to get exposed to high-pressure situations within a secure environment, facilitating the ability to make quick decisions in accordance with best practices. When threats evolve, the effectiveness of an incident response team is reliant on the responsiveness of the team; training provides a foundation upon which staff can modify their practice whenever necessary.

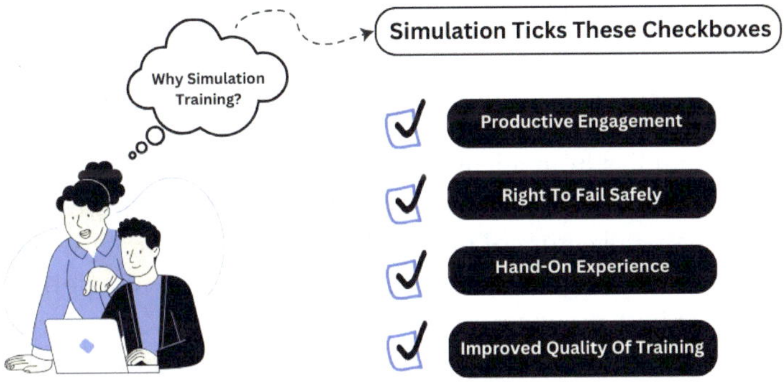

Figure 10.10. Importance of training and simulations.

Source: https://d8it4huxumps7.cloudfront.net/uploads/images/6577d1192509f_simulation_training_benefits.png?d=2000x2000.

Thus, ongoing practice and education are necessary, establishing resilience within groups capable of reducing response times and improving incident resolution. Simulations also present a valuable experience for organizations to review existing procedures and technology. Through simulating potential threats, teams can establish areas of weakness in their incident response systems and plan improvements. This kind of reflection is most important, considering the complexity of modern cyber-attacks that need to be countered by highly evolved and adaptive methods. For example, the use of intelligent anomaly detection through systematic monitoring of networks as outlined by Johnson et al. (2012) portrays the needs for real-time adaptation with an eye on dynamic threat environments. Continuous simulation enables teams to test such automated systems and ensure that they have good correlation with human decision functions. Thus, a training and simulation investment improves not just the near-term capacity for response but a security position of an organization in the longer term as well. Also, since cyber-attacks continue to change, so there is an increased need for an interdisciplinary approach to developing incident response training. The convergence of Crime Science empirical research with information security techniques is at the very heart of Cyber-crime Science, as proposed in Hartel et al. (2010). This new technology fosters evidence-based practices that optimize the effectiveness of training exercises and simulations. Empirical studies allow organizations to discover how various methods impact performance in responding to incidents, thereby continually refining their cycles of training. Last but not least, investing in lengthy training and simulations is not a drill on being ready, but an investment in strategy that enables organizations to be prepared for increasingly sophisticated threats and keeps staff productive, educated, and resistant under pressure (Figure 10.10).

10.5. CASE STUDIES IN INCIDENT RESPONSE

Studying case studies of incident response offers in-depth knowledge of cyber-attack mechanisms and the effectiveness of countermeasures. Case studies represent not only the nature of the incident but also the strategies organizations use to address them. Incident response strategy becomes more effective overall by patterns and best practices that can be drawn out by practitioners through case-by-case analysis. For instance, how an organization responded to a data breach can yield precious lessons, i.e., the importance of real-time monitoring and the implementation of incident response guidelines. These guidelines see to it that companies respond in timely and efficient ways, and therefore avert possible losses. The comprehensive examination of these case studies drives continued progress and serves as the foundation for developing better incident response plans, thereby contributing significantly to an organizations cybersecurity position. Integrating various technologies and methodologies is another important element represented through incident response case studies. One of the most important technological advancements is the development of the Case-Based Reasoning Forensic Triager (CBR-FT), whose application has been shown to significantly increase the efficiency of digital forensic investigations. The application of CBR-FT allows investigators to apply previous knowledge of investigations to identify areas on a suspects operating system most likely to yield evidence, thereby improving the search process. This strengthens the speed and reliability of forensic examination, reducing the potential for missing key evidence. From review of case studies that have made use of CBR-FT, it can be seen that the convergence of technological advancement and incident response policies can yield more efficient results. Such observations push cybersecurity experts to adopt and implement novel approaches that strengthen their response in an ever-increasing threat environment. In the future, lessons from incident response case studies will guide the next generation of cybersecurity practice. Strategy and framework development are necessary to address the evolving nature of threats, as debated in ongoing research community debates. Training, awareness raising, and risk management are required to facilitate organizational preparedness against cyber-attacks, as debated in recent literature. Companies must not only develop good incident response measures, but also a learning and adjustment culture. By integrating lessons learned from previous incidents and with the aid of new technologies, as suggested in recent criticisms, companies can enhance their preparedness against future issues. This cycling back and forth from learning to adaptation to strategy refinement process will be of utmost priority in providing robust defenses in an exponentially expanding cyber environment, as noted by Anglano et al. (2018) and Horsman et al. (2014).

10.5.1. Analysis of Notable Cybersecurity Incidents

Analysis of major cybersecurity incidents is vital in ascertaining how threats are evolving and incident response processes. One of the most significant incidents that depict this is the SolarWinds supply-chain attack that compromised numerous government agencies and businesses globally. The attack highlighted the weaknesses of third-party software and the prolonged dwell time of attackers, which averaged six months before they were detected. Research has determined that responses to such incidents are far from uniform based on the roles of cybersecurity professionals and their frames of reference. Technical staff typically focus on technical lags in attacking, whereas executives tend to emphasize procurement timetables and procurement chain management issues. These different mindsets create gaps in explaining the impact of the event, underlining the need for firms to build a culture of interpretive collaboration during response planning because the convergence to workable solutions came after the event had already happened (Ryan et al., 2025). The Log4Shell vulnerability crisis also clarifies the intricacies in incident response and the diverse perceptions of the stakeholders in cybersecurity. The highly publicized vulnerability hit many organizations and required the issuing of time-consuming responses to counter its impact. Interestingly, implementing practitioners who worked directly on countering Log4Shell understood it as an immediate threat, but other less directly involved practitioners considered it to be a lesser concern, thus minimizing its potential consequences. Research into this incident reveals these perceptions differences are not random but are attuned to the various roles one occupies within an organization. The need for organizational perceptual congruence emerges as a key factor for effective incident management and resilience because various perceptions can lead to response inefficiency and resource misdirection (Ryan et al., 2025). Moreover, the merging of big data application and cyber threats is a new trend in incident analysis, particularly in business sectors such as accounting. As the sophistication of cyber-attacks by malware and ransomware increases, organizations are increasingly turning to big data to enhance decision-making and risk management processes. Studies have established that concern about data breaches has a significant effect on how people view cybersecurity, and that firms have to employ data analytics in order to bolster their protective measures. The findings indicate a close connection between awareness of such threats and heightened fears about cybersecurity, and this underlines the demand for an analysis-driven approach to incident response. Employment of big data not only facilitates anticipating and averting threats but also yields intelligence for enhanced response strategies in future events, and this necessitates immediate adaptation and development of effective policies that reflect evolving cybersecurity threats (Figure 10.11).

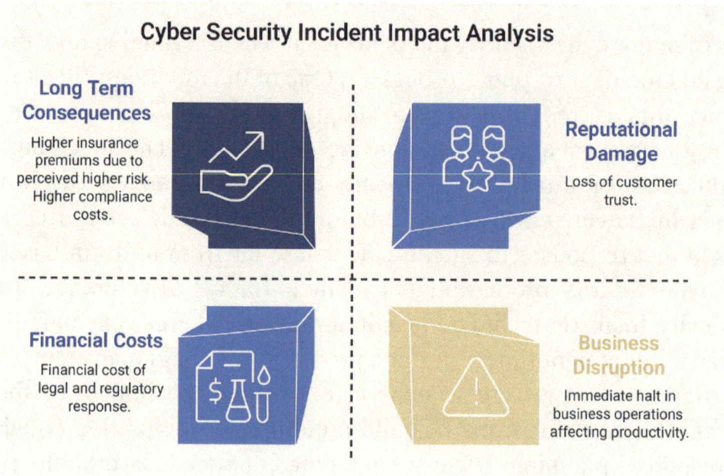

Figure 10.11. The analysis of notable cybersecurity incidents.

Source: https://www.sygnia.co/wp-content/uploads/2025/04/Cyber_Security_Incident_Impact_Analysis.png.

10.5.2. Lessons Learned from Past Incidents

An examination of past cyber incidents offers useful lessons for informing future incident response and forensic processes. A stark example is the SolarWinds supply-chain attack, which revealed stark contrasts in perceptions and response towards cyber threats by various stakeholders. Technical teams were always concerned with the six-month dwell time as the key problem, whereas executives cared more about procurement timeliness and overall organizational implications. These various perceptions highlight the significance of congruence among perception of incidents across various roles in an organization. This systematic mis-understanding in gaps hinders well-coordinated responses and underscores the necessity of developing structures that facilitate common interpretation. Organizations can build cross-mutual resilience and enhance readiness for subsequent attacks by prioritizing incident response convergence, translating differences in perception into structural learning opportunities (Ryan et al., 2025). Alongside perceptual convergence, the lessons from the crisis in the Log4Shell incident highlight the importance of recognizing the human causes of incident response. While technical practitioners witnessed firsthand the sheer work required to contain the incident, executives downplayed its scale. This is a classic case of how overlooked workload and effort across various teams can lead to an asymmetric view of the incidents worth. To close this gap, organizations need to set up systematic communication channels that ensure ongoing feedback and inputs from all levels of personnel during an occurrence. By appreciating the input of various personnel, organizations are able to develop a culture of cooperation that not only improves real-time responses but also promotes future preparedness. This is a shift from analyzing incidents as technical problems to seeing incidents as multifaceted events influenced by various perspectives and experiences (Ryan et al., 2025). Secondly, the use of digital forensics in incident response activities is a rich lesson learned from historical incidents. Effective cyber incident management is based on a good grasp of forensic techniques, which are vital to response and learning processes. The natural elegance of emerging threats necessitates continuous realignment of forensic practices and tools to continue being effective. History has illustrated that the emergent growth of threats precedes existing procedures and has a tendency to render them inadequate. To counter this, organizations must be concerned with developing ideal practices encompassing not only technical competence but also legal and ethical aspects within forensics. This strategy places organizations in a position to counter threats as they move through the intricacies that exist in the digital environment, ultimately enhancing their overall security stance in an era of fast-paced technological evolution.

10.5.3. Impact of Incidents on Organizational Policies

The incident experience significantly impacts organizational policies, and organizations must revisit in depth the prevailing frameworks after incidents. Organizations must analyze their incident response procedures in depth to gain useful lessons after incidents. Feedback is essential; it assists organizations in ascertaining what has gone right, what has gone wrong, and how responses can be enhanced in the future. However, several organizations cannot properly leverage these learning moments. While threats are becoming increasingly sophisticated, organizations can be so technologically oriented that they fail to conduct a full review of their internal policies and procedures. Highlighted by recent research, the focus on merely technical controls hinders organizations from being able to properly revisit their policy frameworks and thus their ability to properly counter emerging threats (Glisson et al., 2017). In addition to this, with the change and development of organizations with new technologies like cloud computing, the insufficiency of current policies is revealed. Cloud services are a challenge unto themselves, in that their design may not have enough forensic features, thereby hindering thorough investigations of incidents. This deficiency can create potential vulnerabilities that malicious parties can use to their advantage, leading to incidents that continue to put pressure on organizational resources.

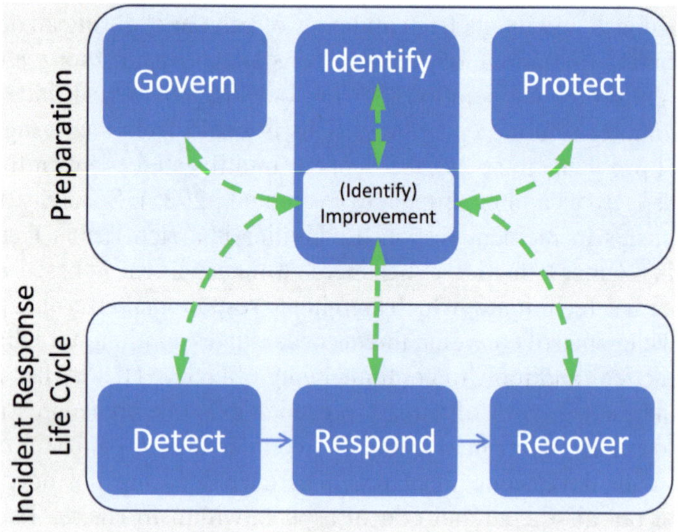

Figure 10.12. Impact of incidents on organizational policies.

Source: https://www.sygnia.co/wp-content/uploads/2025/04/life-cycle.png.

The importance of developing cloud forensic-capable services cannot be overstated because technology development has to be in accordance with incident response naturally. Through a strategy that emphasizes bringing in the aspect of forensics within the services design, organizations can create security issues in advance. This not only enhances incident response capacity but also necessitates reconsidering policies governing service deployment and monitoring (Agarwal et al., 2018). With these limitations considered, organizations are increasingly cognizant of the necessity for incorporating adaptive practices into policy frameworks. Dynamic incidents call for more than mere reactionary measures; organizations must cultivate a culture of continual improvement. It is more than compliance; it is the establishment of a learning environment in which insight obtained from past incidents is used systematically to enhance and revise policies. The inclusion of light agile retrospectives within the incident management process has the potential to facilitate better follow-up activities, allowing businesses to adjust their policies in real time based on experiential evidence. Through the implementation of such practices, organizations can not just enhance their resistance to future incidents but also aid towards a more responsive and dynamic policy environment (Glisson et al., 2017) (Figure 10.12).

10.6. FUTURE TRENDS IN INCIDENT RESPONSE AND FORENSICS

The field of incident response and forensics is on the cusp of drastic change, much of which will be brought about by the accelerating pace of technology development and mounting digital threats. Perhaps the most prominent trend is the increased recognition of mobile forensics because of the pervasive use of smartphones both in business and in life. With mobile phones being the focal point of communication and data storage, forensic examiners are faced with the task of extracting and analyzing data from such devices in an efficient manner. Techniques such as Linux Memory Extraction (LiME) have been discovered to extract volatile memory from Android mobile phones, exemplifying the need for innovative techniques of retrieving incriminating evidence. However, the application of these tools has complexity, such as the necessity for root access, which can hinder the investigation process and may limit what a forensic investigator can access and process effectively (Heriyanto et al., 2014). In addition, advances in artificial intelligence (AI) and machine learning (ML) are shaping the future of incident response and forensic analysis. They facilitate anomaly detection, automate repetitive tasks, and enhance the strength of forensic tools in analyzing data. Merging AI-driven algorithms ensures that sifting through data is more efficient, and analysts can focus on critical evidence and reduce their time investment on mundane work. Furthermore, as forensic tools evolve, they must accommodate new types of data generated by modern devices, including IoT platforms and cloud computing. Emerging studies underscore the importance of keeping oneself abreast with innovations in digital forensics and adapting methodologies to keep up with new platforms and technology, so that forensic specialists can react effectively to emerging threats (Martini et al., 2025). The future of incident response and forensics will also see more collaboration from different sectors, including the government, private businesses, and academia. With cyber threats becoming increasingly sophisticated, a collaborative mechanism is required in the determination of harmonized practices, sharing of best practices, and the creation of a more concerted response mechanism. Cooperation will be important in addressing shared vulnerabilities in the digital space as well as enhancing detection capability and tools. Developing a flexible incident response system, to be tailored based on organizational needs, keeps forensic investigations up to speed in a rapidly changing world. Ultimately, the future of incident response and forensics is all about adaptability, collaboration, and continued improvement of techniques to meet the complexities of the cyber universe.

10.6.1. Artificial Intelligence (AI) and Machine Learning (ML) Applications

The integration of AI and ML in incident response processes is transforming the practice of cybersecurity. With more sophisticated threats to organizational networks, the reliance on traditional security measures is no longer adequate. AI and ML processes offer state-of-the-art functionalities such as real-time threat identification and predictive analytics, enabling organizations to be a step ahead of cyber attackers. Through the analysis of immense amounts of security data, the technologies are able to detect anomalies and possible threats quicker than analyst humans, which is crucial in risk mitigation in an ever-changing digital world. AI-based models in the suggested frameworks are aimed at improving the detection of anomalies and automated response, finally resulting in more secure infrastructures. Sophisticated incident response and forensic functions can facilitate an active security posture, critical in order to address the ever-evolving cyber threat profile (Karim et al., 2024). Moreover, problems faced by security operations, such as sheer volume of security-related information and resource-scarce analysts, require innovative responses that AI and ML are capable of providing. With a Security Operations and Analytics Framework (SOAF) that incorporates machine learning, real-time detection and response are possible, greatly enhancing operational effectiveness. The framework incorporates tools like Wazuh and Elasticsearch, bringing together theoretical advancements in cybersecurity with practical implementation. Among the most significant outcomes of the integration is dramatic reduction in the response time for incidents and increased accuracy in detection, demonstrating the ability of AI-driven systems to handle advanced security environments. Not only does this model facilitate threat detection and assessment with ease, but it also allows security teams to strategically deploy their assets better, reducing the burden on human analysts and creating better organizational resilience (Ayittey et al., 2025). As AI and ML technology continues to evolve, their relevance in cybersecurity will also grow, paving the way for future incident response and forensics trends. The advent of autonomous response technologies, powered by AI-fueled intelligence, can possibly revolutionize the way incidents are managed by organizations. This transition to proactive versus reactive security solutions could significantly lower the impact of cyber incidents, enabling faster recovery and richer forensic analysis following a breach. However, emphasis must also be placed on the threats posed by these technologies, including false positives and adversarial vulnerabilities. Continual research and advancement of AI models will be key to working on these problems, eventually leading to better cybersecurity solutions that keep pace with the accelerated rate of cyber threats. As the virtual environment continues to evolve, so too must our strategies for defending it using innovative applications of artificial intelligence and machine learning (Karim et al., 2024) (Figure 10.13).

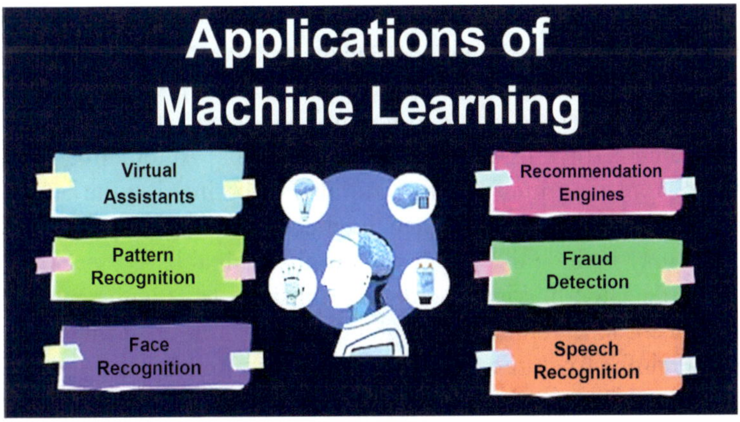

Figure 10.13. Artificial intelligence (AI) and machine learning (ML) applications.

Source: https://daxg39y63pxwu.cloudfront.net/images/blog/10-awesome-machine-learning-applications-of-today/Applications_of_Machine_Learning.webp.

10.6.2. The Rise of Automation in Incident Response

The rise of automation in incident response is a game-changer for how organisations respond to cybersecurity threats. Automation technologies, particularly artificial intelligence (AI) and machine learning, can enable quickly detecting and isolating security breaches by analyzing massive amounts of data in real time. This feature not only improves the productivity of incident response teams but also empowers them with capabilities to respond to threats in a proactive rather than a reactive manner. Through automated processes like log analysis and alert prioritization, security professionals can address more intricate issues that involve human decision. The use of AI on incident response platforms is a component of a broader move to employ technology to address the growing number and sophistication of cyber-attacks that organizations are facing nowadays and reshape the digital security landscape. With a greater extent of automation of incident response being achieved, there are a number of primary controversies surrounding its effectiveness and challenges. While the potential for quicker response times and less error is tremendous, organizations also have issues with data accuracy, biased algorithms, and the potential for false positives. These are credibility threats to automated systems unless well-handled. Adding to this problem is the ever-changing dynamic of cyber threats that can exploit the weaknesses of automated response. With this, effective utilization of automation for incident response calls for not only robust technology but also end-to-end staff training and a formal method of governance. A well-rounded strategy that encompasses constant feedback mechanisms and human oversight guarantees that automation enhances human capabilities rather than replacing them. Future oriented, the rise in automation of incident response corresponds to the growing expectation of organizations to improve their agility in cybersecurity operations. Automating is not only a question of replacing human effort but also of optimizing the collaboration between machine ability and human intellect. Directions into the future in this concern may include the development of explainable AI systems that will explain their decision-making, thereby establishing trust with security teams. Moreover, recent research has highlighted the importance of intense focus on cybersecurity infrastructures and training, as asserted by the Italian Nation research community (Anglano et al., 2018). Overall, automation's arrival in incident response will redefine operation paradigms to new norms, pushing the boundaries on what cybersecurity is conceived and conducted.

10.6.3. Predictions for the Future Landscape of Cybersecurity

The future of cybersecurity is in for a drastic overhaul as emerging technologies continue to redefine operational paradigms across all industries. With more organizations becoming keen on integrating Cyber-Physical Systems (CPS) and Internet of Things (IoT) in their supply chains, efficiency and automation are significantly amplified. However, with this technological transformation comes attendant heightened vulnerabilities because of the sheer volume and variety of devices being networked. Without a sound Security-by-Design philosophy, the possibility of security vulnerabilities is much greater. It is essential for organizations to adopt proactive measures in their digital forensics and incident response procedures to manage such complexity. The emergence of Smart Workplace technology, such as advanced CPS in the form of drones and robots, creates new avenues for insider threats and highlights the need for effective forensic-readiness approaches to watch and counter potential threats in these environments (Ahmadi-Assalemi et al., 2019). Moreover, advances in cybersecurity best practices will be supported by a paradigm shift towards a more scientific approach towards examining security paradigms. Current research frequently grapples with the efficacy and scientific relevance of cybersecurity methodologies.

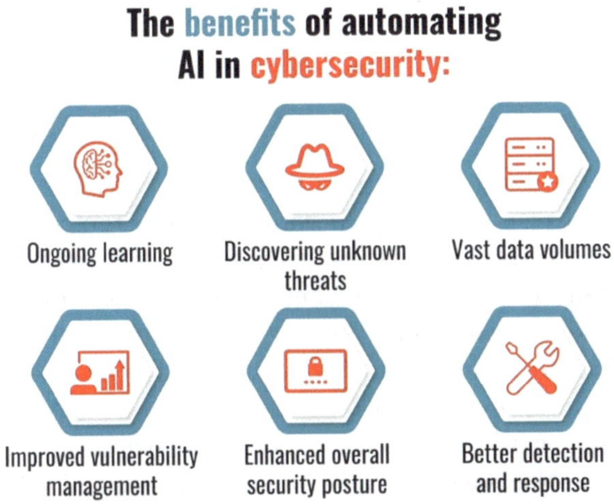

Figure 10.14. Predictions for the future landscape of cybersecurity.

Source: https://marvel-b1-cdn.bc0a.com/f00000000310757/www.fortinet.com/content/dam/fortinet/images/cyberglossary/benefits-of-automating-ai-in-cybersecurity.png.

Through the development of a culture of disciplined observation and evidence-based inquiry, the field can overcome inherent challenges with traditional practices. Increased synergy among specialization disciplines such as engineering and forensics implies a comprehensive platform for validation of cybersecurity phenomena. Most particularly, modern philosophy of science offers sound critical examination of how the methodology of science can be applied to cybersecurity in order to produce meaningful and trustworthy knowledge. Such a change of attitude will further enhance incident response practice and forensic examination, thus building enhanced preparedness and resilience towards future threats (Moore et al., 2017). Finally, the implementation of cutting-edge technologies such as federated Blockchain models will significantly redefine the future of cybersecurity. With the development of a digital Chain-of-Custody (CoC), Blockchain technology has the capability to improve forensic-readiness, providing a secure and open way of tracking digital evidence and ensuring data integrity along investigations. This approach not only facilitates post-incident analysis but also enables stakeholder collaboration in cyber-physical environments. As companies incorporate these advanced frameworks to detect anomalies and behavior patterns in real time, they will better be able to counter insider threats. To build a safe cybersecurity environment that can counter effectively the intricacy of future cyber-attacks, the integration of cutting-edge technologies with robust forensic strategies is needed (Ahmadi-Assalemi et al., 2019) (Figure 10.14).

SUMMARY

In Conclusion, the dynamic nature of the cybersecurity landscape requires an effective incident response program with a defense rooted in state-of-the-art forensic methodologies. The increasing frequency and sophistication of cyber-attacks underscore the necessity for organizations to not only set up but also continuously update their response plans. An effective incident response plan enables organizations to respond to damages instantly, preserve critical evidence, and implement preventive controls against future assaults. Sophisticated tools, such as the argumentation-based reasoner (ABR), demonstrate the potential of technology to enhance the process of assigning cyber-attacks. With its potential to enable forensics experts to elicit knowledge from multivariate data sets, ABR demonstrates how the combination of technical and social proofs can lead to more informed decision-making in cybersecurity responses (Karafili et al., 2020). Moreover, the call for scientific methodology to cybersecurity studies highlights the need for systematic approaches in the evaluation of incidents as well as the creation of responses.

Skeptics have a tendency to argue over the scientific approach to cybersecurity using deficiencies of existing processes. However, as can be illustrated from recent debates, philosophic re-interpretation of scientific approaches can demonstrate that cybersecurity practices do involve scientific fields. Through the use of mathematical modeling and systematic observation, the field of cybersecurity can create more rational frameworks to effectively respond to incidents. This change in attitude not only deepens our knowledge about cyber threats, but it also leads to the creation of evidence-based practices that enhance incident response plans in a quantifiable manner (Moore et al., 2017). Looking ahead, it is important for the cybersecurity profession to embrace not just scientific precision but also emerging technologies to better equip itself to anticipate future incident response and forensic trends. Through the integration of emerging technologies, such as machine learning and artificial intelligence, organizations can enhance their incident detection and response capabilities dynamically. Additionally, as cyber threats evolve to become more complex, there will be greater emphasis on cooperation between different fields—engineering, forensics, and social sciences—to create end-to-end solutions for complex security challenges. Finally, the future of incident response is predicated on adaptive strategies leveraging not only cutting-edge forensic techniques but also collaborative research approaches, enabling organizations to take proactive measures against foreseeable cyber threats.

The evolving cybersecurity landscape requires a holistic approach to incident response and forensics. The forecasting of cyber-attacks has been increasingly underway in industry and government. Existing studies emphasize the importance of leveraging information aggregated from multiple sources such as network telescopes and automated sensors in order to predict attacks. Notably, analyst-verified instances of malware, or cyber event data, reveal structural patterns in cyber events. This approach, founded on rigorous validation methods as described in Barabasi et al. (2018), enhances the precision of forecasts so that organizations can more accurately deploy resources and develop risk management plans with insight. The integration of best forecasting strategies can significantly make a cybersecurity group prepared for future threats, thus transforming incident response for an organization. In addition, the coming together of resources by several parties, e.g., government and private entities, contributes significantly to shaping effective cybersecurity strategies. Seminars such as those organized by powerful personalities in the Australian cybersecurity industry drive the point home about the power of several brains in dealing with urgent issues in cyberspace. Such events enable creative discussion as well as provide an arena for information sharing on best practices and methodologies across industries like defense, finance, and IT. Convergence of diverse perspectives, as the outcomes of forums like the Seoul Conference on Cyberspace reflect, underscores the merit of cross-industry collaboration (Woodall et al., 2025).

Engagement of over one sector in dialogue provides room for the incorporation of a complete comprehension of threats cyber and facilitates the development of comprehensive incident response models that are adaptable to emerging trends and emerging new technology threats. Lastly, the findings of heightened awareness of threat risks cannot be overstated when it comes to incident response and forensics. Scholarship showing systematic behavior in cyber-attacks facilitates heightened thoughts in the future when preparing for cybersecurity. With the application of advanced forecasting models, organizations can gain insights about the likelihood of cyber-attacks, thus preparing them for probable breaches and vulnerabilities. Not only does this risk-reduction strategy make overall cybersecurity infrastructure more resilient but also facilitates valid and focused decision-making. For instance, insights gained through forecasting can effectively channel strategic use of resources, both human capital and security technology, response shifting according to anticipated threats. Ultimately, building such a forward-thinking approach, as described in Barabasi et al. (2018), will be essential to future cybersecurity trends in maintaining effective and responsive incident response systems.

REFERENCES

1. Agarwal, A., Carrier, B., Martini, B., Chang, C., Kalloniatis, C., et al. (2018). A framework for designing cloud forensicenabled services (CFeS). https://core.ac.uk/download/153516639.pdf (accessed on 13th September 2025).
2. Agbanusi, N. C., & Ani, U. P. D. (2014). A comparative assessment of computer security incidence handling. https://core.ac.uk/download/394996780.pdf (accessed on 13th September 2025).
3. Ahmadi-Assalemi, G., Al-Khateeb, H., Cosson, J., et al. (2019). Federated blockchain-based tracking and liability attribution framework for employees and cyber-physical objects in a smart workplace. https://core.ac.uk/download/196263798.pdf (accessed on 13th September 2025).
4. Anglano, C., Aniello, L., Antinori, A., et al. (2018). The future of cybersecurity in Italy: Strategic focus area. https://core.ac.uk/download/188830442.pdf (accessed on 13th September 2025).
5. Angles, R., Curiel, M., Impagliazzo, J., et al. (2016). Latin American perspectives to internationalize undergraduate information technology education. https://core.ac.uk/download/86427281.pdf (accessed on 13th September 2025).
6. Archibald, J., Kapoor, K., & Renaud, K. (2018). Preparing for GDPR: Helping EU SMEs to manage data breaches. https://core.ac.uk/download/228178342.pdf (accessed on 13th September 2025).
7. Augustinos, T. P., Bauer, L., Cappelletti, A., et al. (2016). Cyber insurance: Recent advances, good practices & challenges. https://core.ac.uk/download/82893176.pdf (accessed on 13th September 2025).
8. Ayittey, G., & Ayittey, G. (2025). A security operations and analytics framework: Continuous detection and response. https://core.ac.uk/download/655889301.pdf (accessed on 13th September 2025).
9. Barabasi, B., Bürkner, C., Campbell, C., Carpenter, C., Carter, C., et al. (2018). Malware in the future? Forecasting of analyst detection of cyber events. http://arxiv.org/abs/1707.03243 (accessed on 13th September 2025).
10. Beardall, D. (2023). Unveiling the digital shadows: Cybersecurity and the art of digital forensics. https://core.ac.uk/download/589235599.pdf (accessed on 13th September 2025).
11. Capuano, N., Fenza, G., Loia, V., & Stanzione, C. (2022). Explainable artificial intelligence in cybersecurity: A survey. IEEE Access, 10, 93575–93600. https://doi.org/10.1109/access.2022.3204171 (accessed on 13th September 2025).

12. Carter, D. L., & Carter, J. G. (2015). Effective police homicide investigations: Evidence from seven cities with high clearance rates. https://core.ac.uk/download/46960045.pdf (accessed on 13th September 2025).
13. Glisson, W. B., Grispos, G., & Storer, T. (2015). Security incident response criteria: A practitioner's perspective. https://core.ac.uk/download/46558009.pdf (accessed on 13th September 2025).
14. Glisson, W. B., Grispos, G., & Storer, T. (2017). Enhancing security incident response follow-up efforts with lightweight agile retrospectives. https://core.ac.uk/download/96884346.pdf (accessed on 13th September 2025).
15. Halachev, P. (2024). A big data approach to risk management and control: Cybersecurity in accounting. https://core.ac.uk/download/641543649.pdf (accessed on 13th September 2025).
16. Hartel, P., Junger, M., & Wieringa, R. (2010). Cyber-crime science = Crime science + Information security. https://core.ac.uk/download/pdf/11476794.pdf (accessed on 13th September 2025).
17. Heriyanto, A. P. (2014). Procedures and tools for acquisition and analysis of volatile memory on Android smartphones. https://core.ac.uk/download/41535481.pdf (accessed on 13th September 2025).
18. Horsman, G., Laing, C., & Vickers, P. (2014). A case-based reasoning method for locating evidence during digital forensic device triage. https://core.ac.uk/download/19479046.pdf (accessed on 13th September 2025).
19. Jarocki, S. (2021). Efficacy of incident response certification in the workforce. https://core.ac.uk/download/426946461.pdf (accessed on 13th September 2025).
20. Johnson, C. W., Pezaros, D. P., & White, K. J. S. (2012). Increasing resilience of ATM networks using traffic monitoring and automated anomaly analysis. https://core.ac.uk/download/9400729.pdf (accessed on 13th September 2025).
21. Karafili, E., Lupu, E. C., & Wang, L. (2020). An argumentation-based reasoner to assist digital investigation and attribution of cyber-attacks. http://arxiv.org/abs/1904.13173 (accessed on 13th September 2025).
22. Karim, M. F., Rafi, M. N. A., Rukon, M. I. A., et al. (2024). Enhancing network security: A framework for proactive cyber defense using artificial intelligence and big data. https://core.ac.uk/download/636358637.pdf (accessed on 13th September 2025).

23. Kshetri, N., Rahman, M. M., Sayeed, S. A., et al. (2023). cryptoRAN: A review on cryptojacking and ransomware attacks w.r.t. banking industry–threats, challenges, & problems. http://arxiv.org/abs/2311.14783 (accessed on 13th September 2025).
24. Maple, C., & Phillips, A. (2010). UK security breach investigations report: An analysis of data compromise cases. https://core.ac.uk/download/29820726.pdf (accessed on 13th September 2025).
25. Martini, B., Tassone, C., Slay, J., & Choo, K. K. R. (2025). Mobile device forensics: A snapshot. https://core.ac.uk/download/pdf/30676282.pdf (accessed on 13th September 2025).
26. Mohammed, B. (2024). The impact of artificial intelligence on cyberspace security and market dynamics. https://core.ac.uk/download/630937997.pdf (accessed on 13th September 2025).
27. Moore, T., Pym, D., & Spring, J. M. (2017). Practicing a science of security: A philosophy of science perspective. https://core.ac.uk/download/146052023.pdf (accessed on 13th September 2025).
28. N/A. (2025). Guide to Australia's national security capability. https://core.ac.uk/download/pdf/30676195.pdf (accessed on 13th September 2025).
29. Poe, L. (2018). The development of information assurance and cyber security competencies. https://core.ac.uk/download/511343290.pdf (accessed on 13th September 2025).
30. Rahman, N. (2024). Leveraging large language models for network traffic analysis: Design, implementation, and evaluation of an LLM-powered system for cyber incident reconstruction. https://core.ac.uk/download/645626282.pdf (accessed on 13th September 2025).
31. Razak, S., Al-Dhaqm, A., Ali, A., et al. (2016). A generic database forensic investigation process model. https://core.ac.uk/download/132279410.pdf (accessed on 13th September 2025).
32. Ryan, M. (2025). A mixed methods comparative analysis of organizational incident response to SolarWinds and Log4Shell. https://core.ac.uk/download/658901072.pdf (accessed on 13th September 2025).
33. Wang, Y., Su, Z., Zhang, N., Xing, R., Liu, D., Luan, T. H., & Shen, X. (2022). A survey on Metaverse: Fundamentals, security, and privacy. IEEE Communications Surveys & Tutorials, 25, 319–352. https://doi.org/10.1109/comst.2022.3202047 (accessed on 13th September 2025).
34. Woodall, J., Jennings, P., & Feakin, T. (2025). A shared agenda for the Seoul Conference on Cyberspace, South Korea, 2013. https://core.ac.uk/download/pdf/30675836.pdf (accessed on 13th September 2025).

INDEX

D

E

F

G

H

I

P